Depolymerization of Lignin to Produce Value Added Chemicals

Depolymerization of Lignin to Produce Value Added Chemicals

Pratima Bajpai
Pulp and Paper Consultant, Kanpur, India

First edition

Copyright © 2024 by John Wiley & Sons, Inc. All rights reserved.

Published by John Wiley & Sons, Inc., Hoboken, New Jersey.

Published simultaneously in Canada.

No part of this publication may be reproduced, stored in a retrieval system, or transmitted in any form or by any means, electronic, mechanical, photocopying, recording, scanning, or otherwise, except as permitted under Section 107 or 108 of the 1976 United States Copyright Act, without either the prior written permission of the Publisher, or authorization through payment of the appropriate per-copy fee to the Copyright Clearance Center, Inc., 222 Rosewood Drive, Danvers, MA 01923, (978) 750-8400, fax (978) 750-4470, or on the web at www.copyright.com. Requests to the Publisher for permission should be addressed to the Permissions Department, John Wiley & Sons, Inc., 111 River Street, Hoboken, NJ 07030, (201) 748-6011, fax (201) 748-6008, or online at http://www.wiley.com/go/permission.

Trademarks: Wiley and the Wiley logo are trademarks or registered trademarks of John Wiley & Sons, Inc. and/or its affiliates in the United States and other countries and may not be used without written permission. All other trademarks are the property of their respective owners. John Wiley & Sons, Inc. is not associated with any product or vendor mentioned in this book.

Limit of Liability/Disclaimer of Warranty: While the publisher and author have used their best efforts in preparing this book, they make no representations or warranties with respect to the accuracy or completeness of the contents of this book and specifically disclaim any implied warranties of merchantability or fitness for a particular purpose. No warranty may be created or extended by sales representatives or written sales materials. The advice and strategies contained herein may not be suitable for your situation. You should consult with a professional where appropriate. Further, readers should be aware that websites listed in this work may have changed or disappeared between when this work was written and when it is read. Neither the publisher nor authors shall be liable for any loss of profit or any other commercial damages, including but not limited to special, incidental, consequential, or other damages.

For general information on our other products and services or for technical support, please contact our Customer Care Department within the United States at (800) 762-2974, outside the United States at (317) 572-3993 or fax (317) 572-4002.

Wiley also publishes its books in a variety of electronic formats. Some content that appears in print may not be available in electronic formats. For more information about Wiley products, visit our web site at www.wiley.com.

A catalogue record for this book is available from the Library of Congress

Hardback ISBN: 9781394191635; ePub ISBN: 9781394191659; ePDF ISBN: 9781394191642; oBook ISBN: 9781394191666

Cover Image: © Twenty47studio/Getty Images
Cover Design: Wiley

Set in 9.5/12.5pt STIXTwoText by Integra Software Services Pvt. Ltd., Pondicherry, India

Contents

List of Figures *viii*
List of Tables *x*
Preface *xii*
Acknowledgements *xiii*

1   **General Background and Introduction** *1*
1.1   Structural and Chemical Composition of Lignin *2*
1.2   Major Backbone Units and Representative Linkages in Lignin Molecules *2*
1.3   Types of Lignin *6*

2   **Isolation of Lignin** *12*
2.1   Lignosulfonates *18*
2.2   Kraft Lignin *18*
2.3   Soda Lignin *20*
2.4   Steam Explosion Lignin *20*
2.5   Organosolv Lignins *21*

3   **Lignin Depolymerization Technologies** *29*
3.1   Thermal Depolymerization *30*
3.1.1   Pyrolysis *30*
3.1.2   Hydrothermal Liquefaction *37*
3.2   Biological Depolymerization *53*
3.2.1   Lignin Depolymerization by Fungi *53*
3.2.2   Lignin Depolymerization by Bacteria *56*
3.2.3   Lignin Depolymerization by Enzymes *64*
3.2.3.1 Laccase *65*
3.2.3.2 Lignin Peroxidase *70*
3.2.3.3 Manganese Peroxidase *71*
3.2.3.4 Versatile Peroxidase *72*
3.2.3.5 β-etherase *72*
3.2.3.6 Biphenyl Bond Cleavage Enzyme *73*
3.3   Chemical Depolymerization *87*
3.3.1   Acid-catalyzed Depolymerization *87*
3.3.2   Base-catalyzed Depolymerization *90*
3.3.3   Ionic Liquid-assisted Depolymerization *95*

3.3.4	Supercritical Fluids-assisted Lignin Depolymerization	98
3.3.5	Metallic Catalysis	104
3.4	Oxidative Depolymerization of Lignin	125
3.5	Microwave-aided Depolymerization	133
3.6	Electrochemical Lignin Depolymerization	145
3.7	Reductive De-polymerization of Lignin	151

4	**Lignin-first Biorefining Process**	**156**
4.1	Introduction	156
4.2	The Revolutionary "Lignin-first" Method for Lignocellulosic Catalytic Valorization	157
4.2.1	Reductive Catalytic Fractionation	157
4.2.2	From Phenolic Units to Value-added Products	163
4.3	Future Challenges	165

5	**Lignin Production**	**173**
5.1	Introduction	173
5.2	Pilot-scale	174
5.2.1	Ammonia Fiber Explosion Lignin	174
5.2.2	Steam Explosion Process	175
5.2.3	BioFlex Process	175
5.2.4	German Lignocellulose Feedstock Biorefinery Project	176
5.2.5	Proesa® Lignin	176
5.2.6	FABIOLA™ Lignin	177
5.2.7	Fast Pyrolysis Lignin	178
5.2.8	Sequential Liquid-lignin Recovery and Purification Technology	178
5.3	Commercial scale	179
5.3.1	LignoForce™ Technology	179
5.3.2	LignoBoost™ Technology	180
5.3.3	SunCarbon Lignin	180
5.3.4	Production of Lignosulfonates	181
5.3.5	Kraft Lignin Production	181
5.3.6	Organosolv and Soda Lignin	182
5.3.7	Thermo-mechanical Pulp-bio Lignin	183
5.4	Future Perspectives	183

6	**Applications of Lignin**	**188**
6.1	Introduction	188
6.2	Applications	189
6.2.1	Aromatics, Phenolics and Flavoring Compounds	189
6.2.2	Carbon Materials	191
6.2.3	Lignin-based Nanomaterials	192
6.2.4	Biomedical Application	192
6.2.5	Lignin-based Nanocomposites	193
6.2.6	Urethanes and Epoxy Resins	194
6.2.7	Controlled Release Fertilizer	194
6.2.8	Biosensor and Bioimaging	195
6.2.9	Hydrogen Production	197

6.2.10	Battery Material for Energy Storage	*198*
6.2.11	Dust Control Agent	*200*
6.2.12	Bitumen Modifier in Road Industry	*200*
6.2.13	Cement Additives and Building Material	*201*
6.2.14	Bioplastics	*202*
6.2.15	Use of Lignin as a Binder	*203*
6.2.16	Lignin as Dispersant	*203*
6.2.17	Lignin as Food Additives	*203*
6.2.18	Lignin as Sequestering Agent	*203*
6.2.19	Lignin Bio-oil	*204*

7 Lignin – Business and Market Scenario *212*
7.1 Introduction *212*
7.2 Lignin Market *213*

8 Challenges and Perspectives on Lignin Valorization *219*
8.1 Introduction *219*
8.2 Challenges and Perspectives on Lignin Utilization *220*

Index *225*

List of Figures

1.1	Lignocellulose in biomass and its composition.	2
1.2	Major backbone units and representative linkages in lignin molecules.	3
1.3	The three building blocks of lignin.	4
1.4	Typical linkages present in lignin.	4
1.5	Model lignin structures: (a) softwood, (b) hardwood, and (c) grass.	5
1.6	Dehydrogenation of coniferyl alcohol a aterial.	6
1.7	Structure of lignin in lignocellulosic material.	6
2.1	Typical structural model of lignin.	13
2.2	Monolignol monomer species.	13
2.3	Common linkages found in lignin.	14
2.4	Schematic representation of hardwood lignin.	14
2.5	Schematic representation of softwood lignin.	15
2.6	Simplified and representative structures of common technical lignins.	17
2.7	Photographs showing the differences in physical appearance of a number of typical technical lignins.	23
3.1	Different technologies of lignin depolymerization.	31
3.2	Ranges of reaction temperatures used by different lignin thermal depolymerization methods.	32
3.3	Effect of pyrolysis temperature on the aromatic substitution products of G-type monomers.	32
3.4a	Proposed pyrolytic mechanism of lignin.	36
3.4b	Proposed reaction mechanism pathway of lignin non-catalytic/catalytic fast pyrolysis.	36
3.4c	The purpose mechanism of cleaving the β-O-4 linkages by pyrolysis.	37
3.5	Characteristic pyrolytic behaviors of initial, primary, and charring stages.	37
3.6	Proposed reaction pathways for primary stage of lignin pyrolysis.	38
3.7	Process flow configuration for hydrothermal liquefaction.	40
3.8	Some phenolic products from HTL of lignin.	45
3.9	Some guaiacol derivatives from HTL of lignin.	45
3.10	Structures of some catechols.	46
3.11	Aromatics catabolism from the coniferyl, sinapyl and p-coumaryl branch.	62
3.12	Ligninolytic enzymes and their selective action on lignin components.	66
3.13	Catalytic mechanism of laccase mediated lignin degradation.	66

3.14	Catalytic mechanism of lignin peroxidase (LiP) mediated lignin degradation.	66
3.15	Catalytic mechanism of manganese peroxidase (MnP) mediated lignin degradation.	67
3.16	Catalytic mechanism of versatile peroxidase (VP) mediated lignin degradation.	67
3.17	Three-dimensional structures of ligninolytic enzymes.	68
3.18	Mechanisms of β-O-4 ether and biphenyl linkage degradation.	73
3.19	Cleavage of β-O-4 linkages by acid catalysts.	87
3.20	Mechanism for base catalyzed depolymerization of lignin.	91
3.21	Low-molecular-weight products resulting from the depolymerization of lignin with the use of sodium hydroxide.	94
3.22	The three ionic liquid generations.	95
3.23	Cations and anions of ionic liquids in lignin chemistry.	95
3.24	Plausible reaction pathway of lignin in supercritical water.	102
3.25	The purpose mechanism of cleaving the β-O-4 linkages with the metallic catalyst.	105
3.26	The mechanism of cleaving the β-O-4 linkages with the oxidant.	129
3.27	Oxidative depolymerization mechanism of lignin.	129
3.28	Microwave-assisted catalytic depolymerization of lignin from birch sawdust to produce phenolic monomers utilizing a hydrogen-free strategy.	140
3.29	Understanding lignin depolymerization to phenols via microwave-assisted solvolysis process.	140
3.30	Microwave processing of lignin in green solvents: a high-yield process to narrow-dispersity oligomers.	140
3.31	Ultrasonic and microwave assisted organosolv pretreatment of pine wood for producing pyrolytic sugars and phenols.	141
3.32	Abbreviated reaction scheme showing proposed electrochemical/radical mechanisms during lignin degradation.	145
3.33	A representation of the structure of levulinic acid.	147
3.34	Main levulinic acid-derived products of kraft lignin depolymerization identified by direct injection high-resolution MS.	148
3.35	Reductive depolymerization of kraft lignin to produce aromatics.	154
4.1	The revolution of "lignin-first" approach: from lignocellulosic biomasses to added value.	157
4.2	The chemical-reaction mechanism of lignin-first biorefinery using solvolysis and the catalytic stabilization of reactive intermediates to stable products or protection-group chemistry and subsequent upgrading.	158
4.3	Three lignin-first strategies.	160
4.4	The evolution of reactor configurations for reductive catalytic fractionation.	164
4.5	A schematic overview of potential added value products of lignin biorefinery.	164
6.1	Current and potential applications of technical lignin.	190

List of Tables

1.1	Value-added chemicals formed from lignin through various treatments.	8
2.1	Different types of lignin, their monomer's molecular weight, and lignin content.	15
2.2	Characterization of technical lignins.	16
2.3	Characteristics of the technical lignin.	17
2.4	Some of the major manufacturers of lignins.	17
2.5	Sulfur content and purity of different types of lignins.	23
3.1	Hydrothermal liquefaction of lignin into different products.	40
3.2	Effect of ionic liquids on lignin degradation for different feedstock.	43
3.3	Lignin depolymerization through fungi.	54
3.4	Fungi degradation of lignin in various biomass sources.	55
3.5	Lignin depolymerization through bacteria.	59
3.6	Bacterial degradation of lignin in various biomass sources.	61
3.7	Major ligninolytic enzymes.	65
3.8	Acid-catalyzed depolymerization of lignin.	89
3.9	Reaction conditions and the products of base-catalyzed depolymerization of lignin.	92
3.10	Various phenolic products of sodium hydroxide catalyzed hydrolysis of lignin at 300 °C and 250 bar.	93
3.11	Depolymerization of lignin by catalytic oxidation in ionic liquids.	96
3.12	Reaction conditions and the products of ionic liquid and deep eutectic solvent-based depolymerization.	99
3.13	Reaction conditions and the products of sub- and supercritical depolymerization of lignin.	103
3.14	Metallic catalyzed lignin depolymerization.	107
3.15	Different lignin sources depolymerized with noble metal catalyst.	110
3.16	Nickel catalyst in different support materials to depolymerize lignin.	111
3.17	Reaction conditions and the products of oxidative depolymerization of lignin.	126
3.18	Reaction conditions and the products of microwave-aided depolymerization of lignin.	134
3.19	Depolymerization of lignin under microwave irradiation.	139
4.1	Reductive catalytic fractionation of biomass feedstock.	161

5.1	Different sources of lignin and their current volume.	174
5.2	Lignin production at pilot scale.	174
5.3	Lignin production on commercial scale.	179
6.1	Important results from lignin-related biosensors and applications in bioimaging.	196
7.1	Lignin market key players.	213
7.2	Current industrial applications of lignin.	214
7.3	Lignin market value.	214
7.4	Lignin market share (by product) 2022.	215

Preface

Lignin is a potential feedstock due to its energy content and its abundant availability from pulp and paper mills and biomass based biorefinery. Lignin has great potential for its conversion into value-added products, which could significantly improve the economics of a biorefinery. Emerging opportunities exist in generating high-value small molecules from lignin through depolymerization. This book discusses technologies of lignin depolymerization. Compared with thermal and chemical depolymerization, bioprocessing with microbial and enzymatic catalysis is a clean and efficient method for lignin depolymerization and conversion. Biological methods of lignin depolymerization are gaining significance due to their economic and environmentally benign nature. The feasibility of large-scale implementation of these technologies, including thermal, biological, and chemical depolymerizations is discussed in relation to potential industrial applications. The "lignin-first" biorefining approach and potential applications of lignin-derived monomers and their derivatives as bioactives in food, natural health products, and pharmaceutical sectors; lignin – business and market scenario and challenges and perspectives on lignin valorization are also covered.

This book will help readers to identify the high added value of a biomass residue and support them in its possible use for mass and niche high impact application sectors.

Acknowledgments

I am grateful for the help received from many people and companies/organizations who provided information. I am also thankful to various publishers for allowing me to use their material. My deepest appreciation is extended to Elsevier, Springer, RSC, ACS Publications, John Wiley & Sons, Frontiers Media SA, Hindawi, MDPI, IntechOpen, SpringerOpen, and other open-access journals and publications. My special thanks to Dr. KK Pant IIT Delhi, India and Dr. Weckhuysen, Distinguished Professor, Utrecht University, Netherlands who allowed me to use their material.

1

General Background and Introduction

Abstract

Lignin, a complicated organic polymer, plays a significant structural function in the support tissues of vascular plants. It is particularly prevalent in woody plants and is highly polymerized. Lignin is one of the three crucial elements of wood, along with extractives and carbohydrates. Lignin, a three-dimensional amorphous polymer made of methoxylated phenylpropane structures, is important for the survival of vascular plants. In nature, lignin polymer generally forms ether or ester linkages with hemicellulose which is also connected with cellulose. The structural and chemical composition of lignin; representative linkages in lignin molecules and types of lignin are presented in this chapter.

Keywords   Lignin; Phenylpropane units; Monolignols; Sinapyl alcohol; Coniferyl alcohol; p-coumaryl alcohol; Hardwood lignin; Softwood lignin;

With the rapid growth of populations and rising living standards in developing nations, global energy demand is rapidly rising. To meet this rising energy demand, fossil resources alone will not be sufficient. At the same time, there are significant concerns regarding the impact of climate change, which may be linked to the combustion of fossil fuels that are not renewable. As a result, it is crucial to develop technologies that can use new energy solutions on a large scale and provide more environmentally friendly alternatives to the current economy based on fossil fuels. Because this renewable feedstock can theoretically be incorporated into a carbon dioxide-neutral energy cycle, biomass is an option for the production of sustainable fuels and chemicals. Cellulose, hemicellulose, and lignin are the three main components of biomass.

Aromatic compounds, which can be used as fuel or as intermediate chemicals in the industry, can be obtained from lignin, the organic biopolymer that is found in the second highest concentration anywhere on the planet. Biomass conversion technology's viability can be improved by incorporating lignin into biorefineries. The recalcitrant and complicated nature of the lignin feedstock presents the primary obstacle in this situation. It is a huge challenge to properly convert lignin into functional polymers, but this is a fascinating area of research in both industry and academia (Guvenatam, 2015).

1.1 Structural and Chemical Composition of Lignin

The Swiss botanist Augustin Pyramus de Candolle was the first person to use the term lignin, which comes from the Latin word *lignum*, which means wood (Candolle et al., 1821). Lignin, a complicated organic polymer, plays a significant structural role in the support tissues of vascular plants. It is particularly prevalent in woody plants and is highly polymerized. Lignin is one of the three crucial elements of wood, along with extractives and carbohydrates (Sarkanen and Ludwig, 1971; Sjöström, 1982). Protolignin is the name given to lignin when it is in its natural state, as it is in plants. Lignin, a three-dimensional amorphous polymer made of methoxylated phenylpropane structures, is essential for the survival of vascular plants. In nature, lignin polymer usually forms ether or ester linkages with hemicellulose which is also associated with cellulose. Therefore, these natural polymers construct a complicated and valuable lignocellulose polymer (Figure 1.1).

1.2 Major Backbone Units and Representative Linkages in Lignin Molecules

It is generally acknowledged that the polymerization of three types of phenylpropane units, also known as monolignols, initiates the biosynthesis of lignin (Freudenberg and Neish, 1968; Lewis, 1999; Ralph, 1999; Sarkanen and Ludwig, 1971). These units, sinapyl, coniferyl, and p-coumaryl alcohol, are linked by the chemical bonds of aryl ether (β-O-4), phenylcoumaran (β-5), resinol (β-β), biphenyl ether (5-O-4), and dibenzodioxocin (5–5) (Figure 1.2). In Figure 1.3, the three structures are shown. The most typical linkage among the various typical linkages (β-O-4, β-5, β-1, 5–5, α-O-4, 4-O-5, β-β) (Figure 1.4) is the β-aryl ether (β-O-4), which accounts for more than half of the structure of lignin (Dutta et al., 2014; Rinaldi et al., 2016). Figure 1.5 shows model lignin structures: A softwood, B hardwood, and C grass (Lu and Gu, 2022).

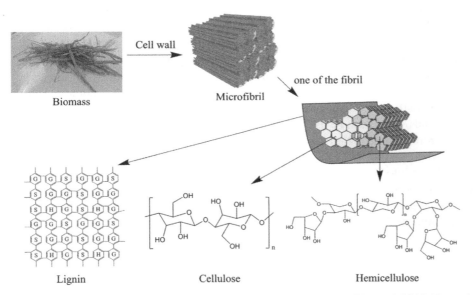

Figure 1.1 Lignocellulose in biomass and its composition. Chonlong Chio et al. 2019 / Reproduced with permission from Elsevier.

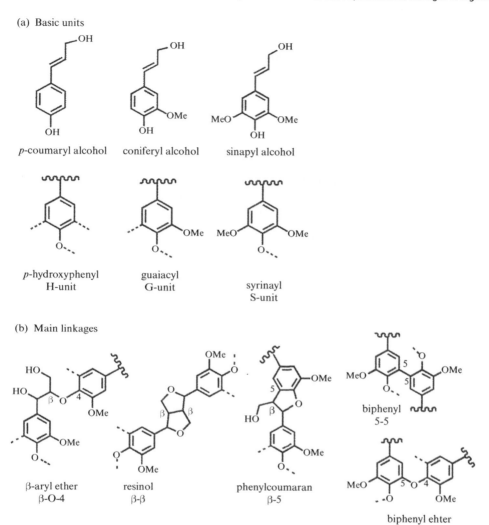

Figure 1.2 Major backbone units and representative linkages in lignin molecules. (a) The building blocks of lignin consist of three primary types of monolignols, namely p-coumaryl alcohol, coniferyl alcohol, and sinapyl alcohol. The alcohols form the corresponding phenylpropanoid units like p-hydroxyphenyl (H), guaiacyl (G), and syringyl (S) in lignin polymer, respectively. (b) Backbone units are conjugated via different chemical bonds (e.g., β-O-4, β-β, 5–5, and β-5) resulting in high resistance to lignin depolymerisation Weng et al. (2021) / Springer Nature / Public Domain CC BY 4.0.

Figure 1.6 demonstrates the phenoxy radicals that are resonance-stabilized and the dehydrogenation of coniferyl alcohol (Chakar and Ragauskas, 2004). The polymerization interaction is set up by the oxidation of the monolignol phenolic hydroxyl groups. It has been demonstrated that an enzymatic pathway catalyzes the oxidation itself. An electron transfer initiates the enzymatic dehydrogenation, resulting in reactive monolignol species and free radicals, which are able to pair with one another. The aromaticity of the benzene ring will be restored by a subsequent nucleophilic attack by water, alcohols, or phenolic hydroxyl groups on the benzyl carbon of the quinone methide intermediate. Polymerization will continue on the produced dilignols.

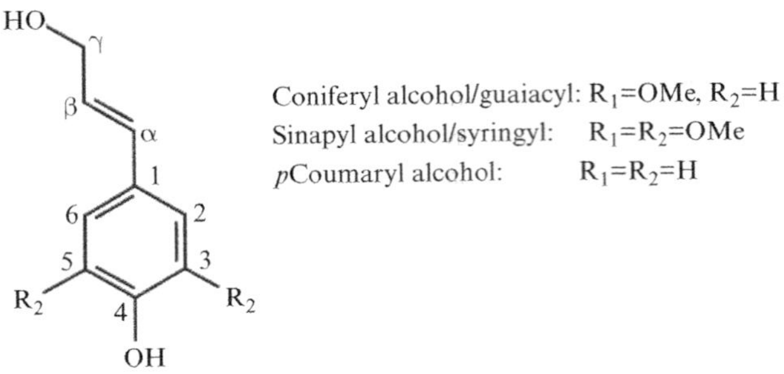

1 General Background and Introduction

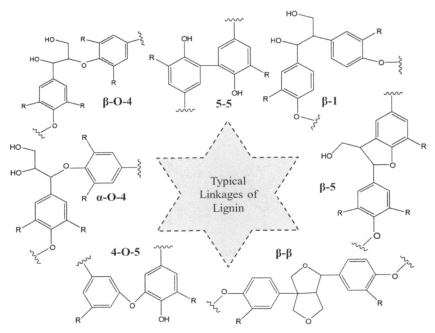

Coniferyl alcohol/guaiacyl: R_1=OMe, R_2=H
Sinapyl alcohol/syringyl: R_1=R_2=OMe
pCoumaryl alcohol: R_1=R_2=H

Figure 1.3 The three building blocks of lignin. Chakar and Ragauskas, 2004 / with permission of ELSEVIER.

Figure 1.4 Typical linkages present in lignin. Agarwal et al. (2018) / with permission of ELSEVIER.

Architecturally, lignin forms a complex, three-dimensional heterogeneous network thanks to the chemical bonds it forms with cellulose and hemicellulose via covalent and non-covalent bonds (Figure 1.7) (Agarwal et al., 2018). Due to lignin's irregular, heterogeneous structure, it remains difficult to produce commodity chemicals with added value.

The primary sources of lignin that can be utilized on a larger scale are spent cooking liquor and the chemical extraction of wood fibers from the pulp and paper industry. Over 50 million tons of lignin-based materials and chemicals are produced annually worldwide. Despite the fact that most lignin in the world is still used as boiler fuel in facilities that process carbohydrates, its low value and abundance indicate that it might be used to create high-value new products.

If converted into chemical compounds, bioproducts based on lignin could lead to a multibillion-dollar industry. Several million tonnes of lignin are produced as a low-value byproduct of industrial cellulosic bioethanol production. It is anticipated that the US bioethanol industry alone will produce up to 60 Mt./year of lignin by the end of 2022 (Holladay et al., 2007a; Joffres et al., 2014).

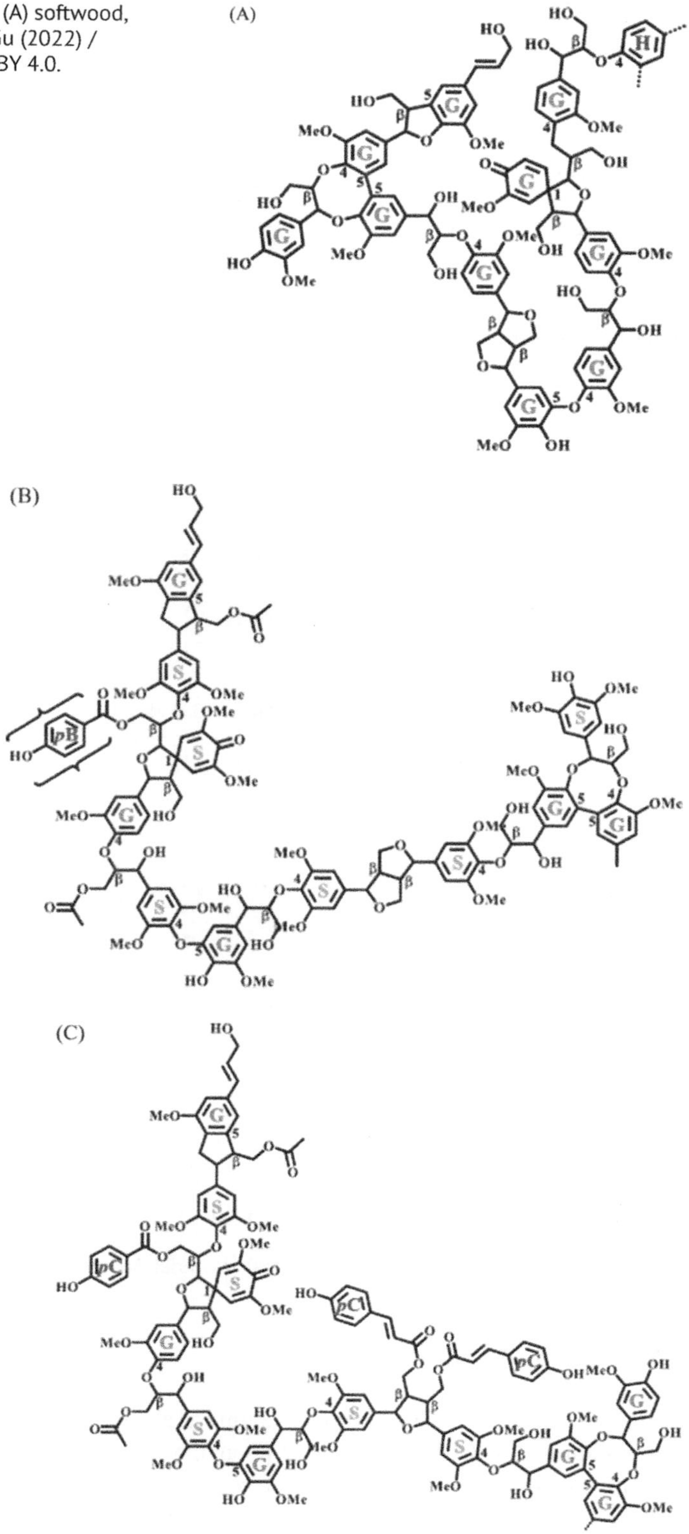

Figure 1.5 Model lignin structures: (A) softwood, (B) hardwood, and (C) grass. Lu and Gu (2022) / Springer Nature / Public Domain CC BY 4.0.

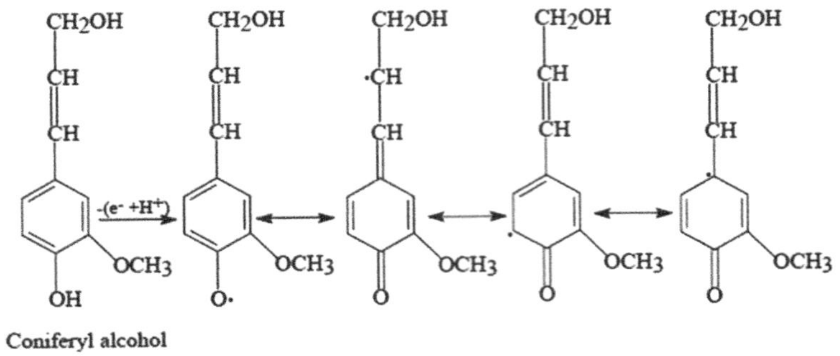

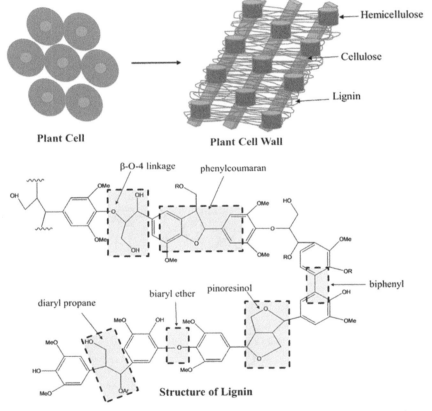

Coniferyl alcohol

Figure 1.6 Dehydrogenation of coniferyl alcohol and the mesomeric radicals. Chakar and Ragauskas, 2004 / with permission of ELSEVIER.

Figure 1.7 Structure of lignin in lignocellulosic material. Agarwal et al. (2018) / with permission of ELSEVIER.

1.3 Types of Lignin

Sinapyl alcohol, coniferyl alcohol and p-coumaryl alcohol, are frequently joined by non-hydrolysable linkages during the dehydrogenation of phenylpropanoid precursors that is carried out by free radicals with the assistance of peroxidase and results in the formation of lignin. The aromatic amorphous heteropolymer lignin lacks any optical activity. These three monoolignols are present in varying amounts in various plant species. For instance, softwood lignin contains a lot of coniferyl alcohol, whereas hardwood lignin contains both sinapyl and coniferyl alcohols and grass lignin contains all the three monolignols (Duval and Lawoko, 2014). The extracted lignin has been divided into four major categories—lignosulfonates, kraft lignin, soda lignin, and organosolv lignin based on the chemical pre-treatment method used: sulfur-free soda lignin is produced when biomass is treated with sodium hydroxide, whereas kraft lignin is produced when biomass is treated with sodium sulfide and sodium hydroxide. The ethanol-water extraction method and the pre-treatment of biomass with aqueous sulfur dioxide produce lignosulfonates and organosolv lignin, respectively. In addition to the four primary types of lignin, ionic liquid lignin, which is produced by treating biomass with ionic liquid is attracting a lot of interest because of its condensed structure and a low β-O-4 content (Wen et al., 2013). Due to the structural changes that take place when lignin is separated from lignocellulose biomass, chemical properties vary between lignin types. Contrary to kraft lignin and lignosulfonates, organosolv lignin, which is practically insoluble in water and natural solvents, contains a higher level of β-O-4 linkages (Bauer et al., 2012).

Even though lignin accounts for 15–40% of a plant's dry weight, it is still not considered a high-value-added product in biorefinery processes (Cao et al., 2017). Indeed, the utilization of lignin has the potential to significantly boost the cost-effectiveness of biorefinery processes based on biomass (Ragauskas et al., 2014). Only 5% of the lignin produced by the paper and pulp industry is used to produce low-quality fuel for use in heat and electricity applications through combustion (Cao et al., 2018). The well-organized valorization of lignin produced by various industrial processes may result in the proliferation of economic and environmental sustainability (Wu et al., 2018).

The highly asymmetrical polymeric structure is the most significant impediment to the lignin conversion process. The fractionation process's effect on product recovery was discussed by a number of researchers (Anderson et al., 2019). Bio-oil yield and quality could be impacted by minor structural changes. By selectively fractionating lignin from other biomass components with fewer structural changes for efficient lignin application, attempts have been made to valorize the lignin conversion process. Compared to extracted lignin whose structure has been altered, the direct hydrogenolysis process yielded aromatic monomers from native lignin at rates of 40–50%, which is five to ten times higher (Shuai et al., 2016).

There have been a number of studies on lignin valorization, in which the monomers and oligomers produced by depolymerizing lignin through thermal, chemical, and biological, pre-treatments can be turned into fuels and other chemicals (Beckham et al., 2016; Ragauskas et al., 2014).

Typically, heterogeneous aromatic compounds are produced following lignin depolymerization based on the feedstock and pre-treatment used (Schutyser et al., 2018). Fine chemicals can only be made with aromatic compounds of high purity. As a result, lignin upgrade is hindered by the heterogeneity of aromatics produced by depolymerization (Liu et al., 2017; Schutyser et al., 2018).

According to Abdelaziz et al. (2016), large quantities of lignin have been produced, estimated at 5–36×10^8 tons annually. The pulp and paper and biomass refinery industries each contribute approximately 6.2×10^7 and 5×10^7 tons of lignin annually, respectively, which includes soda lignin, kraft lignin, and lignosulfonate (Zakzeski, 2010).

Most of the time, lignin is used for energy or thrown away as waste. Due to its rich aromatic skeleton and high carbon-to-oxygen ratio, lignin is a promising feedstock for the production of biofuels and biochemicals (Vishtal and Kraslawski, 2011). In order to take advantage of lignin valorization, it is urgently necessary to acquire an understanding of the degradation procedure and create an efficient metabolic pathway for conversion. Lignin's recalcitrance and complicated structure make it difficult to depolymerize and use effectively. Currently, the most common approaches for lignin depolymerization are thermochemical and biological ones. Pyrolysis (thermolysis), gasification, hydrogenolysis, and chemical oxidation are thermochemical processes that call for extreme conditions, a lot of energy, and costly facilities (Bandounas, 2011). On the other hand, bioprocessing lignin has the advantages of higher specificity, reduced energy consumption, and affordability (Chen and Wan, 2017). In the specific cleavage of lignin linkages, biological depolymerization has demonstrated a number of benefits, and its nature makes this process environmentally friendly (Xu et al., 2019). But there are some disadvantages, such as the difficulty of genetically altering the microbes and their high sensitivity to changes in pH, temperature, and oxygen levels in the reaction system (Chauhan, 2020). Thermochemical methods have been used frequently for a long time, and these can be categorized into various groups based on the catalyst, heating technology, solvents, temperature ranges, and other factors. The liquid that comes out of lignin depolymerization, bio-oil, will have different percentage yields and different kinds of monomers and oligomers because of these factors (Agarwal et al., 2018; Lopez-Camas et al., 2020).

Holladay et al. (2007b) provide an economical evaluation of lignin feedstock-based chemical conversion technologies. Despite its potential, lignin is underutilized by industry as a chemical conversion raw material (Doherty et al., 2011; El Mansouri and Salvadó, 2006).

Biorefining natural feedstocks looks like a good way to use more lignin. The intricate utilization of biomass-derived feedstocks like lignocelluloses, oil and sugar crops, and algae is the foundation of the biorefinery concept (Cherubini, 2010; Demirbas, 2009). The three main components of lignocellulosic materials are: lignin, cellulose, and hemicelluloses (Sjöström, 1982). The majority of biorefineries are presently concentrating on the sugar-based platform for the valorization of hemicelluloses and cellulose (FitzPatrick et al., 2010), whereas lignin is typically regarded as a low-value product (Cherubini et al., 2010; Doherty et al., 2011). In contrast to sugars, which are released as uniformly monomeric carbohydrates, lignin is released as a complex and polydisperse compound. Limited use of lignin in biorefineries is primarily due to its complex structure and uncertain reactivity.

One strategy for realizing the full potential of lignin is through its transformation into useful products. Table 1.1 shows value-added chemicals formed from lignin through various treatments.

Technical lignins, such as kraft lignin, soda lignin, and lignosulphonates, are obtained in processes that deal with treating lignocelluloses, making them an intriguing raw material. Additionally, many technical lignins can be obtained in large quantities and are readily available. However, hydrolysis, organic solvents, and ionic liquids only yield a small fraction of the potentially valuable lignins. These are produced in relatively smaller quantities but may eventually

Table 1.1 Value-added chemicals formed from lignin through various treatments.

Lignin						
Hydrogenation	Pyrolysis	Oxidative hydrolysis	Fast thermolysis	Alkali fusion	Enzymatic oxidation	Microbial conversion
Phenol, cresols, substituted phenols	Phenol, acetic acid, carbon monoxide, methane	Vanillin, dimethyl sulfide dimethyl sulfooxide	Ethylene, acetylene	Catechol and phenolic acid	Oxidized lignin	Lignin with high level of polymerization ferulic, coumaric, vanollic, and other acid

Ullah et al. (2022) / MDPI / Public Domain CC BY 4.0.

develop into products on a commercial scale. Removing lignin from the product streams of small non-wood mills permits the elimination of recovery boiler bottlenecks and offers a solution to some environmental issues but may eventually develop into products on an industrial scale (Gosselink et al., 2004).

Bibliography

Abdelaziz OY, Brink DP, Prothmann J, Ravi K, Sun M, García-Hidalgo J, Sandahl M, Hulteberg CP, Turner C, Lidén G, and Gorwa-Grauslund MF (2016). Biological valorization of low molecular weight lignin. *Biotechnol Adv, 34*: 1318–1346. https://doi.org/10.1016/j.biotechadv.2016.10.001.

Agarwal A, Rana M, and Park JH (2018). Advancement in technologies for the depolymerization of lignin. *Fuel Process Technol, 181*: 115–132.

Anderson EM, Stone ML, Katahira R, Reed M, Muchero W, Ramirez KJ, Beckham GT, and Roman-Leshkov Y (2019). Differences in S/G ratio in natural poplar variants do not predict catalytic depolymerization monomer yields. *Nat Commun, 10*: 2033.

Bandounas L (2011). Isolation and characterization of novel bacterial strains exhibiting ligninolytic potential. *BMC Biotechnol, 11*: 94. https://doi.org/10.1186/1472-6750-11-94.

Bauer S, Sorek H, Mitchell VD, Ibanez AB, and Wemmer DE. (2012). Characterization of Miscanthus giganteus lignin isolated by ethanol organosolv process under reflux condition. *J Agric Food Chem, 60*: 8203–8212.

Beckham GT, Johnson CW, Karp EM, Salvachua D, and Vardon DR (2016). Opportunities and challenges in biological lignin valorization. *Curr Opin Biotech, 42*: 40–53.

Candolle AP (1821). *Regni Vegetabilis Systema Naturale 2*. Argentorati et Londini, Paris, p. 266.

Cao L, Yu IKM, Liu Y, Ruan X, Tsang DCW, Hunt AJ, Ok YS, Song H, and Zhang S (2018). Lignin valorization for the production of renewable chemicals: state-of-the-art review and future prospects. *Bioresour Technol, 269*: 465–475.

Cao L, Zhang C, Chen H, Tsang DCW, Luo G., Zhang S, and Chen J (2017). Hydrothermal liquefaction of agricultural and forestry wastes: State-of-the-art review and future prospects. *Bioresour Technol, 245*: 1184–1193.

Chakar FS and Ragauskas AJ (2004). Review of current and future softwood kraft lignin process chemistry. *Ind Crops Prod, 20*: 131–141.

Chauhan PS (2020). Role of various bacterial enzymes in complete depolymerization of lignin: a review. *Biocatal Agric Biotechnol, 23*: 101498.

Chen Z and Wan C (2017). Biological valorization strategies for converting lignin into fuels and chemicals. *Renew Sust Energ Rev, 73*: 610–621. https://doi.org/10.1016/j.rser.2017.01.166.

Cherubini F (2010). The biorefinery concept: using biomass instead of oil for producing energy and chemicals. *Energy Convers Manag, 51*(7): 1412–1421.

Chio C, Sain MM, and Qin W (2019). Lignin utilization: a review of lignin depolymerization from various aspects. *Renewable Sustainable Energy Rev, 107*: 232–249.

Demirbas A (2009). *Biorefineries: For Biomass Upgrading Facilities*. Springer, Berlin, ISBN: 1848827202.

Doherty W, Mousaviouna P, and Fellows C (2011). Value-adding to cellulosic ethanol: lignin polymers. *Ind Crops Prod, 33*(2): 259–276.

Dutta S, Wu KCW, and Saha B (2014). Emerging strategies for breaking the 3D amorphous network of lignin. *Catal Sci Technol, 4*: 3785–3799.

Duval A and Lawoko M (2014). A review on lignin-based polymeric, micro- and nanostructured materials. *React Funct Polym, 85*: 78–96.

El Mansouri N-E and Salvadó J (2006). Structural characterization of technical lignins for the production of adhesives: application to lignosulphonate, kraft, sodaanthraquinone, organosolv and ethanol process lignins. *Ind Crops Prod*, *24*(1): 8–16.

FitzPatrick M, Champagne P, Cunningham MF, and Whitney RA (2010). A biorefinery processing perspective: treatment of lignocellulosic materials for the production of value-added products. *Bioresour Technol*, *101*(23): 8915–8922.

Freudenberg K and Neish AC (1968). *Constitution and Biosynthesis of Lignin*, Kleinzeller A., Springer G.F., and Whittman H.G. (eds.). Springer-Verlag, New York.

Gosselink RJA, Abächerli A, Semke H, Malherbe R, Käuper P, Nadif A, and van Dam JEG (2004). Analytical protocols for characterization of sulphur-free lignin. *Ind Crops Prod*, *19*(3): 271–281.

Guvenatam B (2015). Catalytic pathways for lignin depolymerization. *PhD Thesis (Research TU/e / Graduation TU/e), Chemical Engineering and Chemistry*. Technische Universiteit Eindhoven.

Holladay JE, Bozell JJ, White JF, and Johnson D (2007a). Top value-added chemicals from biomass. *Vol. 2 Results of Screening for Potential Candidates From Biorefinery Lignin*. U.D.o. Energy, United States of America, p. 79.

Holladay, JE, Bozell, JJ, White, JF, and Johnson, D. (2007b). Top value-added chemicals from biomass. *Volume II- Results of Screening for Potential Candidates from Biorefinery Lignin, A Report*. Available via http://www1.eere.energy.gov/biomass/pdfs/pnnl-16983.pdf.

Joffres B, Lorentz C, Vidalie M, Laurenti D, Quoineaud AA, Charon N, Daudin A, Quignard A, and Geantet C (2014). Catalytic hydroconversion of a wheat straw soda lignin: characterization of the products and the lignin residue. *Appl Catal B Environ*, *145*: 167–176.

Lewis NG (1999). A 20th century roller coaster ride: a short account of lignification. *Curr Opin Plant Biol*, *2*(2): 153–162.

Liu ZH, Olson ML, Shinde S, Wang X, Hao NJ, Yoo CG, Bhagia S, Dunlap JR, Pu Y, Kao KC, and Ragauskas AJ (2017). Synergistic maximization of the carbohydrate output and lignin processability by combinatorial pretreatment. *Green Chem*, *19*: 4939–4955.

Lopez-Camas K, Arshad M, and Ullah A (2020). Chemical modification of lignin by polymerization and depolymerization. *Lignin: Biosynthesis and Transformation for Industrial Applications*, Sharma S. and *Kumar A.* (eds.). Springer, Cham, p. 139–180.

Lu X and Gu X (2022). A review on lignin pyrolysis: pyrolytic behavior, mechanism, and relevant upgrading for improving process efficiency. *Biotechnol Biofuels*, *15*: 106.

Ragauskas AJ, Beckham GT, Biddy MJ, Chandra R, Chen F, Davis MF, Davison BH, Dixon RA, Gilna P, and Keller M (2014). Lignin valorization: improving lignin processing in the biorefinery. *Science*, *344*: 1246843.

Ralph J (1999). Lignin structure: recent developments. *Proceedings of the 6th Brazilian Symposium Chemistry of Lignins and Other Wood Components*, Guaratingueta, Brazil, October, p. 97–112.

Rinaldi R, Jastrzebski R, Clough MT, Ralph J, Kennema M, Bruijnincx PCA, and Weckhuysen BM (2016). Paving the way for lignin valorisation: recent advances in bioengineering, biorefining and catalysis. *Angew Chem Int Ed*, *55*: 8164–8215.

Sarkanen KV and Ludwig CH (1971). *Lignin, Occurrence, Formation, Structure and Reactions*. Wiley/Interscience, New York, p. 95–240.

Schutyser W, Renders T, Van den Bosch S, Koelewijn SF, Beckham GT, and Sels BF (2018). Chemicals from lignin: an interplay of lignocellulose fractionation, depolymerisation, and upgrading. *Chem Soc Rev*, *47*: 852–908.

Shuai L, Amiri MT, Questell-Santiago YM, Héroguel F, Li Y, Kim H, Meilan R, Chapple C, Ralph J, and Luterbacher JS (2016). Formaldehyde stabilization facilitates lignin monomer production during biomass depolymerization. *Science*, *354*: 329–333.

Sjöström E (1982). *Wood Chemistry: Fundamentals and Applications.* Academic Press, p. 223. ISBN 0-12-647480-x.

Ullah M, Liu P, Xie S, and Sun S (2022). Recent advancements and challenges in lignin valorization: green routes towards sustainable bioproducts. *Molecules, 27*: 6055.

Vishtal AG and Kraslawski A (2011). Challenges in industrial applications of technical lignins. *Bioresources, 6*: 3547–3568.

Wen JL, Sun SL, Xue BL, and Sun RC (2013). Quantitative structures and thermal properties of birch lignins after ionic liquid pretreatment. *J Agric Food Chem, 61*: 635–645.

Weng C, Peng X, and Han Y (2021). Depolymerization and conversion of lignin to value-added bioproducts by microbial and enzymatic catalysis. *Biotechnol Biofuels, 14*: 84. https://doi.org/10.1186/s13068-021-01934-w.

Wu X, Fan X, Xie S, Lin J, Cheng J, Zhang Q, Chen L, and Wang Y (2018). Solar energy-driven lignin-first approach to full utilization of lignocellulosic biomass under mild conditions. *Nat Catal, 1*: 772–780.

Xu Z, Lei P, Zhai R, Wen Z, and Jin M (2019). Recent advances in lignin valorization with bacterial cultures: microorganisms, metabolic pathways, and bio-products. *Biotechnol Biofuels, 12*: 32. https://doi.org/10.1186/s13068-019-1376-0.

Zakzeski J (2010). The catalytic valorization of lignin for the production of renewable chemicals. *Chem Rev, 110*: 3552–3599. https://doi.org/10.1021/cr900354u.

2

Isolation of Lignin

Abstract

Lignin is produced by the cell walls of plants, agricultural crops, and wood. Between 15 and 40% of the dry matter in woody plants is made up of lignin, which is mainly a structural material that gives cell walls more strength and rigidity. Compared to cellulose and other structural polysaccharides, lignin is more resistant to the majority of biological attacks. Currently, chemical pulping of wood yields lignin. However, a number of biomass refineries are now operational. P-coumaryl, coniferyl, and sinapyl alcohols are the three basic phenylpropane units that make up lignin's chemical structure. The pulping process produces the technical lignins as a byproduct. The most commonly used commercial lignins are kraft lignin and lignosulfonates. Organosolv lignins are another type of lignin. Isolation and characterization of different types of lignin are presented in this chapter.

Keywords   Lignin; Technical lignin; Kraft lignin; Lignosulfonates; Organosolv lignins; Soda lignin; Steam explosion lignin; Phenylpropane units; Monolignols; Sinapyl alcohol; Coniferyl alcohol; p-coumaryl alcohol

Currently, lignin is produced through the chemical pulping of wood. Due to the fact that numerous biomass refineries are currently in operation, the lignin that is produced as a byproduct of the production of cellulosic ethanol would be an excellent feedstock for the production of products with added value. Melt-spinnable lignins made by oganosolv pulping are simple to make. Compared to lignins obtained through chemical pulping, these lignins are purer. The biosphere contains more than 300 billion tonnes of lignin, which grows by approximately 20 billion tonnes per year.

Lignin is mostly found in the cell walls of club mosses, ferns, and vascular plants (Akin and Benner, 1988; Baurhoo et al., 2008; Gregorováa et al., 2006; Kirk, 1971; Matsushita, 2015; McCrady, 1991; Miidla, 1980; Piló-Veloso et al., 1993; Rosas et al., 2014; Souto et al., 2018). In contrast to polysaccharides, which have a clear structure, lignin has unique properties like being aromatic and having less oxygen. Because of these properties, lignin is a desirable feedstock for the transformation of it into useful chemicals or materials, as well as renewable chemical building blocks. However, the potential applications of lignin are highly dependent on its availability, a thorough understanding of the source of lignin, method of isolation, and the anticipated applications' technical requirements.

Lignin is an organic polymer with three dimensions. It creates crucial structural components for vascular plants' support tissues. It is primarily found in woody plants, where it is more complex and heavily polymerized. Lignin is absorbed into the wood's cellulose walls. The term for this process is lignification. It gives trees more rigidity and significantly increases the cell's strength and hardness. According to Rouhi and Washington (2001), this is crucial to enable woody plants to stand straight and vertical. According to Nordström (2012), the complex structure of natural lignin, which is found in a variety of plants, includes both aromatic and aliphatic components. Even though information about lignin has been available for more than a century, its significance has generally been recognized since the early 1900s (Glasser et al., 2000).

The complex structure of lignin limits our understanding of it. The field of lignin has seen significant growth in recent years as new chemical analysis techniques have been implemented. Because of this, we now know about the structure of lignin and how it can be used. Lignin is a random, three-dimensional network polymer made of phenylpropane units that are linked in different ways.

Plant fibers are held together by mechanical supports, which is made possible by lignin. Lignin is also important in the transportation of water and nutrients because it reduces the amount of water that penetrates the xylem's cell walls. Lastly, because it prevents destructive enzymes from passing through the cell wall, lignin is a key component of a plant's natural defense mechanism against deterioration (Bajpai, 2017; Sarkanen and Ludwig, 1971; Sjöström, 1993).

Figure 2.1 shows a typical structural model of lignin (Lu et al., 2017). Sinapyl, p-coumaryl, and coniferyl alcohols give lignin its three fundamental phenylpropane units (Figure 2.2). Figure 2.3 shows common linkages found in lignin.

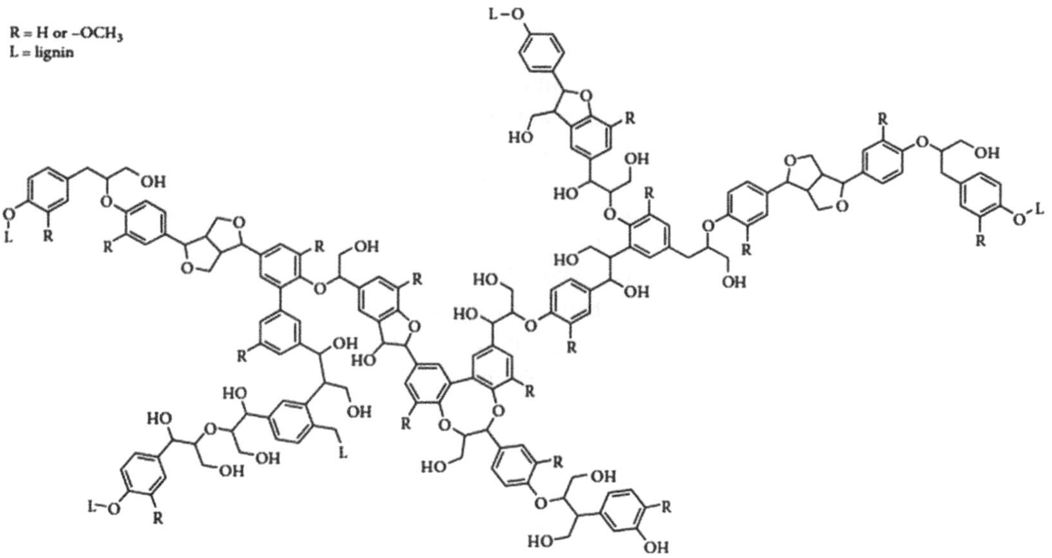

Figure 2.1 Typical structural model of lignin. Lu et al., 2017 / John Wiley & Sons / Public Domain CC BY 4.0.

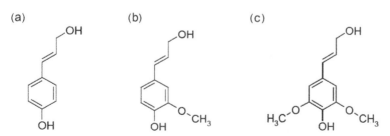

Figure 2.2 Monolignol monomer species. (a) *p*-coumaryl alcohol (4-hydroxyl phenyl, H), (b) coniferyl alcohol (guaiacyl, G), (c) sinapyl alcohol (syringyl, S). Based on Bajpai (2017); Ekielski and Mishra (2020).

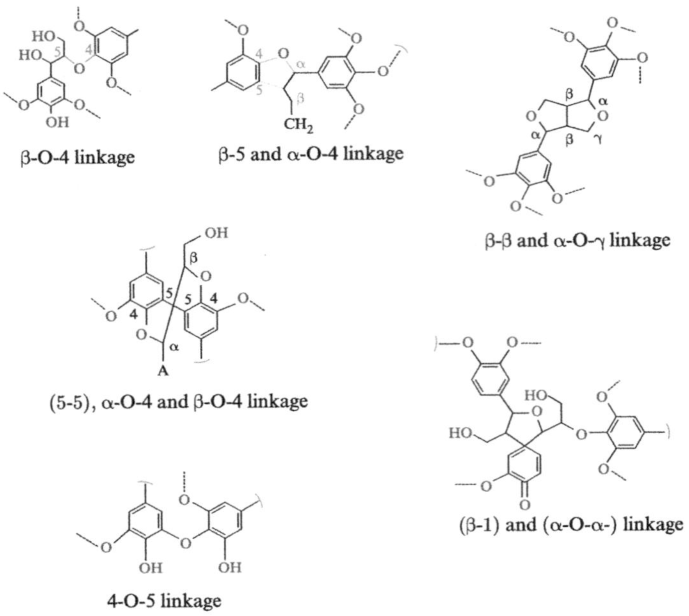

Figure 2.3 Common linkages found in lignin. Ekielski and Mishra (2020) / MDPI / Public Domain CC BY 4.0.

During biological lignification, radical coupling reactions join these to form a complex three-dimensional macromolecule. The important linkages include β-O-4, β-5, β-b, and 5–5 linkages among others. The structure and quantity of lignin found in different species vary. Coniferyl and sinapyl alcohols, for instance, can be found in small amounts in hardwood lignin; coniferyl alcohol is the primary component of softwood lignin. Sinapyl, p-coumaryl, and coniferyl can be found in grass lignin (Matsushita, 2015). Hardwood and softwood lignin's structures are depicted in Figures 2.4 and 2.5 (Gargulak and Lebo, 1999; Nimz, 1974; Zakzeski et al., 2010).

Lignin is produced by the cell walls of plants, agricultural crops, and wood. In plants, cellulose and lignin work together to serve a structural purpose. Additionally, it serves as a strong defence against insect and fungal assaults. The structure and composition of lignin is found to vary. It

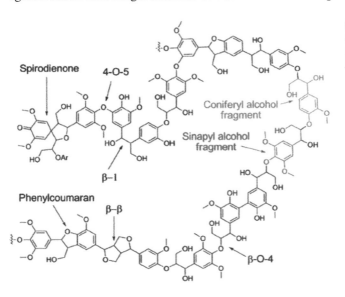

Figure 2.4 Schematic representation of hardwood lignin. Reproduced with permission Zakzeski et al., 2010 / American Chemical Society.

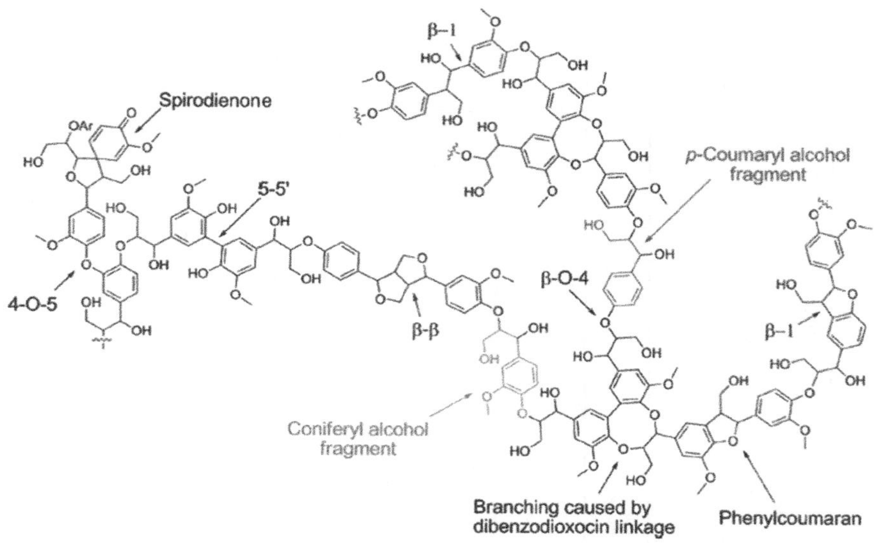

Figure 2.5 Schematic representation of softwood lignin. Reproduced with permission Zakzeski et al., 2010 / American Chemical Society.

depends upon the species of tree or plant, the time of year, the climate, and the age of the plant. In terms of p-hydroxyphenyl (H), guaiacyl (G), and syringyl (S) composition, as well as the relative abundance of chemical linkages in the polymer, the composition of lignin is highly dependent on the two plant species. Table 2.1 shows different types of lignin, their monomer's molecular weight, and lignin content.

Almost always, the process of separating biomass into its component parts will have a significant impact upon the molecular structure of lignin, resulting in the release of lignin (technical lignins). Consequently, the technical lignins have distinct characteristics (Table 2.2). Figure 2.6 shows simplified and representative structures of common technical lignins.

Table 2.1 Different types of lignin, their monomer's molecular weight, and lignin content.

	Types of Lignin	Source	Monomer Molecular Weight (g mol^{-1})	(mmol g^{-1}) Lignin Content	Chemicals/catalysts
Sulfur process	Kraft lignin	Wood chips, softwoods, hardwoods	2000–3000	1.25	NaOH, Na$_2$S
	Lignosulfonates	Softwoods, hardwoods, annual plants	20 000–50 000	1.25–2.5	Ca(HSO$_3$)$_2$ or Mg(HSO$_3$)$_2$
Sulfur free process	Organosolv lignin	Hardwoodm, Softwood, and wheat straw	2000–5000	0	Methanol, ethanol, various bronsted acid catalysts (H$_2$SO$_4$)
	Alkali/soda lignin	Hardwood, bagasse, wheat straw, and flax	5000–6000	0	NaOH, NH$_4$OH, Ca(OH)$_2$

Based on Aro and Fatehi, 2017; Basakçılardan Kabakcı and Tanis, 2021; Kim et al., 2016; Ullah et al., 2022; Xu et al., 2020; Yoo et al., 2020.

Table 2.2 Characterization of technical lignins.

Characterization of kraft lignin

Species	Hydroxyl group		Molecular weight			References
	Total (mmol g^{-1})	Phenolic (mmol g^{-1})	Mw (×103)	Mn (×103)	Mw Mn^{-1}	
Softwood	6.5–8.6	2.7–3.5	1.1–45.7	0.5–7.7	2.2–13.4	(El Mansouri and Salvadó, 2006, 2007; Ekeberg et al., 2006; Mansson, 1983; Ponomarenko et al., 2014)
Hardwood	6.5–8.4	4.3–4.7	2.4–4.8	0.4–1.3	1.8–12.0	(Mansson, 1983; Pan and Saddler, 2013; Ponomarenko et al., 2014)

Characterization of Lignosulfonate

Species	Hydroxyl group		Molecular weight			References
	Total (mmol g^{-1})	Phenolic (mmol g^{-1})	Mw (×103)	Mn (×103)	Mw Mn^{-1}	
Softwood	Not available	1.2–1.9	10.5–60.2	2.7–6.5	6.7–22.3	(Alonso et al., 2001; El Mansouri and Salvadó, 2007; Ekeberg et al., 2006)
Hardwood	Not available	1.4–1.5	6.9–7.8	2.4–4.6	1.7–3.0	(Alonso et al., 2001; Ekeberg et al., 2006; Ye et al., 2013; Zhou et al., 2013)

Characterization of ogranosolv lignin

	Species	Hydroxyl group		Molecular weight			References
		Total (mmol g^{-1})	Phenolic (mmol g^{-1})	Mw (×103)	Mn (×103)	Mw Mn^{-1}	
(ethanol)	Softwood	6.3–10	2.7–3.1	2.9–5.4	1.8–3.1	1.6–1.8	(Pan et al., 2005; Sannigrahi et al., 2010)
	Hardwood	5.7	2.8	2.0–2.6	1.3–1.6	1.5–1.6	(Pan et al., 2005; Pan and Saddler, 2013)
(Formic acid)	Miscanthus	3.7–4.9	1.6–2.4	2.8	1.1	2.5	(El Mansouri and Salvadó, 2006, 2007; El Mansouri et al., 2012)
(Acetic acid)	Hardwood	5.5–5.8	3.5–4.0	0.9	Not available	Not available	(Benar et al., 1999; Kin, 1990)
(Acetic acid/ formic acid)	Wheat straw	3.4	1.0	2.2	1.6	1.3	(Delmas et al., 2011)

Bajpai (2021) / with permission of ELSEVIER.

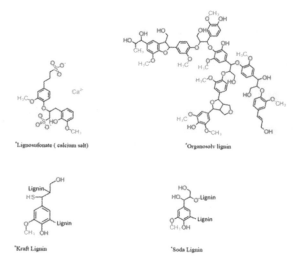

Figure 2.6 Simplified and representative structures of common technical lignins. Ekielski and Mishra (2020) / MDPI / Public Domain CC BY 4.0.

The various pre-treatment-derived lignins differ in terms of physicochemical properties and their chemical structure.

Presently, lignin is available in large quantities and can be transformed into a variety of other raw materials through fractionation, purification, and chemical alterations (functionalized).

Table 2.3 presents the characteristics of the technical lignin. Table 2.4 lists some of the biggest companies that make these products.

Table 2.3 Characteristics of the technical lignin.

Molecular weight
Water solubility
Degree of contamination (example remaining covalently bound sugar residues or incorporation of non-native elements, such as sulfur)
Extent of condensation
Functional group decoration of the macromolecule.

Bajpai (2021) / with permission of ELSEVIER.

Table 2.4 Some of the major manufacturers of lignins.

Alberta Pacific
Borregaard LignoTech,
CIMV
Domtar
Domsjö
Tembec
UPM
Weyerhaeuser

Bajpai (2021) / with permission of ELSEVIER.

The pulping process produces the technical lignins as a byproduct. The most commonly used commercial lignins are kraft lignin and lignosulfonates. Organosolv lignins are another type of lignin. These lignins are made by pulping materials with organic solvents like formic acid, acetic acid, and ethanol among others. The pulping method determines how the lignin is structured. Additionally, different lignins have distinct functional groups and molecular weights. Industrial uses of lignin are constrained by type of phenylpropane units, its functional groups, molecular weight distributions, and linkage between structural units (Matsushita, 2015). The technical lignins differ greatly in physical characteristics such as solubility, hydrophilicity, and hydrophobicity, as well as in molecular structure, weight, and chemical composition (including impurities). The strategies that are viable for further valorization will largely be determined by these characteristics (Bruijnincx et al., 2016).

2.1 Lignosulfonates

Lignosulfonates are produced by sulfite pulping processes. The liquor used to cook the wood is made by combining sulfur dioxide and an aqueous base. Sulfuric acid is produced when sulfur dioxide reacts with water to form sulfur dioxide, which then breaks down and sulfonates the lignin by substituting a hydroxyl group for a sulfonate one. Because of this, the lignin can be solubilized and removed from the cellulose without getting precipitated. The sugars in the spent sulfite liquor—mostly monosaccharides—must be destroyed before the lignosulfonate can be used as a concrete additive to reduce water content (Niaounakis, 2015).

In the lignosulfonate process, sulfite with either magnesium or calcium as the counterion is used. The process takes place over a broad pH range, from 2 to 12. Water and some organics and amines with high polarities dissolve the product. Due to the inclusion of sulfonate groups on the arenes, the lignosulfonate produced by the sulfite process has a higher average molecular weight and monomer molecular weight than kraft lignin. Lignosulfonates are distinct due to the sulfonate groups that are primarily introduced in the α-position of the propyl side chain. Kraft lignins, on the other hand, have a sulfonate on the aromatic ring and are sulfonated.

Consequently, the resulting lignin is water-soluble, distinguishing it from other technical lignins. Lignosulfonates are used as adhesives, stabilizers, dispersants, and surfactants due to the unique colloidal properties they possess thanks to their high density of functional groups. The lignosulfonates have a generally higher sub-atomic weight and higher ash content and still contain a significant amount of carbohydrates. Hardwoods and softwoods both contain lignosulfonates, which can be purchased commercially (www.dutchbiorefinerycluster.nl).

2.2 Kraft Lignin

Sodium sulfide and sodium hydroxide are used in the kraft pulping method of alkaline pulping. It has control of 96% of the market. Phenols and lignins with high and low molecular weights that are bound to carbohydrate residues give soda and kraft process lignin its highly polydisperse nature. The native lignin structure is substantially destroyed during kraft pulping. In contrast to soda lignin, kraft lignin contains sulfur. Since the black liquor is burned to produce energy and recover chemicals, it is not free.

The lignin that is produced in the soda and kraft processes is characterized by greater lignin fragmentation. The production of carbon fiber from kraft lignin in the past was unsuccessful; particularly the softwood kraft lignin, which only produces char on heating (Kubo et al., 1977, 1996, 1997; Kubo and Kadla, 1987). The lack of a primary lignin fraction with plasticizing and softening properties could be the cause of this behaviour. Utilizing highly purified hardwood lignin, carbon fiber was successfully produced from industrial kraft lignin (Kadla et al., 2002). The lignin was first treated with heat for 60 minutes at 145 °C in a vacuum to make fibers. The molecular weight was increased and the volatile components of the lignin were removed.

The lignin's spinnability increased when a small amount of poly(ethylene oxide) was added as a plasticizer. In addition, spinning ought to be feasible at a lower temperature than when using only lignin. When poly(ethylene oxide) was added at a concentration greater than 10%, the lignin fibers self-fused. Due to carefully monitored thermostabilization conditions, such as a very slow increase ($12\,°C\,h^{-1}$) in temperature of the lignin fiber to 250 °C, the ensuing carbonization enhanced the strength characteristics as shown in Table 2.2. Lignin was thermostabilized at 250 °C for 60 minutes in air before being carbonized at 1000 °C to produce carbon fiber. Before thermo-stabilization and carbonization, 5% poly(ethyleneterephtalate) could be added to the lignin to further enhance its strength properties (Kubo and Kadla, 2005; www.benthamscience.com).

Technical kraft lignin was first extracted from black liquor in 1942 by the MeadWestvaco Corporation, the largest kraft lignin producer in the world. It was the only factory in the world by 2011 that offered technical lignin for sale to businesses. Later, Innventia and Chalmers University Technology created the LignoBoost® process, which reduces the amount of ash and carbohydrate in the recovered lignin to produce technical kraft lignin with fewer impurities. Metso Co. bought the technology. In Canada FP Innovation also developed a low-ash, low-carbohydrate, and low-sulfur recovery process called Lignoforce®. GreenValue in Switzerland also offers commercially available, low impurity, non-wood, technical soda lignin (Gosselink, 2011; Smolarski, 2012; Zhu, 2013; iopscience.iop.org).

Extraction of highly pure lignin from a kraft pulp mill forms the basis of the LignoBoost technology, which was developed by Innventia and Chalmers University of Technology in Sweden. Carbon dioxide is used for lowering the pH of the black liquor in order to precipitate the lignin. A filter press is used to extract water from the precipitate.

Problems with sodium separation and conventional filtration are addressed by redissolving the lignin in spent wash water and acid. In order to produce pure lignin, water is removed from the resulting slurry once more and washed with acidified water. When acidified, carboxylic acids and all phenols undergo protonation. The result is pure lignin containing between 2 and 3 weight % sulfur and little to no contamination from carbohydrates or ash. The lignin is chemically linked to about half of the sulfur.

Metso expanded the technology after acquiring it in its entirety in 2008. At the end of 2011, Metso sold Domtar its first commercial LignoBoost technology plant. The plant is located in a pulp mill in Plymouth, North Carolina, USA. It produced 25 000 tons of kraft lignin in 2013. A second plant in Sunila, Finland, which was acquired by Stora Enso was established and started producing 50 000 tonnes of dried lignin annually in the third quarter of 2015. Mead-Westvaco produces a similar lignin for commercial use in Charleston, USA. Approximately 30 000 tonnes are produced annually (www.dutchbiorefinerycluster.nl).

There is a growing number of commercial lignin suppliers. However, some purity is required for high-value applications, so numerous technologies are being developed. Commercially available lignosulphonates are also available. However, due to the presence of impurities, they are utilized for low-value tasks (Chen, 2014; Gosselink, 2011). Kraft lignin has a very condensed structure, many C–C bond links that will not break (like biphenyl and methylene-bridged ones), and a few easy-to-break ether bonds.

Covalently, sulfur species, especially thiols, are also incorporated into the structure making them significant impurities that may prevent subsequent valorization (for sulfur-based catalytic depolymerization, a known poison for many metal catalysts or in material applications). Softwoods and hardwoods are used in the commercial production of kraft lignins. These lignins possess a high hydroxyl content.

2.3 Soda Lignin

Soda lignins are distinct from lignosulfonates and kraft lignins in that they do not contain any sulfur-containing reagents. Lignin becomes resistant and condensed as a result of the pulping process's relatively harsh conditions. Soda lignins, like kraft lignins, have a low to moderate purity and only a small amount of ash and carbohydrates. Vinyl ethers are present in soda lignins, in contrast to kraft lignins and lignins produced under acidic conditions. Both hardwoods and annual crops are used for the commercial production of soda lignins.

2.4 Steam Explosion Lignin

Steam explosion involves pre-treatment of woody biomass for a brief period of time with steam at high pressure and a temperature of 200 °C or higher, followed by rapid decompression (Krutov et al., 2017; Palmqvist et al., 1996; Soederstroem et al., 2004; Stenberg et al., 1998). The material obtained after explosion is extracted using either an organic solvent or an aqueous alkali, and the lignin, a byproduct with few impurities (carbohydrates and wood extractives), is obtained. Steam explosion lignin is more like native lignin in terms of the quantity and composition of functional groups than any other produced technical lignin. However, the molecular weight of the lignin decreases significantly.

The biomass is rapidly decompressed after treatment with steam to temperatures of 200–220 °C during the steam explosion process. After the explosion of wood, individual fibers and fiber bundles are produced. Extraction with aqueous alkali or an organic solvent can be used for isolation of lignin in large quantities from hardwoods (Josefsson et al., 2002; Robert et al., 1988). A green fiber was produced through the hydrogenation of the lignin, chloroform and carbon disulfide extractions, heat treatment, and melt spinning. The fibers were thermostabilized in air at 210 °C and carbonized at 1000 °C to produce carbon fiber for general purpose (Sudo and Shimizu, 1992). Table 2.2 displays the mechanical characteristics of these fibers. Similar types of lignin were also melt spun, subjected to phenolysis, purification, heat treatment at 280 °C in a vacuum, stabilized in air at 300 °C, and carbonized once more at 1000 °C, but there was no change in strength properties (Table 2.2). However, the yield of purified lignin was much higher (Sudo et al., 1993).

Shimizu et al. (1998) carried out additional research employing the steam explosion method. They investigated a wide variety of softwoods and hardwoods and studied the effects of the process parameters on the production of lignin that is suitable for carbon fiber conversion. During the steam treatment of biomass, the lignin undergoes degradation and condensation reactions due to the acidic environment. It continues via a common carbonium ion intermediate.

Along these lines, the internal nucleophilic centers in the lignin will oppose the expansion of a low sub-nuclear weight nucleophile such as phenol. Degradation emerges as the primary structural change as a result (Li et al., 2007).

The distribution of molecular weights in the organosolv process is comparable to this one. Steam explosion with sulfur dioxide pre-impregnation yields alkaline extractable lignin from hardwood with a high fractionation efficiency and greater efficacy than steam explosion without pre-impregnation (Li et al., 2009). Steam explosion could not produce high efficiency for softwoods, regardless of preimpregnation. The two-step steam explosion method is comparable to the one-step method in terms of efficiency.

2.5 Organosolv Lignins

Organosolv pulping is a type of treatment for lignocellulosic biomass in which water and an organic solvent are combined, frequently at high temperatures. This fractionation technology makes it relatively simple to isolate the lignin. Organosolv processing frequently makes use of methanol, ethanol, acetone, organic acids such as formic acid and acetic acid, or combinations of these acids and cyclic ethers. As a result, the cellulose is delignified and the lignin is dissolved in the extraction solvent. The hemicellulose fraction is either intentionally added or produced in situ as the acid catalyzes the process (www.dutchbiorefinerycluster.nl).

In 1893, the first study of fractionating wood with organic solvents was done. Hydrochloric acid and ethanol were used in this sudy (Sidiras and Salapa, 2015). The process was the subject of extensive research at the end of the 1960s (Rinaldi et al., 2016).Two commercially viable plants were in operation in 1992. These were Organocell, Acetosolv, ASAM, and Milox (Muurinen et al., 2000).

Cosolvents such as THF-water (tetrahydrofuran) and GVL-water (γ-valerolactone) are the subject of numerous studies right now. The ability of these two solvents to partially dissolve and decrystallize crystalline cellulose into sugar platform vaporization is the reason for their increasing use (Smith et al., 2017). The application of organosolv is anticipated to increase in the coming years as biorefineries emerge, as its activity has significantly increased over the past 50 years (Rinaldi et al., 2016).

Aqueous ethanol is used in the Alcell process. It was created by Repap Ventures and marketed in 1989. Aqueous ethanol is used for delignification in this procedure. The cooking liquor has a low pH because hemicellulose produces organic acids. Buildup responses between the alpha-position of the side chain and the 6-position of another aromatic ring can happen in protolignin (Pye and Lora, 1991; Sarkanen, 1990; Shimada et al., 1997).

31P NMR was used to investigate the condensed phenolic groups in solubilized and residual lignin during the Alcell process. Alcell lignin (solubilized lignin) had a lower condensed phenolic hydroxyl content than kraft lignin because some of the protolignin was condensed during the Alcell process, which caused problems with solubility and pulp retention.

Using acetic acid, acetosolv lignins from grass, softwood, and hardwood can be obtained. Sulfuric acid, for example, is a powerful acid that catalyzes extensive and selective delignification (Davis et al., 1986; Kin, 1990; Liu et al., 2000; Parajó et al., 1993; Sano et al., 1989, 1990; Shukry et al., 2008; Young and Davis, 1986).

Utilizing model compounds, the structure and delignification mechanism of acetic acid lignin were established (Davis et al., 1987; Yasuda and Ito, 1987; Yasuda, 1988). Hydrolysis, homolytic cleavage of β-aryl ethers, acetylation of hydroxyl groups, acidolysis, and formaldehyde elimination at the c-position are among the reactions. During acetic acid pulping, there are also condensations between and within molecules. A method for delignification making use of formic and acetic acid was developed at the beginning of the 1900s (Erismann et al., 1994; Freudenberg 1959).

A mixture of water, formic acid, and acetic acid was utilized in the pilot-scale CIMV process to produce Biolignin™ from wheat (Ekielski et al., 2021). This lignin had a few free hydroxyl groups, a low polydispersity, and a low molecular weight. The processing conditions under which organosolv lignins are produced have a significant impact on their structure. Organosolv lignins are crystal clear, typically containing very few carbohydrates and ash impurities. It is generally incorrect to assume that organosolv lignins are also the most structurally similar to native lignin and still contain a higher percentage of aryl ether linkages that are simple to break. However, this only applies to organosolv lignins produced in a controlled environment in the laboratory. Organosolv lignins, like kraft or soda lignins, can be just as chemically resistant despite having a structure that is more homogenous overall and having a lower molecular weight as well as polydispersity. Hardwoods and softwoods are used in the pilot and demonstration plants that produce organosolv lignins (www.dutchbiorefinerycluster.nl; Nimz and Casten, 1986; Williamson, 1987).

Alcell lignin's spinnability was improved when polyethylene oxide was used as a plasticizer. However, the thermo-stabilization in air had to be performed at a very low heating rate in order to prevent the filaments from self-fusing. Compared to other kinds of carbon fiber made with organosolv lignins, the mechanical properties of the carbon fiber improved after it was carbonized (Table 2.2) (Kadla et al., 2002).

After being subjected to thermal treatment at lower pressure to alter the structure of the lignin, the spinnability of the acetic acid lignin produced by acetic acid pulping of birch wood improved. The average molecular weight increased, but the amount of methoxyl and acetyl groups stayed the same. Thermo-stabilization in air at 250 °C and carbonization at 1000 °C produced carbon fiber with the strength properties shown in Table 2.2 (Uraki et al., 1995; Gellerstedt et al., 2010). On the other hand, fusible lignin, was not produced by a comparable method employing softwood. Carbon fiber could only be made by fractionating lignin to get rid of materials with higher molecular weights. However, the carbon fiber that was produced had a lower strength than the fibers that were made from hardwoods (Table 2.2) (Kubo et al., 1977, 1996, 1998).

Hydrotropic agents are salts that, when present in high concentration, significantly increase the aqueous solubility of poorly soluble substances. Separating lignin also requires high-temperature concentrated aqueous solutions of hydrotropic agents (Gabov et al., 2014). From the 1950s to the 1980s, this process was studied as an alternative to the traditional sulfite and kraft pulping processes (Gabov et al., 2013). According to Willför and Gustafsson (2010), this method is simpler than kraft pulping, has a higher cellulose yield, lower capital costs, and heat savings. Hydrotropic pulping produces precipitated lignin that is more pure and can be used to make other chemical products, in contrast to kraft and sulfite pulping, which use contaminating inorganic chemicals (U.S.Congress, 1989). According to Gabov et al. (2014), hydrotropic lignin resembles organosolv lignin, which is of high quality.

However, the primary reason for the industry's lack of interest is that the hydrotropic process is not suitable for softwoods. Because of the rising interest in bio-based items, this option may be reconsidered in light of ongoing endeavors to develop woody yields like poplar and willow for energy. Additionally, efforts in this direction are ongoing, and successful research has been conducted on the application of ionic liquids to non-woody biomass, softwoods, and hardwoods (Cláudio et al., 2015; Mäki-Arvela et al., 2010; Muhammad et al., 2012; Tan et al., 2009).

Table 2.5 (Mäki-Arvela et al., 2010) demonstrates the purity and sulfur content of various lignins. The distinct physical appearances of a number of common technical lignins are depicted in Figure 2.7.

Table 2.5 Sulfur content and purity of different types of lignins.

Type of lignin	Scale of production	Separation method	Pre-treatment chemistry	Sulfur content	Purity
Kraft	Industrial	Precipitation (pH change) or ultrafiltration	Alkaline	Moderate	Moderate
Soda	Industrial	Precipitation (pH change) or ultrafiltration	Alkaline	Free	Moderate–low
Lignosulfonate	Industrial	ultrafiltration	Acid	High	Low
Organosolv	Pilot/demo	Dissolved air flotation, precipitation (addition of non-solvent)	Acid	Free	High
Hydrolysis	Industrial/pilot	Acid		Low-free	Moderate–low
Steam explosion	Demo/pilot	Acid		Low-free	Moderate–low
AFEX	Pilot	Alkaline		Free	Moderate–low

Based on (Bruijnincx et al., 2016; Mood et al., 2013)

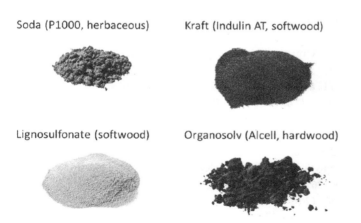

Figure 2.7 Photographs showing the differences in physical appearance of a number of typical technical lignins (Bruijnincx et al., 2016). Reproduced with permission.

Bibliography

Akin DE and Benner R (1988). Degradation of polysaccharides and lignin by ruminal bacteria and fungi. *Appl Environ Microbiol*, 54: 1117–1125.

Alonso MV, Rodrı́guez JJ, Oliet M, Rodrı́guez F, Garcı́a J, and Gilarranz MA (2001). Characterization and structural modification of ammonic lignosulfonate by methylolation. *J Appl Polym Sci*, 82: 2661–2668.

Aro T and Fatehi P (2017). Production and application of lignosulfonates and sulfonated lignin. *ChemSusChem*, 10: 1861–1877.

Bajpai (2021). *Carbon fiber, Chapter 3-Types of lignins and characteristics*. Elsevier, p. 51–66.

Bajpai P (2017). Carbon fibre from Lignin. *Springer Briefs in Material Science*, Springer (Springer Nature).

Basakçılardan Kabakcı S and Tanis MH (2021). Pretreatment of lignocellulosic biomass at atmospheric conditions by using different organosolv liquors: a comparison of lignins. *Biomass Convers Biorefinery*, 11: 2869–2880.

Baurhoo B, Ruiz-Feria CA, and Zhao X (2008). Purified lignin: nutritional and health impacts on farm animals–A review. *Anim Feed Sci Technol*, 144: 175–184.

Benar P, Gonccalves AR, Mandelli D, and Schuchardt U (1999). Eucalyptus organosolv lignins: study of the hydroxymethylation and use in resol. *Bioresour Technol*, 68: 11–16.

Bruijnincx P, Weckhuysen B, Gruter G, and Engelen-Smeets E (2016). *Lignin Valorisation: The Importance of a Full Value Chain Approach*. p. 22. https://www.dutchbiorefinerycluster.nl/.../Lignin_valorisation_-_APC_June_2016.pdf (accessed April, 2022).

Chen MCW (2014). *Commercial viability analysis of lignin based carbon fibre*. Master Dissertation. Simon Fraser University, Burnaby, Canada.

Cláudio A, Neves M, Shimizu K, Lopes J, Freire M, and Coutinho J (2015). The magic of aqueous solutions of ionic liquids: ionic liquids as a powerful class of Catanionic hydrotropes. *Green Chem*, 17: 3948–3963.

Davis JL, Nakatsubo F, Murakami K, and Umezawa T (1987). Organic acid pulping of wood IV: reactions of arylglycerol- β-guaiacyl ethers. *Mokuzai Gakkaishi*, 33: 478–486.

Davis JL, Young RA, and Deodhar SS (1986). Organic acid pulping of wood III. Acetic acid pulping of spruce. *Mokuzai Gakkaishi*, 32: 905–914.

Delmas GH, Benjelloun-Mlayah B, Bigot YL, and Delmas M (2011). Functionality of wheat straw lignin extracted in organic acid media. *J Appl Polym Sci*, 121: 491–501.

El Mansouri NE and Salvadó J (2006). Structural characterization of technical lignins for the production of adhesives: application to lignosulfonate, kraft, soda-anthraquinone, organosolv and ethanol process lignin. *Ind Crop Prod*, 24: 8–16.

El Mansouri NE and Salvadó J (2007). Analytical methods for determining functional groups in various technical lignins. *Ind Crop Prod*, 26: 116–124.

El Mansouri NE, Vilaseca J F, and Salvadó J (2012). Structural changes in organosolv lignin during its reaction in an alkaline medium. *J Appl Polym Sci*, 126: E213–E220.

Ekeberg D, Gretland KS, Gustafsson J, Bra°ten SM, and Fredheim GE (2006). Characterisation of lignosulphonates and kraft lignin by hydrophobic interaction chromatography. *Anal Chim Acta*, 565: 121–128.

Ekielski A and Mishra PK (2020). Lignin for bioeconomy: the present and future role of technical lignin. *Int J Mol Sci* Dec 23, 22(1): 63. https://doi.org/10.3390/ijms22010063.

Ekielski A and Mishra PK (2021). Lignin for bioeconomy: the present and future role of technical lignin. *Int J Mol Sci*, 22(1): 63. https://doi.org/10.3390/ijms22010063.

Erismann NM, Freer J, Baeza J, and Durán N (1994). Organosolv pulping VII: delignification selectivity of formic acid pulping of Eucalyptus grandis. *Bioresour Technol*, 47: 247–256.

Freudenberg K (1959). Biosynthesis and constitution of lignin. *Nature*, 183: 1152–1155.

Gabov K, Fardim P, and da Silva Ju´nior FG (2013). Hydrotropic fractionation of birch wood into cellulose and lignin: a new step towards green biorefinery. *BioResources*, 8(3): 3518–3531.

Gabov K, Gosselink RJ, Smeds AI, and Fardim P (2014). Characterization of lignin extracted from birch wood by a modified hydrotropic process. *J Agric Food Chem*, 62(44): 10759–10767.

Gargulak JD and Lebo E (1999). *Commercial use of lignin-based materials, volume 742 of ACS symposium series.* American Chemical Society, pp. 304–320.

Gellerstedt G, Sjöholm E, and Brodin I (2010). The wood-based biorefinery: a source of carbon fiber? *Open Agr J*, 3: 119–124 OA.

Glasser WG, Northey RA, and Schultz TP (2000). *Lignin: Historical, Biological, and Materials Perspective.* American Chemical Society, Washington, DC.

Gosselink RJA (2011). *Lignin as a renewable aromatic resource for the chemical industry.* Doctoral Thesis Wegeningen University, Wegeningen, Netherlands.

Gregorováa A, Košíkováa B, and Moravčíkb R (2006). Stabilization effect of lignin in natural rubber. *Polym Degrad Stab*, 31: 229–233.

Josefsson T, Lennholm H, and Gellerstedt G (2002). Steam explosion of aspen wood. Characterisation of reaction products. *Holzforschung*, 56: 289–297.

Kadla JF, Kubo S, Venditti RA, Gilbert RD, Compere A, and Griffith W (2002). Lignin-based carbon fibers for composite fiber applications. *Carbon*, 40: 2913–2920.

Kim JS, Lee YY, and Kim TH (2016). A review on alkaline pretreatment technology for bioconversion of lignocellulosic biomass. *Bioresour Technol*, 199: 42–48.

Kin Z (1990). The acetolysis of beech wood. *Tappi J*, 73: 237–238.

Kirk TK (1971). Effects of microorganisms on lignin. *Annu Rev Phytopathol*, 9: 185–210.

Krutov S, Ipatova E, and Vasilyev A (2017). Steam explosion treatments of technical hydrolysis lignin. *Holzforschung*, 71(7–8): 571–574.

Kubo S, Ishikawa N, Uraki Y, and Sano Y (1977). Preparation of lignin fibers from softwood acetic acid lignin. Relationship between fusibility and the thermal structure of lignin. *Mokuzai Gakkaishi*, 43: 655–662.

Kubo S, Ishikawa N, Uraki Y, and Sano Y (1997). Preparation of lignin fibres from softwood acetic acid lignin: by atmospheric acetic acid pulping. *Mokuzai Gakkaishi*, 43: 655–662.

Kubo S and Kadla JF (1987). Lignin-based carbon fibers: effect of synthetic polymer blending on fiber properties. *J Polym Environ* 2005, 13: 97–105.

Kubo S and Kadla JF (2005). Lignin-based carbon fibres: effect of synthetic polymer blending on fibre properties. *J Polym Environ*, 13(2): 97–105.

Kubo S, Uraki Y, and Sano Y (1996). Thermomechanical analysis of isolated lignins. *Holzforschung*, 50: 144–150.

Kubo S, Uraki Y, and Sano Y (1998). Preparation of carbon fibers from softwood lignin by atmospheric acetic acid pulping. *Carbon*, 36: 1119–1124.

Li J, Gellerstedt G, and Toven K (2009). Steam explosion lignins; their extraction, structure and potential as feedstock for biodiesel and chemicals. *Bioresour Technol.*, 100: 2556–2561.

Li J, Henriksson G, and Gellerstedt G (2007). Lignin depolymerization and its critical role for delignification of aspen wood by steam explosion. *Biores Technol*, 98: 3061–3068.

Liu Y, Carriero S, Pye K, and Argyropoulos DS (2000) A comparison of the structural changes occurring in lignin during Alcell and kraft pulping of hardwoods and softwoods. *ACS Symposium Series 742 Lignin: Historical, Biological, and Materials Perspectives.* Grasser WG, Norhey RA, and Schultz TP (eds.). American Chemical Society, Washington, DC, pp 447–464.

Lu Y, Lu Y, Hu H, Xie F, Wei X, and Fan X (2017). Structural characterization of lignin and its degradation products with spectroscopic methods. *Spectroscopy*, 2017: 1–15.

Mäki-Arvela P, Anugwom I, Virtanen P, Sjöholm R, and Mikkola JP (2010). Dissolution of lignocellulosic materials and its constituents using ionic liquids: a review. *Industrial Crops and Products*, 32(3): 175–201.

Mansson P (1983). Quantitative determination of phenolic and total hydroxy groups in lignins. *Holzforschung*, 37: 143–146.

Matsushita Y (2015). Conversion of technical lignins to functional materials with retained polymeric properties. *Wood Sci*, 61: 230–250. https://doi.org/10.1007/s10086-015-1470-2.

McCrady E (1991). The nature of lignin. *Alkaline Paper Advocate*, 4.

Miidla H (1980). Lignification in plants and methods for its study. *Regul Rosta Pitan Rast*, 87.

Mood SH, Golfeshan AH, Tabatabaei M, Jouzani GS, Najafi GH, Gholami M, and Ardjm M (2013). Lignocellulosic biomass to bioethanol, a comprehensive review with a focus on pretreatment. *Renewable and Sustainable Energy Rev*, 27: 77–93.

Muhammad, N., Omar, W.N., Man, Z., Bustam, M.A., Rafiq, S., and Uemura, Y. (2012). Effect of ionic liquid treatment on pyrolysis products from bamboo. *Ind Eng Chem Res*, 51: 2280–2289.

Muurinen E (2000). *Organsolv pulping—a review and distillation study related to peroxyacid pulping.* Master Dissertation University of Oulu, Oulu, Finland.

Niaounakis M (2015). *Biopolymers: Applications and Trends*, 1st ed., Elsevier.

Nimz HH (1974). Beech lignin—proposal of a constitutional scheme. *Angew Chem Int Ed Engl*, 13: 313–321.

Nimz HH and Casten R (1986). Chemical processing of lignocellulosics. *Holz Roh- Werkst*, 44: 207–212.

Nordström Y (2012). Development of softwood kraft lignin based carbon fibres. *Licentiate Thesis, Division of Material Science Department of Engineering Sciences and Mathematics*, Luleå University of Technology.

Palmqvist E, Hahn-Hagerdal H, Galbe M, Larsson M, Stenberg K, Szengyel Z, Tengborg C, and Zacchi G (1996). Design and operation of a bench-scale process development unit for the production of ethanol from lignocellulosics. *Bioresour Technol*, 58(2): 171–179.

Pan X, Arato C, Gilkes N, Gregg D, Mabee W, Pye K, Xiao Z, Zhang X, and Saddler J (2005). Biorefining of softwoods using ethanol organosolv pulping: preliminary evaluation of process streams for manufacture of fuel-grade ethanol and co-products. *Biotech Bioeng*, 90: 473–481.

Pan X and Saddler JN (2013). Effect of replacing polyol by organosolv and kraft lignin on the property and structure of rigid polyurethane foam. *Biotechnol Biofuel*, 6: 12–21.

Parajo´ JC, Alonso JL, and Va´zquez D (1993). On the behavior of lignin and hemicelluloses during the acetosolv processing of wood. *Bioresour Technol*, 46: 233–240.

Piló-Veloso D, Nascimento E A, and Morais SAL (1993). Isolamento e análise estrutural de ligninas. *Química Nova*, 16: 435–448.

Ponomarenko J, Dizhbite T, Lauberts M, Viksna A, Dobele G, Bikovens O, and Telysheva G (2014). Characterization of softwood and hardwood LignoBoost kraft lignins with emphasis on their antioxidant activity. *Bioresources*, 9: 2051–2068.

Pye EK and Lora JH (1991). The AlcellTM process. A proven alternative to kraft pulping. *Tappi J*, 74: 113–118.

Rinaldi R, Jastrzebski R, Clough MT, Raplh J, Kennema M, Bruijnincx PCA, and Weckhuysen BM (2016). Paving the way for lignin valorisation: recent advances in bioengineering, biorefining, catalysis. *Angew Chem Int Ed*, 55(29): 8164–8215.

Robert D, Bardet M, Lapierre C, and Gellerstedt G (1988). Structural changes in aspen lignin during steam explosion treatment. *Cell Chem Technol*, 22: 221–230.

Rosas JM, Berenguer R, Valero-Romero MJ, Rodríguez-Misarol J, and Cordero T (2014). Preparation of different carbon materials by thermochemical conversion of lignin. *Frontiers in Materials*, 1: 1–17.

Rouhi AM and Washington C (2001). Only facts will end lignin war. *Sci Technol*, 79(14): 52–56.

Sannigrahi P, Ragauskas AJ, and Miller SJ (2010). Lignin structural modifications resulting from ethanol organosolv treatment of loblolly pine. *Energy Fuel*, 24: 683–689.

Sano Y, Maeda H, and Sakashita Y (1989). Pulping of wood at atmospheric pressure I: pulping of hardwoods with aqueous acetic acid containing a small amount of organic sulfonic acid. *Mokuzai Gakkaishi*, 35: 991–995.

Sano Y, Nakamura M, and Shimamoto S (1990). Pulping of wood at atmospheric pressure II: pulping of birch of wood with aqueous acetic acid containing a small amount of sulfuric acid. *Mokuzai Gakkaishi*, 36: 207–211.

Sarkanen KV (1990). Chemistry of solvent pulping. *Tappi J*, 73(10): 215–219.

Sarkanen KV and Ludwig CH (1971). *Wood Chemistry Lignin: Occurrence, Formation, Structure and Reactions*. Sarkanen KV and Ludwig CH (ed.). Wiley-Interscience, New York, p. 916.

Shimada K, Hosoya S, and Ikeda T (1997). Condensation reactions of softwood and hardwood lignin model compounds under organic acid cooking conditions. *J Wood Chem Technol*, 17: 57–72.

Shimizu K, Sudo K, Ono H, Ishihara M, Fujii T, and Hishiyama S (1998). Integrated process for total utilization of wood components by steam explosion pre-treatment. *Biomass Bioenergy*, 14: 195–203.

Shukry N, Fadel SM, Agblevor FA, and EI-Kalyoubi SF (2008). Some physical properties of acetosolv lignins from bagasse. *J Appl Poly Sci*, 109: 434–444.

Sidiras DK and Salapa I S (2015). *Engineering conferences international*. Available from: http://dc.engconfintl.org/cgi/viewcontent.cgi?article=1013&context=biorefinery_I (accessed February 20, 2018).

Sjöström E (1993). *Wood Chemistry: Fundamentals and Application*. Academic Press, Orlando, p. 293.

Smith MD, Cheng X, Petridis L P, Mostofian B, and Smith JC (2017). Organosolv-water cosolvent phase separation on cellulose and its influence on the physical deconstruction of cellulose: a molecular dynamics analysis. *Sci Rep*, 7: 14494.

Smolarski N (2012). *High-vallue opportunities for lignin: unlocking its potential*. Available from http://greenmaterials.fr/wpcontent/uploads/2013/01/High-value-Opportunities-for-Lignin-Unlocking-its-Potential-Market-Insights.pdf (accessed December 20, 2022).

Soederstroem J, Galbe M, and Zacchi G (2004). Effect of washing on yield in one- and two-step steam pretreatment of softwood for production of ethanol. *Biotechnol Progress*, 20(3): 744–749.

Souto F, Calado V, and Pereira N (2018). Lignin-based carbon fiber: a current overview. *Mater Res Express*, 5: 072001.

Stenberg K, Tengborg C, Galbe M, and Zacchi G (1998). Optimization of steam pretreatment of SO2-impregnated mixed softwoods for ethanol production. *J Chem Technol Biotechnol*, 71(4): 299–308.

Sudo K and Shimizu K (1992). A new carbon fiber from lignin. *J Appl Polym Sci*, 1992(44): 127–134.

Sudo K, Shimizu K, Nakashima N, and Yokoyama A (1993). A new modification method of exploded lignin for the preparation of a carbon fiber precursor. *J Appl Polym Sci*, 48: 1485–1491.

Tan SSY, MacFarlane DR, Upfal J, Edye LA, Doherty WOS, Patti AF, Pringle JM, and Scott JL (2009). Extraction of lignin from lignocellulose at atmospheric pressure using alkylbenzenesulfonate ionic liquid. *Green Chem*, 11: 339–345.

U.S. Congress, Office of Technology Assessment. (1989). *Technologies for Reducing Dioxin in the Manufacture of Bleached Wood Pulp*. OTA-BP-O-54. Washington, D.C.: U.S. Government Printing Office.

Ullah M, Liu P, Xie S, and Sun S (2022). Recent advancements and challenges in lignin valorization: green routes towards sustainable bioproducts. *Molecules*, 27: 6055.

Uraki Y, Kubo S, Nigo N, Sano Y, and Sasaya T (1995). Preparation of carbon fibers from organosolv lignin obtained by aqueous acetic acid pulping. *Holzforschung*, 49: 343–350.

Willför S, and Gustafsson J (2010). *The Forest Based Biorefinery: Chemical and Engineering Challenges and Opportunities*. http://web.abo.fi/instut/pcc/presentations_pdf/Willf%C3%B6r_Gustafsson_Lignin.pdf.

Williamson PN (1987). Repap's ALCELL process: how it works and what it offers. *Pulp Pap Can*, 88(12): 47.

Xu L, Zhang S-J, Zhong C, Li B-Z, and Yuan Y-J (2020). Alkali-based pretreatment-facilitated lignin valorization: a review. *Ind Eng Chem Res*, 59: 16923–16938.

Yasuda S (1988). Behavior of lignin in organic acid pulping II: reaction of phenylcoumaran and 1,2-diaryl-1,3-propanediol with acetic acid. *J Wood Chem Technol*, 8: 155–164.

Yasuda S and Ito N (1987). Behavior of lignin in organic acid pulping I: reaction of arylglycerol-b-aryl ethers with acetic acid. *Mokuzai Gakkaishi*, 33: 708–715.

Ye DZ, Zhang MH, Gan LL, Li QL, and Zhang X (2013). The influence of hydrogen peroxide initiator concentration on the structure of eucalyptus lignosulfonate. *Int J Biol Macromole*, 60: 77–82.

Yoo CG, Meng X, Pu Y, and Ragauskas AJ (2020). The critical role of lignin in lignocellulosic biomass conversion and recent pretreatment strategies: a comprehensive review. *Bioresour Technol*, 301: 122784.

Young RA and Davis JL (1986). Organic acid pulping of wood. Part II. Acetic acid pulping of aspen. *Holzforschung*, 40: 99–108.

Zakzeski J, Bruijnincx PC, Jongerius AL, and Weckhuysen BM (2010). The catalytic valorization of lignin for the production of renewable chemicals. *Chem Rev* Jun 9, 110(6): 3552–3599.

Zhou H, Yang D, Qiu X, Wu X, and Li Y (2013). A novel and efficient polymerization of lignosulfonates by horseradish peroxidase/ H2O2 incubation. *Appl Microbiol Biotechnol*, 97: 10309–10320.

Zhu W (2013). Equilibrium of lignin precipitation—the effects of pH, temperature, ion strength and wood origins. *Bachelor Degree Chalmers University of Technology Goethenburg*, Sweden.

Relevant Websites

www.dutchbiorefinerycluster.nl
www.benthamscience.com
www.iopscience.iop.org

3

Lignin Depolymerization Technologies

Abstract

The natural aromatic polymer source lignin is the most abundant on earth. P-coumaryl alcohol, coniferyl alcohol, and sinapyl alcohol are the three primary monolignols in the lignin structure. The fact that these components make the structure of lignin a suitable raw material for a variety of products with added value has attracted the attention of the scientific community. Even though most of the world's lignin is still used as boiler fuel in carbohydrate processing plants, its low value and high volume have shown it has a lot of potential to make high-value new products. Lignin-based bioproducts could generate a multibillion-dollar industry if it were converted into chemical compounds. Numerous reports have discussed lignin fragmentation as an alternative method for producing chemicals that are presently extracted from fossil fuel sources. The thermal, chemical, and biological methods for depolymerizing lignin are discussed in this chapter. Due to their economic viability and low impact on the environment, biological methods of lignin depolymerization are becoming increasingly popular. There are a number of commercially viable methods for lignin depolymerization that are currently used on an industrial scale. Each approach has its own set of benefits and drawbacks, as well as obstacles to overcome and opportunities for further development.

Keywords *Lignin; Depolymerization; Thermal depolymerization; Chemical depolymerization; Biological depolymerization; Fungi; Bacteria; Enzymes*

By 2050, there will be roughly ten billion people living in the world, which will increase energy consumption and require more fuels and chemicals to support modern lifestyles.

Considering the primary sources of fossil fuels, this increase could result in a number of financial and environmental issues. There will be an imminent and already underway depletion of this non-renewable resource (Arpia et al., 2021; Azadi et al., 2013). In light of this, increasing the use of renewable energy is absolutely necessary. Lignocelluloses have emerged as an appealing and promising material for product manufacture. Lignocelluloses have attracted attention mainly because they are plentiful, do not emit carbon dioxide, and, most importantly, do not jeopardize the safety of the food supply (Wang et al., 2019). Cellulose, hemicellulose, and lignin make up lignocellulosic materials.

Carbohydrate fraction has long driven the use of lignocelluloses, leaving a low-value product from one-third of the lignocellulosic material. According to Kleinert and Barth (2008), this strategy reduces biorefineries' profitability and effectiveness. Presently, the same industries that process carbohydrate fractions use lignin as an energy source at a low cost. The biorefinery makes a little money from this activity. However, if lignin is transformed into various chemicals, its potential increases. As a result, incomes will rise and a new alternative fuel source will emerge.

Depolymerization of Lignin to Produce Value Added Chemicals, First Edition. Pratima Bajpai.
© 2024 John Wiley & Sons, Inc. Published 2024 by John Wiley & Sons, Inc.

Numerous approaches to depolymerize the structure of lignin have been vigorously pursued, for example, biological and thermochemical depolymerization processes. Specific cleavage of lignin linkages by biological means has demonstrated a number of advantages and is environmentally friendly by nature (Xu et al., 2019, 2021). There are, however, some problems such as the difficulty of genetically altering the microbes and their high sensitivity to changes in pH, temperature, and oxygen levels in the reaction system (Chauhan, 2020). Depending on the driving force, warming technology, solvents, temperature ranges, and other variables, different groups of thermochemical processes can be distinguished. There will be a variety of monomers and oligomers produced by these groups, as well as bio-oil yields at various rates which results from depolymerization of lignin (Agarwal et al., 2018; Lopez-Camas et al., 2020).

3.1 Thermal Depolymerization

This section focuses on the use of thermochemical processes to fragment lignin. Thermochemical techniques can be used, either with or without catalysts. Depolymerization can be further developed using a few new approaches, such as microwave-aided innovation. These techniques are better suited for use in industrial settings due to their shorter reaction time, and higher bio-oil and monomer yields. Some of the remaining obstacles include the severe reaction conditions, higher char yield, and lower depolymerization selectivity, resulting in a more challenging separation of the final products (Chio et al., 2019; Nguyen et al., 2021).

A number of methods for lignin depolymerization are commercially viable (Figure 3.1) and are currently used on an industrial scale. Combustion, pyrolysis, hydrothermal liquefaction (HTL), and gasification are the primary subtypes of thermal depolymerization. Pyrolysis, in particular fast/flash pyrolysis, and HTL have attracted a lot of interest as the fundamental technologies for the thermal depolymerization of lignin in comparison to combustion and gasification at very high temperatures (Figure 3.2).

3.1.1 Pyrolysis

One of the most studied processes for fractionating lignin is pyrolysis. The absence of oxygen and higher response temperatures are the fundamental characteristics of pyrolysis. As the temperature rises, lignin's structure breaks down a number of chemical bonds.

Because these reactions can take place in a variety of ways, the stages of pyrolysis are difficult to categorize. For better comprehension, pyrolysis has been divided into two stages (Chio et al., 2019; Kawamoto, 2017):

- Primary pyrolysis (200–400 °C)
- Secondary pyrolysis (400–800 °C)

Pyrolysis is a common method of thermal depolymerization in which feedstocks are thermally broken down into platform chemicals and fuels in an inert atmosphere. Numerous types of thermal conversion begin with pyrolysis (Hoang et al., 2021; Wang et al., 2017; Zhang et al., 2007). Lignin will be converted into bio-oil, gaseous products, and char through pyrolysis. Bio-oil is thought to be the primary product of depolymerization.

The term "fast pyrolysis" refers to a process of anoxic thermal degradation. It involves the rapid breakdown of materials inside a reactor into volatile compounds. In order to carry out rapid pyrolysis, the reactor is heated to a temperature of 400–600 °C, at a higher heating rate. Compared to traditional thermochemical methods, quick pyrolysis of lignin is more successful and achievable in light of the fact that the reaction conditions can be changed without difficulty.

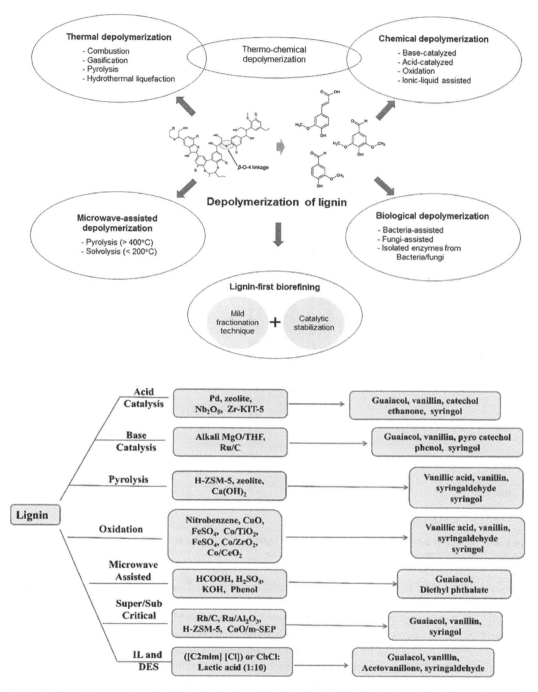

Figure 3.1 Different technologies of lignin depolymerization. A. Zhou et al. (2022) / Frontiers Media / CC BY 4.0. B. Roy et al. (2022) / MDPI / CC BY 4.0.

Monomeric phenolic compounds, a significant quantity of char, and acetic acid are produced when lignin is pyrolyzed (Patwardhan et al., 2011).

Figure 3.3 demonstrates the connection between the temperature of pyrolysis and the pattern of aromatic substitution in lignin degradation products. Most of the aromatic compounds generated

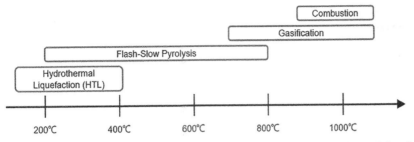

Figure 3.2 Ranges of reaction temperatures used by different lignin thermal depolymerization methods. Zhou et al. (2022) / Frontiers Media / CC BY 4.0.

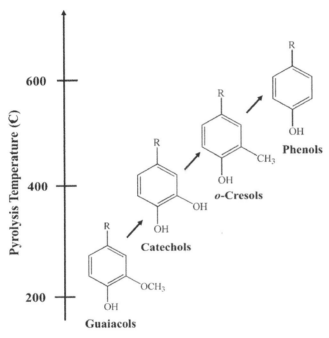

Figure 3.3 Effect of pyrolysis temperature on the aromatic substitution products of G-type monomers. Reproduced with Permission Agarwal, 2018 / ELSEVIER.

during primary lignin pyrolysis are syringols from S-lignins and four-substituted guaiacols from G-lignins. Direct homolysis of α-O-4 links occurs at this point but there is no direct homolysis of the C–C and C–O bonds, making the aromatic methoxyl group very stable. The primary lignin pyrolysis byproducts serve as hydrogen donors and stabilize the radicals produced by lignin (Kotake et al., 2015). As a result, the yield of monomeric products increases. Guaiacols/syringols undergo a rapid transformation into phenols, o-cresols and catechols during the secondary pyrolysis stage of the reaction when the pyrolysis temperature reaches 400–450 °C (Kawamoto, 2017). Within this temperature range, side chain C–C bonds are also cracked, increasing monomer yield. Additionally, around 450 °C, homolysis of O–CH_3 bonds occur, and the aromatic methoxyl group turns out to be very responsive. As the temperature rises to 550 °C, CO, a particular non-condensable gas, is significantly produced more, catechols and pyrogallols typically vanish and become coke (Asmadi et al., 2011). O-cresols and phenols remain fairly stable during pyrolysis at temperatures higher than 700 °C, while PAH production rises. During pyrolysis, the greatest amount of residual char is produced by lignin, which is lignocellulose's most heat-resistant component (Bu et al., 2014).

Fast pyrolysis of lignin occurs between 200 and 400 °C, breaking β-O-4 and α-O-4 bonds. At about 300 °C, side chain aliphatic groups are released, forming smaller oxygenated compounds (Huang et al., 2010). Free radicals are frequently produced when a C–O bond is broken, facilitating the breakage of C–C bonds. Between 550 and 650 °C is the ideal temperature for producing bio-oil, at which point 20% phenolic products and 40% bio-oil can be produced. Demethoxylation reactions are favored by high pyrolysis temperatures, resulting in increased catechol, phenol, alkyl phenols, and pyrogallol production while inhibiting phenol methoxylation (Jiang et al., 2010; Trinh et al., 2013). Bio-oil is transformed into gaseous products when the temperature is raised further than the ideal value (Fan et al., 2017a, 2017b). The production of undesirable pyrolytic products is frequently accelerated by high temperature. Fast pyrolysis of lignin occurs between 200 and 400 °C, breaking β-O-4 and α-O-4 bonds. At about 300 °C, side chain aliphatic groups are released, forming small oxygenated compounds (Mukkamala et al., 2012). Additionally, the presence of sulfur, which encourages char formation and polymerization (de Wild et al., 2017; Li et al., 2012) significantly reduces the yield of bio-oil during lignin pyrolysis. In comparison to lignocellulosic biomass pyrolysis, lignin pyrolysis contains about 80% phenol-containing compounds. This is because lignin has a unique phenyl structure (Fan et al., 2017a, 2017b; Borges et al., 2014). However, the three major biomass components' yields of aromatic hydrocarbons decline in the following order under the similar catalytic fast pyrolysis conditions: cellulose > hemicellulose≫lignin (Borges et al., 2014). Compared to biomass-pyrolyzed biodiesel, which has a heating value of 16 to 19 MJ Kg^{-1}, lignin-pyrolyzed biodiesel has a higher heating value (30 MJ Kg^{-1}) because of the lower total acid content (de Wild et al., 2017; Li et al., 2015; Qi et al., 2007).

Depolymerization performance is affected by the choice of catalytic system, size of feedstock, retention time, heating rate, and pyrolysis temperature (Chen et al., 2019; Hoang et al., 2021). Lignin undergoes thermal decomposition at a variety of temperatures due to its extensive variety of linkages (Shen et al., 2015a, 2015b; Zhou et al., 2017). By connecting multiple reactors directly, including thermogravimetric reactors, reactors with fixed beds, reactors with fluidized beds, etc., to photoionization (PI) MS by means of a molecular beam nozzle, heated transfer lines, or the incorporation of various furnaces into the ionization source, online MS detection of volatiles from pyrolysis has been discovered, and researchers have intuitively observed how the distinct products change over time and temperature (Dufour et al., 2013; Hurt et al., 2013; Jia et al., 2015; Le Brech et al., 2016; Zhou et al., 2017; Zhu et al., 2020).

In general, there are two stages to lignin pyrolysis (Chen et al., 2019; Le Brech et al., 2016; Zhu et al., 2022):

1) Using weak ether bonds like β-O-4 and α-O-4, feedstocks melt into a liquid and undergo heterolytic cleavage at low temperatures to create lignin monomers.
2) As the temperature rises, several C–O and C–C bonds get broken to form additional monomers, and additional reactions like demethoxylation and demethylation will take place to form a variety of phenols.

In the meantime, the homolysis of chemical bonds will produce a lot of free radicals, which will then be rearranged and repolymerized to result in oligomers and even coke (Li and Takkellapati, 2018; Zhu et al., 2020). The radical and concerted mechanisms, which may differ based on the structure of the feedstock and the conditions and stages of the reaction, have been the primary focus of discussions regarding lignin's primary pyrolysis mechanism (Chen et al., 2019; Jarvis et al., 2011). To simplify the reaction system, model compounds that concentrate on the cleavage mechanisms of α-O-4 and β-O-4 are frequently used. A hyperthermal spout directed the continuous pyrolysis of 2-phenethyl phenyl ether through PIMS (Jarvis et al., 2011). The main reaction pathways for the breakage of α-O-4 were Maccoll and/or retro-ene eliminations at atmospheric pressure and pyroprobe-integrated chemical ionization MS at temperatures below 1000 °C, according to studies using in situ synchrotron vacuum ultraviolet PIMS at different temperatures (He et al., 2016; Sheng et al., 2017). The homolysis reaction, which generates free radicals, will slowly take over at high temperatures and for prolonged periods. This leads to the growth of aromatic hydrocarbons and the production of coke (Zhu et al., 2020). The homolysis of C–O caused the breakage of α-O-4, as shown by the SVUV PIMS detection of the benzyl radical and the phenoxy radical in the pyrolysis studies on α-O-4 model compounds (Dai et al., 2019; He et al., 2016). Because the essential pyrolysis of lignin primarily results in the production of monomers like guaiacol, the system of the second-stage lignin pyrolysis response was the focus of the web-based analysis of guaiacol pyrolysis (Dai et al., 2019; Liu et al., 2018).

Polycyclic aromatic hydrocarbon formation and elimination of methoxyl, methyl, and other groups were both discovered. These findings have direct implications for reducing repolymerization and increasing product selectivity (Zhu et al., 2022). Due to repolymerization and the low selectivity of thermal pyrolysis, the pyrolytic oil typically has a more complex composition and a lower monomer yield (Schutyser et al., 2018). As a result, the pyrolysis of lignin requires a suitable catalytic system. Due to their distinctive shape selectivity and pore structure, zeolites have attracted the most research because they are able to stabilize reactive intermediates and further transform monomers that have been depolymerized into desired products (like aromatics) (Chen et al., 2019; Jae et al., 2011; Schutyser et al., 2018). The study of guaiacylglycerol-guaiacyl ether pyrolysis using HZSM-5 catalysis has recently been used for an online MS method and in situ atmospheric pressure photoionization (APPI) monitoring of the primary products of biomass pyrolysis (Chen et al., 2020; Liu et al., 2021). The previously proposed phenolic pool mechanism was made clear through the successful detection of several heavy phenolic oligomers (Stanton et al., 2018). An online investigation into the rapid and catalytic pyrolysis of biomass using a microfluidized bed found that hierarchical zeolites had a higher selectivity for aromatics (Jia et al., 2017). Isomer-selectively detecting intermediates in the catalytic pyrolysis of guaiacol was accomplished through the utilization of photoelectron photoion coincidence spectroscopy with SVUV and a temporal analysis of the products reactor in order to better comprehend the mechanism. In addition, the fulvenone ketene, the core species of the reaction network, was found for the first time (Hemberger et al., 2017). Carbon-based catalysts, metal oxides, and other comparable substances likewise can upgrade the pyrolysis of lignin (Chen et al., 2019). For instance, it was discovered that the guaiacol hydrodeoxygenation catalyzed by Pd/NC was stable and had a high selectivity for benzene (a yield of up to 85.1%). (Liu et al., 2021).

Both fixed bed as well as fluidized bed reactors are frequently utilized for pyrolysis of lignin (Lee et al., 2016; Zhang et al., 2015c). Heat is transferred from the external surface of lignin to its internal surface in these reactors. Bio-oil and aromatic compounds are frequently produced in high quantities in fixed bed as well as fluidized bed pyrolysis reactors.

In a fixed bed bioreactor, the substrate is heated unevenly. This results in lower bio-oil yield in contrast to the fluidized bed reactor, where thorough mixing between the substrate and solid particles maximizes convective heat transfer efficiency which is favorable for the demethoxylation of lignin (Jiang et al., 2010). Furthermore, the melting point of lignin is decreased when functional groups, mostly hydroxyl and methoxyl groups, are present, which produces char (Mukkamala et al., 2012). Additionally, sulfur favors char formation and reduces bio-oil yield through polymerization (de Wild et al., 2017). During the rapid pyrolysis of lignin, a higher H/Ceff ratio, which indicates a reduced oxygen content, encourages the substantial production of aromatic compounds (Li et al., 2012). During pyrolysis, long retention times frequently result in a reduction in the yield of bio-oil but an increase in the yields of light organic compounds and non-condensable gases. However, a shorter retention time is advantageous for obtaining a lot of bio-oil but the quality is poor (Trinh et al., 2013). Pyrolysis vapors' secondary decomposition rate is sped up by long retention times. Therefore, for attaining the desired quality and quantity of bio-oil, an optimal retention time, amount of catalyst, and heating rate are frequently required (Kalogiannis et al., 2015; Kim et al., 2015). Lignin pyrolytic product yield and selectivity are significantly enhanced by catalysts. Zeolites and metal chlorides have been extensively utilized for pyrolysis of lignin (Bi et al., 2018; Junior et al., 2018; Luo et al., 2012). Metal catalysts reduce the production of undesirable chemicals while simultaneously increasing the yield of pyrolysis products (Maldhure and Ekhe, 2013; Wang et al., 2015a, 2015b, 2015c). The order of the synergist capacity of metal chlorides for lignin pyrolysis is as follows:

$FeCl_3 > CaCl_2 > KCl$.

Even though noble metals have a lot of catalytic activity, they are still expensive to use for pyrolysis of lignin. The catalytic pyrolysis of lignin has been studied for a variety of high aluminum zeolites, HZSM-5 (Custodis et al., 2016) including H-mordenite, H-β, H-ferrierite, HZSM-5(25), HZSM-5(50), and H-USY, and also macropore materials like SBA-15, MSU-J and MCM-41, silica, and—γ Al2O3 (Custodis et al., 2016; Kurnia et al., 2017). At 650 °C, HZSM-5 helped lignin pyrolysis produce aromatic monomers effectively. Two hardwood lignins were pyrolyzed by a number of NiO-containing zeolites (NiO/HY, NiO/HBETA; NiO/HZSM-5) (Milovanovic et al., 2017). It was discovered in this study that NiO/HY reduced the formation of coke while increasing the quantity of gaseous and liquid products. Despite the fact that NiO/HZSM-5 yielded the least bio-oil, the bio-oil had the most aromatic compounds but fewer oxygenates than other oils.

Xu et al. (2017) investigated the ex situ catalytic rapid pyrolysis of lignin using ammonia on a variety of zeolites (ZnO/HZSM-5, HZSM-5, HY, MCM-41, β-zeolite, and ZnO/HY). Bio-oil and aromatic amines had total carbon yields of 9.8 and 5.6 %, respectively, and aromatic amines had selectivity of up to 87.3 % for aniline. Due to the fact that the Si/Al molar ratio is determined by the Si/Al molar ratio, ZSM-5 zeolite possesses greater pyrolytic activity toward various lignins. Bronsted acid catalysts like ZSM-5 have the desirable porous structure, tunable acidity, higher thermal stability, and water tolerance for producing aromatic hydrocarbons (Engtrakul et al., 2016). HZSM-5 zeolite has a higher catalytic activity toward the pyrolysis of different lignins because the number of Bronsted acid sites is determined by the Si/Al molar ratio (Kim et al., 2015; Ma et al., 2012; Shen et al., 2016). During pyrolysis of lignin, zeolite catalysts increase the yield of aromatic products, reduce the formation of char, and encourage the cracking of oxygenated compounds (Li et al., 2014). A moderately high pyrolysis temperature facilitates both zeolite's higher desorption of coke precursors and the large production of oxygenates with low molecular weights. Thermal decomposition primarily contributes to the formation of coke/char, which obstructs the pores of zeolite, during the rapid pyrolysis of lignin. The majority of thermal char and catalytic coke deposits are found on the exterior of the zeolite surface and within its internal pores, respectively. The zeolite catalyst's carbon conversion efficiency decreases as a result of this rapid, irreversible deactivation (Wang et al., 2014). In addition, phenolic compounds derived from lignin serve as a precursor for coke and are firmly adhered to the zeolite acid sites, deactivating the catalyst. As a result, for the pyrolysis of lignin, pure zeolites are not thought to be good catalysts (Rezaei et al., 2016). The lignin pyrolysis products are significantly influenced by zeolite's shape and pore size. For instance, microporous ZSM-5 rapidly pyrolyzes lignin to produce a significant quantity of aromatic compounds, followed by beta, mordenite, and Y zeolites. Due to pore blockage, however, microporous ZSM-5 and mordenite zeolites are unable to efficiently cleave bulky syringyl monolignols; only beta and γ zeolites are able to effectively catalyze the deoxygenation reactions of oxygenates obtained from lignin (Yu et al., 2012).

Bulky molecules' resistance to diffusion and mass transfer into and out of the zeolite micropores must, therefore, be reduced. It was discovered that suppressing mass transfer resistance over hierarchical HZSM-5 effectively enhanced oxygenated compound deoxygenation while reducing the coke formation (Bi et al., 2018). In contrast to microporous HZSM-5 zeolite, the use of mesoporous

3 Lignin Depolymerization Technologies

aluminosilicates had no impact on the yield of aromatic compounds or the formation of coke during lignin pyrolysis (Custodis et al., 2016). For treating bulky monolignols obtained from the catalytic rapid pyrolysis of lignin, hierarchical zeolites with interconnected mesopores and micropores offer promising options (Bi et al., 2018; Li et al., 2017; Ma and Zhao, 2015).

While maintaining Bronsted acid sites for acid catalysis, hierarchical zeolites guarantee the entry of bulky molecules. In hierarchical zeolites, mesoporous structures enable large molecules to diffuse to the Bronsted acid sites (Hertzog et al., 2018; Jia et al., 2017). Selective upgradation of crude bio-oil obtained from biomass pyrolysis is made possible by the size and mesopore surface density of Bronsted acid sites.

The main pyrolysis pathways that lead to pyrolysis products are depicted in Figure 3.4a, b, c (Chio et al., 2019; Ma et al., 2012; Yang et al., 2020a).

Characteristic pyrolytic behaviors of the initial, primary, and charring stages are depicted in Figure 3.5 (Lu and Gu, 2022). There are three stages to the lignin pyrolysis process: the initial

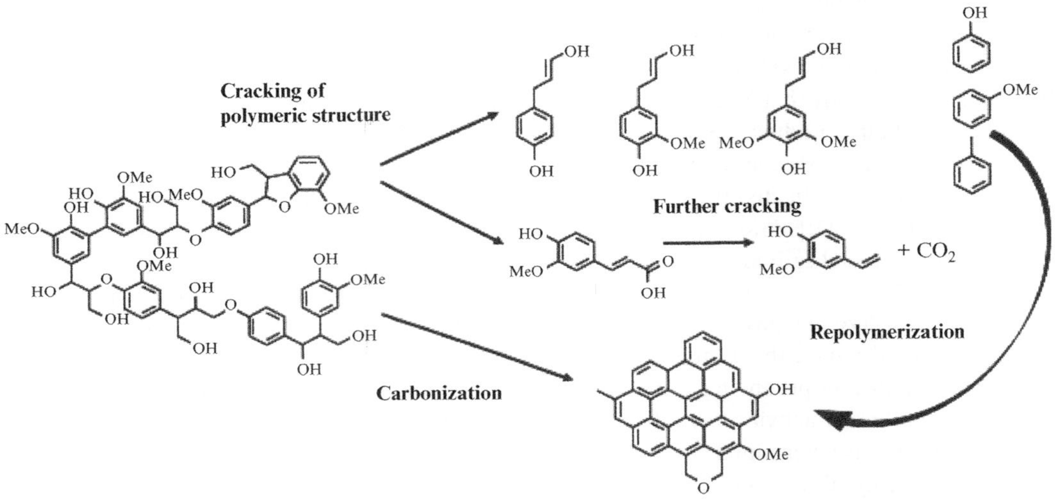

Figure 3.4a Proposed pyrolytic mechanism of lignin. Reproduced with permission Yang et al. (2016) / ELSEVIER.

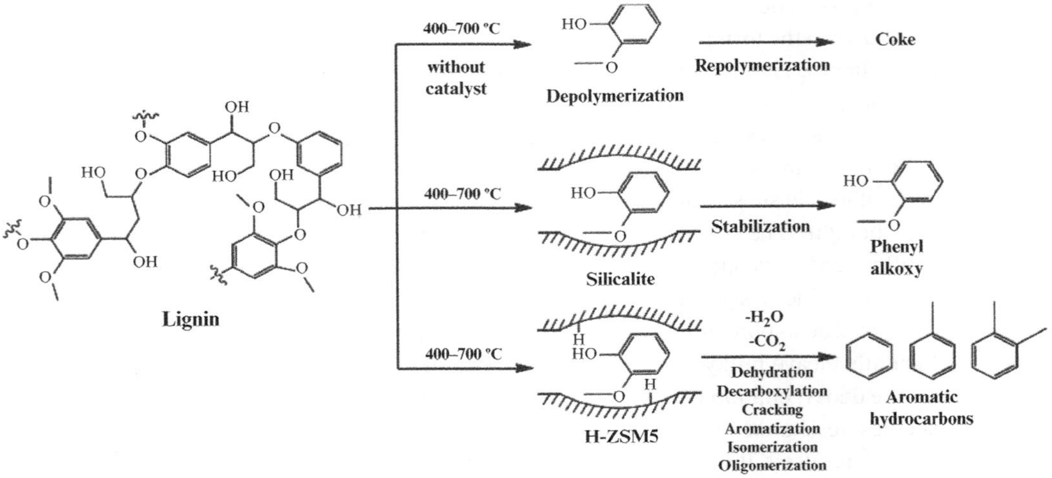

Figure 3.4b Proposed reaction mechanism pathway of lignin non-catalytic/catalytic fast pyrolysis. (Ma et al., 2012). Reproduced with permission.

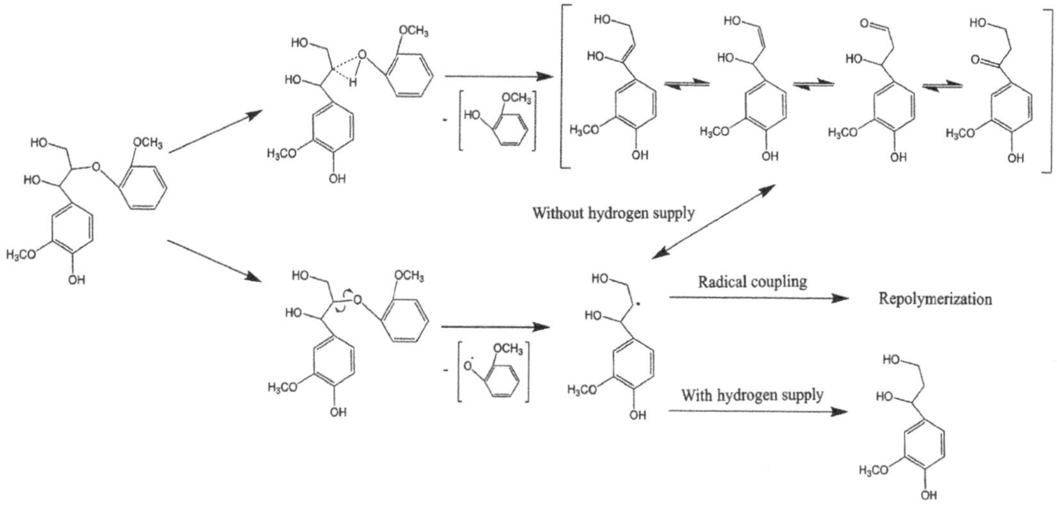

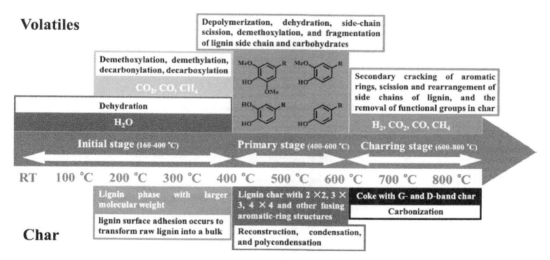

Figure 3.4c The purpose mechanism of cleaving the β-O-4 linkages by pyrolysis. Reproduced with permission from Chio et al. (2019) / ELSEVIER.

Figure 3.5 Characteristic pyrolytic behaviors of initial, primary, and charring stages. Lu and Gu, (2022) / Springer Nature / CC BY 4.0.

stage, the primary stage, and the charring stage. Mechanisms and behaviors of pyrolysis are unique to each stage.

Figure 3.6 shows the proposed reaction pathways for the primary stage of lignin pyrolysis (Supriyanto et al., 2020).

3.1.2 Hydrothermal Liquefaction

Hydrothermal liquefaction (HTL), a fundamental method for lignin depolymerization, has recently received a lot of attention. For biomass with a high water content, HTL is a better option than other thermal depolymerization methods like pyrolysis and gasification because it does not need any initial drying procedures, which are expensive and consume a lot of energy (Kumar et al., 2017).

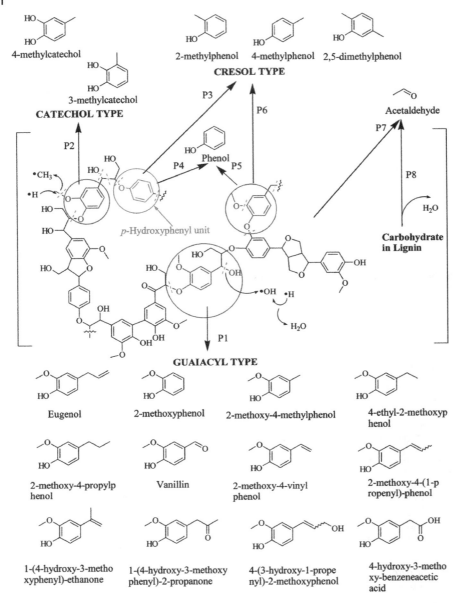

Figure 3.6 Proposed reaction pathways for primary stage of lignin pyrolysis (P1: depolymerization and dehydration; P2: side-chain scission and demethylation; P3, P4: depolymerization; P5: demethoxylation; P6: side-chain scission and demethoxylation; P7: fragmentation of lignin side chains; P8: dehydration and fragmentation of carbohydrate). Supriyanto et al. (2020) / ELSEVIER / CC BY 4.0.

Typically, phenolics (like syringol, vanillin, and guaiacol) or low-oxygen liquid bio-oil are extracted from lignin using a solvent with or without catalysts at a high pressure (5–28 MPa) and moderate temperature (200–400 °C).

According to Li and Takkellapati (2018), hot-compressed water is frequently used for HTL of lignin. Subcritical water (≤374 °C and 22 MPa) is a good reaction medium for biomass conversion due to its high organic compound solubility and low viscosity, which set it apart from supercritical water and water at room temperature (≥374 °C and 22 MPa).To put it another way, water is both a green solvent and a catalyst and reactant in the lignin HTL process (Yang et al., 2020b). Consequently, it is thought that HTL can depolymerize lignin. This method is both sustainable and kind to the environment, has a high conversion rate, and produces less secondary pollution (Cao et al., 2018).

The depolymerization mechanism of lignin through HTL for producing aromatic derivatives with added value has been the subject of numerous investigations (Zhou et al., 2022). Typically, lignin hydrolysis into methoxy phenolics occurs first because breaking ether bonds needs less energy than breaking C bonds.

The proposed three-step HTL of lignin depolymerization included the hydrolysis of lignin, the cleavage of ether bonds and C–C bonds between monomers, the degradation of methoxy groups on the benzene ring, and the alkylation of functional groups on the benzene ring (Cao et al., 2018). The distribution and yield of lignin HTL-derived products are significantly influenced by reaction time and temperature. However, other reaction variables like reactor pressure and stirring rate have less of an impact. The simultaneous promotion of lignin depolymerization and solid residue accumulation will result in an increase in residue and low-molecular- weight compounds (Cao et al., 2018; Singh et al., 2014).

The lignin HTL's reaction time is an important factor in determining the nature of the final products (Zhou et al., 2022). As the temperature rises, both the repolymerization of intermediate products and the depolymerization of lignin will accelerate simultaneously, resulting in an increase in compounds with a low molecular weight and solid residue (Singh et al., 2014). By lignin HTL, longer reaction times can make low molecular end products more complex, and even longer reaction times can make low molecular end products condense into coke (Xu and Li, 2021). Due to the fact that the HTL process is controlled kinetically, a significant factor has also been identified as the heating rate (Li and Takkellapati, 2018).

Because water is used as a solvent in HTL, the route is less harmful to the environment. Water also serves as a catalyst for numerous reactions and as a solvent (Kruse and Dahmen, 2015). This is due to the improved polarity, solubility, and transportation properties of water as it approaches its supercritical point (above 375 °C) and shows good solvent ability (Haarlemmer et al., 2016; Toor et al., 2011). Due to its suitability for the transformation of feedstock with a high moisture content into products of high quality and stability, HTL reduces the amount of energy required for biomass drying from a financial perspective (Haarlemmer et al., 2016).

As a result of these factors, HTL has emerged as a highly specialized method that can be used to convert a variety of feedstocks.

Cheng et al. (2012) investigated alkali lignin's hydrothermal co-solvent degradation into bi-phenolic compounds indicating that the water–ethanol cosolvent system yields more efficiently under subcritical and supercritical conditions. Bio-crude oil yield and quality were significantly influenced by temperature and the co-solvent ratio, whereas liquefaction products yield and quality were unaffected by time.

Singh et al. (2014) used a co-solvent system of ethanol and methanol (1:10) to make aromatic ethers and substituted phenols. A flow chart for the semi-continuous process of liquefying biomass is shown in Figure 3.7. Therefore, lignin HTL is an appropriate method for rapidly transforming it into high-quality, useful products. Table 3.1 provides a summary of the published research on the HTL of lignin into reported products, focusing on the most important bifunctional chemicals produced and the reported yields.

Lignin hydrothermal liquefaction is affected by a number of factors. Effects of temperature and heating rate, catalyst concentration and type, and reaction time are all examples of these. In

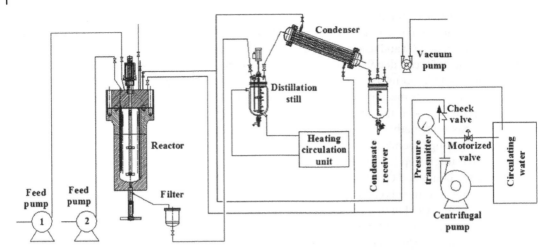

Figure 3.7 Process flow configuration for hydrothermal liquefaction. Adapted from Yang et al. (2020).

Table 3.1 Hydrothermal liquefaction of lignin into different products.

Lignin source
Organosolv hardwood and wheat straw lignin

Liquefaction preview
The liquefaction involved the pre-heating of lignin and formic acid in a supercritical fluid consisted of CO_2/acetone/water in molar ratios of 2.7/1/1. The reaction was completed within 3.5 h at 300 °C and 10 bar pressure.

Yield
10–12% monomeric aromatic compounds

Products
Phenolic oil was obtained consisting monomeric phenol and oligomeric aromatic compounds Schwarz et al. (2016).

Lignin source
Alkaline lignin

Liquefaction preview
HTL was conducted in a 250 mL autoclave with heating power of 1.5 kW. An alkaline lignin (8.0 g) and 100 mL of water are added and reactor was purged with nitrogen and agitated using a stirrer (200 rpm). The temperature was varied for 30 min.

Yield
Not reported

Products
Isolated products include benzenediols, monophenolic hydroxyl products, weak-polar products, and water-soluble products (low-molecular weight organic acids, alcohols, etc.) Konduri et al. (2015).

Table 3.1 (Continued)

Lignin source
Kraft lignin

Liquefaction preview
Kraft lignin, heterogeneous catalyst (K_2CO_3), phenol and deionized water, is uninterruptedly pumped by high pressure diaphragm pump at a flow rate of 1 kg h^{-1}. This was a continuous reactor set-up. The reaction temperature was 350 °C and 25 MPa.

Yield
With respect to dry lignin fed into the reactor, approximately 70 weight % of lignin oil was obtained.

Products
Phenol, alkylphhenols, phenolic dimers, catechol and guaiacol
Nguyen et al. (2014)

Lignin source
Enzymatic hydrolysis lignin

Liquefaction preview
The lignin, water and RANEY nickel (dry matter) were placed in a 5 mL autoclave. The autoclave was purged with nitrogen. The reaction mixture was heated at different temperatures for different reaction times.

Yield
Varying yields ranging from 3 to 15 mg g^{-1} lignin.

Products
Phenol, catechol and guaiacol
Beauchet et al. (2012)

Lignin source
Organosolv lignin

Liquefaction preview
HTL tests were completed in a batch autoclave. The reactor was heated with an electric heating jacket. Temperature and pressure were checked online. A 700 mL chilled condenser was connected to the autoclave exit port. Lignin, catalyst, and deionized water with and without catalysts.

Yield
Approximately 25 weight % of monomeric phenol.

Products
A complex mixtures of monomeric catechol and methoxylated and alkylated phenol
Struven and Meier D (2016)

Lignin source
Kraft lignin

Liquefaction preview
The treated kraft lignin, ionic liquid catalyst and, and Pd/C catalyst were placed in a 75 mL autoclave reactor. The reactor was sealed and purged with H2. The reaction was conducted at 200 °C at a stirring speed of 800 rpm.

Yield
Maximum liquids products yield was 50 weight %, while based on Kraft lignin was 13 weight % yield.

Products
Phenol, catechol and guaiacol
Chang et al. (2016).

(Continued)

Table 3.1 (Continued)

Lignin source
Kraft lignin

Liquefaction preview
Kraft lignin and deionized water were reassigned into the reactor. The system was vented with nitrogen three times. For each experiment the reactor was heated in 30 min from ambient to a working temperature of 130 °C or 180 °C or 230 °C, after which the experiment was continued for 15 or 60 min.

Yield
Phenolic oil content ranges from 5.4 to 10.6 weight %, with 78 weight % guaiacol.

Products
Guaiacol
Kruse and Dahmen (2015)

Lignin source
Wheat straw biomass

Liquefaction preview
Sequential biomass pretreatment and followed by room temperature extraction with an activated resin. The biorefinery approach also included supercritical CO_2 extraction of phenolic compounds.

Yield
Not reported

Products
High molecular weight phenolic compounds, targeting Tricin
Yan et al. (2015)

Alhassan et al. (2020) / IntechOpen / CC BY 4.0.

thermo-synthetic cycles like liquefaction, the most important factor is the response temperature. Because of the fact that the temperature of the reaction affects the degree of liquefaction, the lignin HTL process is temperature-dependent. By carefully selecting the heating rate, the rate can be slowed down at which unsaturated oligomeric phenol condensation occurs (Chen and Long, 2016). According to temperature, HTL is divided into subcritical liquefaction and supercritical liquefaction. Lignin feedstock undergoes degradation at temperatures between 200 and 300 °C. Biocrude oil yield tends to decline at temperatures above 300 °C.

Hu et al. (2014) claim that temperature has an impact on lignin degradation in such a way that an increase in temperature in the degradation of black liquor lignin favors bond cleavage, the removal of functional groups, and carbonization. Yang et al. (2015) discovered that temperature greatly impacted product distribution than reaction time.

During lignin degradation, various categories of catalysts have already demonstrated positive effects. Suitable homogenous catalysts, such as sodium hydroxide were found (Konduri et al., 2015; Murciano Martinez et al., 2016). Mixed oxide catalysts are also getting more attention in research (Cheng et al., 2012). On the other hand, a recent trend in the literature suggested a paradigm in favor of ionic liquids (ILs). The use of ILs in biomass refineries has received more attention in recent years. Particularly, lignin degradation catalysis by ILs has received significant attention. Various classes of ILs, including protonic, biocompatible, and biorenewable ILs have been studied (Zhang et al., 2016b). The selectivity of ILs has the greatest impact on lignin HTL when they are utilized as catalysts and/or cosolvents. The length of the cationic alkyl chain, temperature, anionic hydrophobicity, and type of solvent, among other factors, contribute to ILs' selectivity. For instance, the Lewis/Bronsted acidity of the alkyl chain length greatly influences the acidic ILs' catalytic activity.

Inter-molecular bonding interactions give ILs catalysts a lot of acidity. This contrasts with the protonation of the compounds that results in the acidity of conventional acid catalysts. As a result, the first one displayed a tendency to eliminate reactor corrosion; a significant response designing issue that has been presenting significant difficulties. It was proposed that, lignin disintegration was supported by means of a π–π connection between an alkyl imidazolium chloride impetus and the π-bond in the fragrant rings design of lignin, recommending the extra disintegration possibilities of alkyl-based ILs (Zakzeski, 2010). As per Zhuo et al. (2015), it was discovered that 2-phenyl-2-Imidazoline-based ILs with shorter side chains at C-1 had higher acidity than those with longer side chains and were lower than ILs that had been—SO3H functionalized. The acidity of dual-functionalized ILs is even higher. The distribution of the products is also linked to the acidity of the ILs that are used as catalysts and co-solvent medium. The hydrolysis reaction was favored by a highly acidic medium, resulting in products that were water-soluble whereas basic medium encourages reactions of liquefaction that result in organic products (Liu et al., 2016).

Table 3.2 shows the effect of ionic liquids on lignin degradation for different feedstock.

Table 3.2 Effect of ionic liquids on lignin degradation for different feedstock.

Chemical composition.	Feedstock used	Efficiency/findings	Ref.
1-ethyl-3-methylimidazolium acetate ([emim][OAc]).	Wheat straw	The use of the ionic liquid was effective in selective fractionation of cellulose, hemicellulose, and lignin into relatively high purity fractions. The catalyst was also effective in the valorization of the phenolic fraction	Yan et al. (2015)
The ILs including 1-(4-sulfobutyl)-3-methyl imidazolium hydrosulfate ([C_4H_8-SO_3Hmim]HSO_4), N-methyl imidazolium hydrosulfate. (HSO_4), 1-butyl-3-methyl imidazolium hydrosulfate ([bmim] HSO4), and 1-(2-carboxyethyl)-3-methyl imidazolium chloride ([C_2H_4COOHmim]Cl).	Sugarcane bagasse.	All ILs studied were very effective towards total degradation of lignin components, showing excellent recyclability up to five times. However, this results into numerous products which caused characteristic separation difficulty	Celikbag et al. (2016)
Dialkylimidazolium-based e.g. ([C4mim]MeSO3); ([C4mim]OAc); ([C4mim]Cl).	Regenerated lignin	The pH, IL composition, and IL content were established to significantly affect the degradation and chemical conversion of lignin structure. It was concluded that low pH helped lignin depolymerization nevertheless destroyed the substructure of lignin	Hu et al. (2014).

Alhassan et al. (2020) / IntechOpen / CC BY 4.0.

Deep eutectic solvents (DESs), are the most appealing catalyst in this field (Hayyan et al., 2014; Zhang et al., 2016). DESs also have additional properties like low volatility, high selectivity, chemical and thermal stability, and green characteristics. One of the additional benefits of using DESs as catalysts is their ease of preparation (Alhassan et al., 2016; Peleteiro et al., 2016). The chemistry of these catalysts suggests that either the acidic quantity, strength, or chemical composition of the quaternary salt affected their delignification efficiency (Alhassan and Kumar, 2016; Hayyan et al., 2014). Choline-derived DESs were used by Liu et al. (2016) to catalyze kraft lignin's selective hydrogenolysis into mono-phenol. Due to its strong acidity and superior thermal stability, [Ch] $MeSO_3$ demonstrated excellent lignin dissolution. The feasibility of extracting DESs based on monocarboxylic, dicarboxylic, and polycarboxylic acids from lignin were investigated. Monocarboxylic acids with a higher acidic strength had better lignin extractability than low acid monocarboxylic acids and dicarboxylic acids because of the release of carbon dioxide when dicaroxylic acids were used. Controversially, the hydroxyl group of polycarboxylic acids is made available for interaction with the lignin's etherified hydroxyl components due to their low viscosity resulting in a high extraction of lignin (Zhang et al., 2016).

In addition to their usual applications in biomass refining, such as lignin degradation, ILs have reportedly been used in the selective production and isolation of biorefinery products like total reducing sugars, furfural and glycerin separation (De Andrade Neto et al., 2016; Hayyan et al., 2010; Peleteiro et al., 2016; Zhuo et al., 2015). As an alternative to microwave assisted methylation, it has been reported that these benzylic alcohols in lignin are selectively oxidized into benzylic ketones before the β-O-4 hydrogenolysis treatment (Zhu et al., 2016).

Yuan et al. (2010) investigated how reaction time affected lignin degradation. The researchers found that for all of the ether bonds in lignin to completely degrade and for the stable C–C bonds to degrade gradually afterward, a long reaction time was required. Secondary reactions like re-polymerization, rearrangement, and cross-linking, were made possible by the intermediates' prolonged reaction induction period, resulting in slightly higher product yields. Response time has fundamentally impacted the properties of individual lignin liquefaction items. Chen et al. (2012) claim that product yield was unaffected by reaction time. Another significant factor in lignin degradation is its composition.

Zhou (2014) found that kraft lignin's HTL yield of water-soluble hydrocarbons was lower than sawdust's liquefaction yield of the same components. Their finding was attributed to kraft lignin's low carbohydrate content. The composition of the solvent is another significant factor in the degradation of lignin.

Yuan et al. (2010) found that the distribution of the products was significantly influenced by the solvent's composition. They noticed that low residue formation was the result of the addition of phenol to the reaction medium, which inhibited side reactions such as the re-polymerization of intermediate products. When comparison was made to individual mono-solvent systems, co-solvent systems, particularly water-ethanol co-solvent systems, demonstrated superior lignin degradation (Cheng et al., 2012).

The HTL process can theoretically and practically produce a large number of chemical compounds. These substances are divided into three groups: phenols, guaiacols, and catechols. However, these key group derivatives have received a lot of attention from the media. For instance, various research teams have produced lignin phenolic compounds. HTL products contain a lot of ortho-methoxyphenolic compounds, which can be identified in their NMR spectra by the presence of $–OCH_3$. Chemical and biochemical uses for this high-volume product include drug manufacturing and clinical diagnosis. For instance, measuring the ortho-methoxyphenols was used in the urinary assay monitoring of wood smoke exposure. Due to its antioxidant properties, perfumes, disinfectants, and other products are produced. It has been reported to be used as a starting point for the synthesis of guaiacol compounds. In a similar manner, 3-methoxyphenol serves as both a catalyst and a building block in the production of antioxidants. The destructive distillation of the phenol portion of coal tar results in the production of guaiacol.

Figure 3.8 shows some phenolic products from HTL of lignin. Figure 3.9 shows some guaiacol derivatives from HTL of lignin. Figure 3.10 presents some selected catechol derivatives.

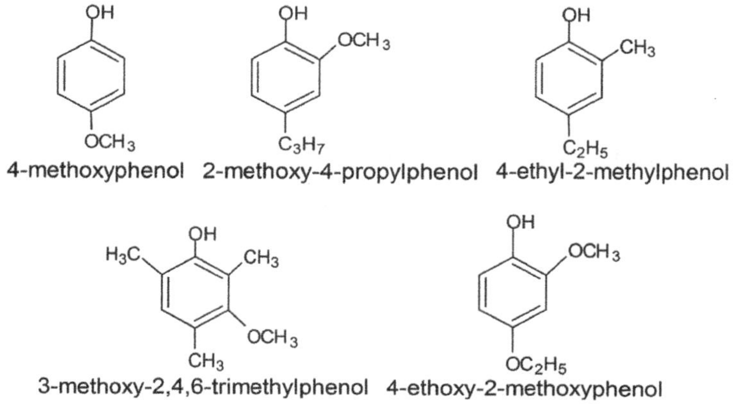

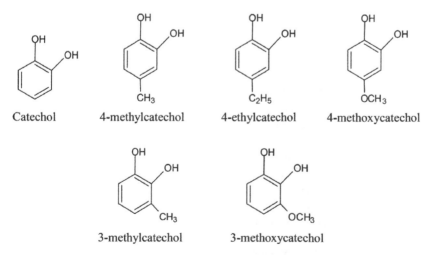

Figure 3.8 Some phenolic products from HTL of lignin. Alhassan et al. (2020) / IntechOpen / CC BY 4.0.

Figure 3.9 Some guaiacol derivatives from HTL of lignin. Alhassan et al. (2020) / IntechOpen / CC BY 4.0.

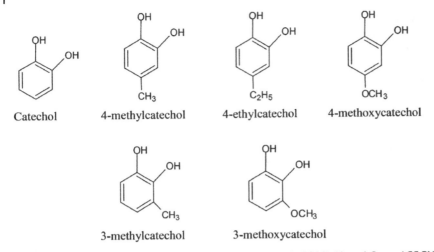

Figure 3.10 Structures of some catechols. Alhassan et al. (2020) / IntechOpen / CC BY 4.0.

Bibliography

Agarwal A, Rana M, and Park JH (2018). Advancement in technologies for the depolymerization of lignin. *Fuel Process Technol*, 181: 115–132.

Alhassan Y, Hornung U, and Bugaje IM (2020). Lignin hydrothermal liquefaction into bifunctional chemicals: a concise review. *Biorefinery Concepts, Energy and Products*, Beschkov V (ed). IntechOpen. https://doi.org/10.5772/intechopen.90860.

Alhassan Y and Kumar N (2016). Single step biodiesel production from Pongamiapinnata (Karanja) seed oil using deep eutectic solvent (DESs) catalysts. *Waste Biomass Valorization*, 7(5): 1055–1065.

Alhassan Y, Kumar N, and Bugaje IM (2016). Hydrothermal liquefaction of de-oiled *Jatropha curcas* cake using deep eutectic solvents (DESs) as catalysts and co-solvents. *Bioresour Technol*, 199: 375–381.

Arpia AA, Chen WH, Lam SS, Rousset P, and de Luna MDG (2021). Sustainable biofuel and bioenergy production from biomass waste residues using microwave-assisted heating: a comprehensive review. *Chem Eng J*, 403: 126233.

Asmadi M, Kawamoto H, and Saka S (2011). Thermal reactivities of catechols/pyrogallols and cresols/xylenols as lignin pyrolysis intermediates. *J Anal Appl Pyrolysis*, 92: 76–87.

Azadi P, Inderwildi OR, Farnood R, and King DA (2013). Liquid fuels, hydrogen and chemicals from lignin: a critical review. *Renew Sustain Energy Rev*, 21: 506–523.

Beauchet R, Monteil-Rivera F, and Lavoie JM (2012). Conversion of lignin to aromatic-based chemicals (L-chems) and biofuels (L-fuels). *Bioresour Technol*, 121: 328–334.

Bi YD, Lei XJ, Xu GH, Chen H, and Hu JL (2018). Catalytic fast pyrolysis of kraft lignin over hierarchical HZSM-5 and Hβ zeolites. *Catalysts*, 8: 82.

Borges FC, Du Z, Xie Q, Trierweiler JO, Cheng Y, Wan Y, Liu Y, Zhu R, Lin X, Chen P, and Ruan R (2014). Fast microwave assisted pyrolysis of biomass using microwave absorbent. *Bioresour Technol*, 2014 Mar; 156: 267–274.

Bu Q, Lei H, Wang L, Wei Y, Zhu L, Zhang X, Liu Y, Yadavalli G, and Tang J (2014). Bio-based phenols and fuel production from catalytic microwave pyrolysis of lignin by activated carbons. *Bioresour Technol*, 2014 Jun; 162: 142–147.

Cao L, Yu IK, Liu Y, Ruan X, Tsang DC, Hunt AJ, Ok YS, Song H, and Zhang S (2018). Lignin valorization for the production of renewable chemicals: state-of-the-art review and future prospects. *Bioresour Technol*, 269: 465–475.

Celikbag Y, Via BK, Adhikari S, Buschle-Diller G, and Auad ML (2016). The effect of ethanol on hydroxyl and carbonyl groups in biopolyol produced by hydrothermal liquefaction of loblolly pine: 31P-NMR and 19F-NMR analysis. *Bioresour Technol*, 214: 37–44.

Chang G, Huang Y, Xie J, Yang H, Liu H, Yin X, and Wu C (2016). The lignin pyrolysis composition and pyrolysis products of palm kernel shell, wheat straw, and pine sawdust. *Energy Convers Manag*, 124: 587–597.

Chauhan PS (2020). Role of various bacterial enzymes in complete depolymerization of lignin: a review. *Biocatal Agric Biotechnol*, 23: 101498.

Chen X, Che QF, Li SJ, Liu ZH, Yang HP, Chen YQ et al. (2019). Recent developments in lignocellulosic biomass catalytic fast pyrolysis: strategies for the optimization of bio-oil quality and yield. *Fuel Process Technol*, 196: 106180.

Chen Y, Chai L, Tang C, Yang Z, Zheng Y, Shi Y, and Zhang H (2012). Kraft lignin biodegradation by *Novosphingobium* sp. B-7 and analysis of the degradation process. *Bioresour Technol*, Nov; 123: 682–685.

Chen XM, Zhu LY, Cui CH, Zhu YN, Zhou ZY, and Qi F (2020). In situ atmospheric pressure photoionization mass spectrometric monitoring of initial pyrolysis products of biomass in real time. *Anal Chem*, 92(1): 603–606.

Chen Z and Long J (2016). Organosolv liquefaction of sugarcane bagasse catalyzed by acidic ionic liquids. *Bioresour Technol*, 214: 16–23.

Cheng S, Wilks C, Yuan Z, Leitch M, and Xu C (2012). Hydrothermal degradation of alkali lignin to bio-phenolic compounds in sub/supercritical ethanol and water-ethanol co-solvent. *Polym Degrad Stab*, 97(6): 839–848.

Chio C, Sain M, and Qin W (2019). Lignin utilization: a review of lignin depolymerization from various aspects. *Renew Sustain Energy Rev*, 107: 232–249.

Custodis VB, Karakoulia SA, Triantafyllidis KS, and van Bokhoven JA (2016). Catalytic fast pyrolysis of lignin over high-surface-area mesoporous aluminosilicates: effect of porosity and acidity. *ChemSusChem*, 23; 9(10): 1134–1145.

Dai G, Zhu Y, Yang J, Pan Y, Wang G, Reubroycharoen P, and Wang S (2019). Mechanism study on the pyrolysis of the typical ether linkages in biomass. *Fuel*, 249: 146–153.

De Andrade Neto JC, De Souza Cabral A, De Oliveira LRD, Torres RB, and Morandim-Giannetti ADA (2016). Synthesis and characterization of new low-cost ILs based on butylammonium cation and application to lignocelluloses hydrolysis. *Carbohydr Polym*, 143: 279–287.

de Wild PJ, Huijgen WJJ, Kloekhorst A, Chowdari RK, and Heeres HJ (2017). Biobased alkylphenols from lignins via a two-step pyrolysis – hydrodeoxygenation approach. *Bioresour Technol*, 229: 160–168.

Dufour A, Weng JJ, Jia LY, Tang XF, Sirjean B, Fournet R et al. (2013). Revealing the chemistry of biomass pyrolysis by means of tunable synchrotron photoionisation-mass spectrometry. *RSC Adv*, 3(14): 4786–4792.

Engtrakul C, Mukarakate C, Starace AK, Magrini KA, Rogers AK, and Yung MM (2016). Effect of ZSM-5 acidity on aromatic product selectivity during upgrading of pine pyrolysis vapors. *Catal Today*, 269: 175–181.

Fan L, Chen P, Zhang Y, Liu S, Liu Y, Wang Y, Dai L, and Ruan R (2017a). Fast microwave-assisted catalytic co-pyrolysis of lignin and low-density polyethylene with HZSM-5 and MgO for improved bio-oil yield and quality. *Bioresour Technol*, Feb; 225: 199–205.

Fan L, Zhang Y, Liu S, Zhou N, Chen P, Cheng Y, Addy M, Lu Q, Omar MM, Liu Y, Wang Y, Dai L, Anderson E, Peng P, Lei H, and Ruan R (2017b). Bio-oil from fast pyrolysis of lignin: effects of process and upgrading parameters. *Bioresour Technol*, Oct; 241: 1118–1126.

Haarlemmer G, Guizani C, Anouti S, Deniel M, Roubaud A, and Valin S (2016). Analysis and comparison of bio-oils obtained by hydrothermal liquefaction and fast pyrolysis of beech wood. *Fuel*, 174: 180–188.

Hayyan A, Hashim MA, Hayyan M, Mjalli FS, and Alnashef IM (2014). A new processing route for cleaner production of biodiesel fuel using a choline chloride based deep eutectic solvent. *J Clean Prod*, 65: 246–251.

Hayyan M, Mjalli FS, Hashim MA, and AlNashef IM (2010). A novel technique for separating glycerine from palm oil-based biodiesel using ionic liquids. *Fuel Process Technol*, 91(1): 116–120.

He T, Zhang YM, Zhu YA, Wen W, Pan Y, Wu JL et al. (2016). Pyrolysis mechanism study of lignin model compounds by synchrotron vacuum ultraviolet photoionization mass spectrometry. *Energy Fuels*, 30(3): 2204–2208.

Hemberger P, Custodis VBF, Bodi A, Gerber T, and van Bokhoven JA (2017). Understanding the mechanism of catalytic fast pyrolysis by unveiling reactive intermediates in heterogeneous catalysis. *Nat Commun*, 8: 15946.

Hertzog J, Carré V, Jia L, Mackay CL, Pinard L, Dufour A, Mašek O, and Aubriet F (2018). Catalytic fast pyrolysis of biomass over microporous and hierarchical zeolites: characterization of heavy products. *ACS Sustain Chem Eng*, 6: 4717–4728.

Hoang AT, Ong HC, Fattah IMR, Chong CT, Cheng CK, Sakthivel R, and Ok YS (2021). Progress on the lignocellulosic biomass pyrolysis for biofuel production toward environmental sustainability. *Fuel Process Technol*, 223: 106997.

Hu J, Shen D, Wu S, Zhang H, and Xiao R (2014). Effect of temperature on structure evolution in char from hydrothermal degradation of lignin. *J Anal Appl Pyrolysis*, 106: 118–124.

Huang J, Liu C, Wei S, Huang X, and Li H (2010). Density functional theory studies on pyrolysis mechanism of β-d-glucopyranose. *J Mol Struct*, 958: 64–70.

Hurt MR, Degenstein JC, Gawecki P, Borton DJ, Vinueza NR, Yang L, Agrawal R, Delgass WN, Ribeiro FH, and Kenttämaa HI (2013). On-line mass spectrometric methods for the determination of the primary products of fast pyrolysis of carbohydrates and for their gas-phase manipulation. *Anal Chem*, 85(22): 10927–10934.

Jae J, Tompsett GA, Foster AJ, Hammond KD, Auerbach SM, Lobo RF, and Huber GW (2011). Investigation into the shape selectivity of zeolite catalysts for biomass conversion. *J Catal*, 279: 257–268.

Jarvis MW, Daily JW, Carstensen HH, Dean AM, Sharma S, Dayton DC, Robichaud DJ, and Nimlos MR (2011). Direct detection of products from the pyrolysis of 2-phenethyl phenyl ether. *J Phys Chem A*, Feb 3; 115(4): 428–438.

Jia LY, Le-Brech Y, Shrestha B, Bente-von Frowein M, Ehlert S, Mauviel G, Zimmermann R, and Dufour A (2015). Fast pyrolysis in a microfluidized bed reactor: effect of biomass properties and operating conditions on volatiles composition as analyzed by online single photoionization mass spectrometry. *Energy Fuels*, 29(11): 7364–7374.

Jia LY, Raad M, Hamieh S, Toufaily J, Hamieh T, Bettahar MM, Mauviel G, Tarrighi M, Pinard L, and Dufour A (2017). Catalytic fast pyrolysis of biomass: superior selectivity of hierarchical zeolites to aromatics. *Green Chem*, 19: 5442–5459.

Jiang G, Nowakowski DJ, and Bridgwater AV (2010). Effect of the temperature on the composition of lignin pyrolysis products. *Energy Fuel*, 24: 4470–4475.

Junior JS, Carvalho W, and Ataíde C (2018). Catalytic effect of ZSM-5 zeolite and HY-340 niobic acid on the pyrolysis of industrial Kraft lignins. *Ind Crop Prod*, 111: 126–132.

Kalogiannis KG, Stefanidis SD, Michailof CM, Lappas AA, and Sjöholm E (2015). Pyrolysis of lignin with 2DGC quantification of lignin oil: effect of lignin type, process temperature and ZSM-5 in situ upgrading. *J Anal Appl Pyrolysis*, 115: 410–418.

Kawamoto H (2017). Lignin pyrolysis reactions. *J Wood Sci*, 63: 117–132.

Kim JY, Lee JH, Park J, Kim JK, An D, Song LK, and Choi JW (2015). Catalytic pyrolysis of lignin over HZSM-5 catalysts: effect of various parameters on the production of aromatic hydrocarbon. *J Anal Appl Pyrolysis*, 114: 273–280.

Kleinert M and Barth T (2008). Towards a lignincellulosic biorefinery: direct one-step conversion of lignin to hydrogen-enriched biofuel. *Energy Fuels*, 22(2): 1371–1379.

Konduri MK, Kong F, and Fatehi P (2015). Production of carboxymethylated lignin and its application as a dispersant. *Eur Polym J*, 70: 371–383.

Kotake T, Kawamoto H, and Saka S (2015). Pyrolytic formation of monomers from hardwood lignin as studied from the reactivities of the primary products. *J Anal Appl Pyrolysis*, 113: 57–64.

Kruse A and Dahmen N (2015). Water—A magic solvent for biomass conversion. *J Supercrit Fluids*, 96: 36–45.

Kumar G, Shobana S, Chen WH, Bach QV, Kim SH, Atabani AE, and Chang JS (2017). A review of thermochemical conversion of microalgal biomass for biofuels: chemistry and processes. *Green Chem*, 19(1). https://doi.org/10.1039/c6gc01937d.

Kurnia I, Karnjanakom S, Bayu A, Yoshida A, Rizkiana J, Prakoso T, Abudula A, and Guan G (2017). In-situ catalytic upgrading of bio-oil derived from fast pyrolysis of lignin over high aluminum zeolites. *Fuel Process Technol*, 167: 730–737.

Le Brech Y, Jia LY, Cisse S, Mauviel G, Brosse N, and Dufour A (2016). Mechanisms of biomass pyrolysis studied by combining a fixed bed reactor with advanced gas analysis. *J Anal Appl Pyrolysis*, 117: 334–346.

Lee HW, Kim Y, Jae J, Sung BH, Jung S, Kim SC, Jeon J, and Park Y (2016). Catalytic pyrolysis of lignin using a two-stage fixed bed reactor comprised of in-situ natural zeolite and ex-situ HZSM-5. *J Anal Appl Pyrolysis*, 122: 282–288.

Li B, Lv W, Zhang Q, Wang T, and Ma L (2014). Pyrolysis and catalytic pyrolysis of industrial lignins by TG-FTIR: kinetics and products. *J Anal Appl Pyrolysis*, 108: 295–300.

Li D, Briens C, and Berruti F (2015). Improved lignin pyrolysis for phenolics production in a bubbling bed reactor – effect of bed materials. *Bioresour Technol*, 189: 7–14.

Li N, Bi Y, Xia X, Chen H, and Hu J (2017). Hydrodeoxygenation of methyl laurate over Ni catalysts supported on hierarchical HZSM-5 zeolite. *Catalysts*, 7(12): 383.

Li T and Takkellapati S (2018). The current and emerging sources of technical lignins and their applications. *Biofuel Bioprod Biorefin*, 12(5): 756–787.

Li X, Su L, Wang Y, Yu Y, Wang C, Li X, and Wang Z (2012). Catalytic fast pyrolysis of Kraft lignin with HZSM-5 zeolite for producing aromatic hydrocarbons. *Front Envir Sci Eng*, 6(3): 295–303.

Liu CJ, Ye LL, Yuan WH, Zhang Y, Zou JB, Yang JZ, Wang Y, Qi F, and Zhou Z (2018). Investigation on pyrolysis mechanism of guaiacol as lignin model compound at atmospheric pressure. *Fuel*, 232: 632–638.

Liu CJ, Zhou CQ, Wang Y, Liu XH, Zhu LY, Ma H, Zhou Z, and Qi F (2021). Gasphase hydrodeoxygenation of bio-oil model compound over nitrogen-doped carbon-supported palladium catalyst. *Proc Combust Inst*, 38(3): 4345–4353.

Liu F, Liu Q, Wang A, and Zhang T (2016). Direct catalytic hydrogenolysis of kraft lignin to phenols in choline derived ionic liquids. *ACS Sustain Chem Eng*, Acssuschemeng. 6b00620. 4(7): 3850–3856.

Lopez-Camas K, Arshad M, and Ullah A (2020). Chemical modification of lignin by polymerization and depolymerization. *Lignin: Biosynthesis and Transformation for Industrial Applications*, Sharma S and Kumar A. (ed.). Springer, Cham, pp. 139–180.

Lu X and Gu X (2022). A review on lignin pyrolysis: pyrolytic behavior, mechanism, and relevant upgrading for improving process efficiency. *Biotechnol Biofuels*, 15: 106.

Luo Z, Wang S, and Guo X (2012). Selective pyrolysis of Organosolv lignin over zeolites with product analysis by TG-FTIR. *J Anal Appl Pyrolysis*, 95(2012): 112–117.

Ma B and Zhao C (2015). High-grade diesel production by hydrodeoxygenation of palm oil over a hierarchically structured Ni/HBEA catalyst. *Green Chem*, 17: 1692–1701.

Ma Z, Troussard E, and van Bokhoven JA (2012). Controlling the selectivity to chemicals from lignin via catalytic fast pyrolysis. *Appl Catal A Gen*, 423–424: 130–136.

Maldhure AV and Ekhe JD (2013). Pyrolysis of purified kraft lignin in the presence of AlCl3 and ZnCl2. *J Environ Chem Eng*, 1: 844–849.

Milovanovic J, Luque R, Tschentscher R, Romero AA, Li H, Shih K, and Rajic N (2017). Study on the pyrolysis products of two different hardwood lignins in the presence of NiO contained-zeolites. *Biomass Bioenergy*, 103: 29–34.

Mukkamala S, Wheeler MC, van Heiningen ARP, and DeSisto WJ (2012). Formate-assisted fast pyrolysis of lignin. *Energy Fuel*, 26: 1380–1384.

Murciano Martinez P, Punt AM, Kabel MA, and Gruppen H (2016). Deconstruction of lignin linked p-coumarates, ferulates and xylan by NaOH enhances the enzymatic conversion of glucan. *Bioresour Technol*, 216: 44–51.

Nguyen LT, Phan DP, Sarwar A, Tran MH, Lee OK, and Lee EY (2021). Valorization of industrial lignin to value-added chemicals by chemical depolymerization and biological conversion. *Ind Crop Prod*, 161: 113219.

Nguyen TD, Maschietti M, Belkheiri T, Åmand L, Theliander H, Vamling L, Olausson L, and Andersson S (2014). Catalytic depolymerisation and conversion of Kraft lignin into liquid products using near-critical water. *J Supercrit Fluids*, 86: 67–75.

Patwardhan PR, Brown RC, and Shanks BH (2011). Understanding the fast pyrolysis of lignin. *ChemSusChem*, 4(11): 1629–1636.

Peleteiro S, Santos V, Garrote G, and Paraj JC (2016). Furfural production from eucalyptus wood using an acidic ionic liquid. *Carbohydr Polym*, 146: 20–25.

Qi Z, Jie C, Tiejun W, and Ying X (2007). Review of biomass pyrolysis oil properties and upgrading research. *Energy Convers Manag*, 48: 87–92.

Rezaei PS, Shafaghat H, and Daud WMAW (2016). Aromatic hydrocarbon production by catalytic pyrolysis of palm kernel shell waste using a bifunctional Fe/HBeta catalyst: effect of lignin-derived phenolics on zeolite deactivation. *Green Chem*, 18: 1684–1693.

Roy R, Rahman MS, Amit TA, and Jadhav B (2022). Recent advances in lignin depolymerization techniques: a comparative overview of traditional and greener approaches. *Biomass*, 2: 130–154. https://doi.org/10.3390/biomass2030009.

Schutyser W, Renders T, Van den Bosch S, Koelewijn SF, Beckham GT, and Sels BF (2018). Chemicals from lignin: an interplay of lignocelluloses fractionation, depolymerisation, and upgrading. *Chem Soc Rev*, 47(3): 852–908.

Schwarz D, Dorrstein J, Kugler S, Schieder D, Zollfrank C, and Sieber V (2016). Integrated biorefinery concept for grass silage using a combination of adapted pulping methods for advanced saccharification and extraction of lignin. *Bioresour Technol*, 216: 462–470.

Shen D, Zhao J, and Xiao R (2016). Catalytic transformation of lignin to aromatic hydrocarbons over solid-acid catalyst: effect of lignin sources and catalyst species. *Energy Convers Manag*, 124: 61–72.

Shen D, Zhao J, Xiao R, and Gu S (2015a). Production of aromatic monomers from catalytic pyrolysis of black-liquor lignin. *J Anal Appl Pyrolysis*, 111: 47–54.

Shen DK, Liu GF, Zhao J, Xue JT, Guan SP, and Xiao R (2015b). Thermo-chemical conversion of lignin to aromatic compounds: effect of lignin source and reaction temperature. *J Anal Appl Pyrolysis*, 112: 56–65.

Sheng H, Murria P, Degenstein JC, Tang W, Riedeman JS, Hurt MR, Dow A, Klein I, Zhu H, Nash JJ, Abu-Omar M, Agrawal R, Delgass WN, Ribeiro FH, and Kenttämaa HI (2017). Initial products and reaction mechanisms for fast pyrolysis of synthetic G-lignin oligomers with β-O-4 linkages via on-line mass spectrometry and quantum chemical calculations. *Chemistryselect*, 2: 7185–7193.

Singh R, Prakash A, Dhiman SK, Balagurumurthy B, Arora AK, Puri SK, and Bhaskar T (2014). Hydrothermal conversion of lignin to substituted phenols and aromatic ethers. *Bioresour Technol*, Aug; 165: 319–322.

Stanton AR, Iisa K, Mukarakate C, and Nimlos MR (2018). Role of biopolymers in the deactivation of ZSM-5 during catalytic fast pyrolysis of biomass. *ACS Sustain Chem Eng*, 6(8): 10030–10038.

Struven JO and Meier D (2016). Hydrocracking of organosolv lignin in subcritical water to useful phenols employing various Raney nickel catalysts. *ACS Sustain Chem Eng*, 7(3): Acssuschemeng.6b00342.

Supriyanto, Usino DO, Ylitervo P, Dou J, Sipponen MH, and Richards T (2020). Identifying the primary reactions and products of fast pyrolysis of alkali lignin. *J Anal Appl Pyrol*, 151: 104917.

Toor SS, Rosendahl L, and Rudolf A (2011). Hydrothermal liquefaction of biomass: a review of subcritical water technologies. *Energy*, 36(5): 2328–2342.

Trinh NT, Jensen PA, Sárossy Z, Dam-Johansen K, Knudsen NO, Sørensen HR, and Egsgaard H (2013). Fast pyrolysis of lignin using a pyrolysis centrifuge reactor. *Energy Fuels*, 27: 3802–3810.

Wang H, Pu Y, Ragauskas A, and Yang B (2019). From lignin to valuable products–strategies, challenges, and prospects. *Bioresour Technol*, 271: 449–461.

Wang K, Kim KH, and Brown RC (2014). Catalytic pyrolysis of individual components of lignocellulosic biomass. *Green Chem*, 16: 727–735.

Wang S, Ru B, Lin H, Sun W, and Luo Z (2015c). Pyrolysis behaviors of four lignin polymers isolated from the same pine wood. *Bioresour Technol*, Apr; 182: 120–127.

Wang SR, Dai GX, Yang HP, and Luo ZY (2017). Lignocellulosic biomass pyrolysis mechanism: a state-of-the-art review. *Prog Energy Combust Sci*, 62: 33–86.

Wang WL, Ren XY, Chang JM, Cai LP, and Shi SQ (2015b). Characterization of bio-oils and bio-chars obtained from the catalytic pyrolysis of alkali lignin with metal chlorides. *Fuel Process Technol*, 138: 605–611.

Wang WL, Ren XY, Li LF, Chang JM, Cai LP, and Geng J (2015a). Catalytic effect of metal chlorides on analytical pyrolysis of alkali lignin. *Fuel Process Technol*, 134: 345–351.

Xu L, Yao Q, Zhang Y, and Fu Y (2017). Integrated production of aromatic amines and N-doped carbon from lignin via ex situ catalytic fast pyrolysis in the presence of ammonia over zeolites. *ACS Sustain Chem Eng*, 5: 2960–2969.

Xu YH and Li MF (2021). Hydrothermal liquefaction of lignocellulose for value-added products: mechanism, parameter and production application. *Bioresour Technol*, 342: 126035.

Xu Z, Lei P, Zhai R, Wen Z, and Jin M (2019). Recent advances in lignin valorization with bacterial cultures: microorganisms, metabolic pathways, and bio-products. *Biotechnol Biofuels*, 12: 32.

Yan B, Li K, Wei L, Ma Y, Shao G, Zhao D, Wan W, and Song L (2015). Understanding lignin treatment in dialkylimidazolium-based ionic liquid-water mixtures. *Bioresour Technol*, 196: 509–517.

Yang C, Wang S, Yang J, Xu D, Li Y, Li J, and Zhang Y (2020b). Hydrothermal liquefaction and gasification of biomass and model compounds: a review. *Green Chem*, 22: 8210–8232.

Yang J, Wang X, Shen B, Hu Z, Xu L, and Yang S (2020a). Lignin from energy plant (Arundo donax): pyrolysis kinetics, mechanism and pathway evaluation. *Renew Energy*, 161: 963–971.

Yang S, Yuan TQ, Li MF, and Sun RC (2015). Hydrothermal degradation of lignin: products analysis for phenol formaldehyde adhesive synthesis. *Int J Biol Macromol*, 72: 54–62.

Yang T, Jie Y, Li B, Kai X, Yan Z, and Li R (2016). Catalytic hydrodeoxygenation of crude bio-oil over an unsupported bimetallic dispersed catalyst in supercritical ethanol. *Fuel Process Technol*, 148: 19–27.

Yu Y, Li X, Su L, Zhang Y, Wang Y, and Zhang H (2012). The role of shape selectivity in catalytic fast pyrolysis of lignin with zeolite catalysts. *Appl Catal A Gen*, 447–448: 115–123.

Yuan Z, Cheng S, Leitch M, and Xu CC (2010). Hydrolytic degradation of alkaline lignin in hot-compressed water and ethanol. *Bioresour Technol*, 01(23): 9308–9313.

Zakzeski J, Bruijnincx PCA, Jongerius AL, and Weckhuysen BM (2010). The catalytic valorization of ligning for the production of renewable chemicals. *Chem Rev*, 110: 3552–3599.

Zhang C-W, Xia S-Q, and Ma P-S (2016b). Facile pretreatment of lignocellulosic biomass using deep eutectic solvents. *Bioresour Technol*, 219: 1–5.

Zhang H, Xiao R, Nie J, Jin B, Shao S, and Xiao G (2015c). Catalytic pyrolysis of black-liquor lignin by co-feeding with different plastics in a fluidized bed reactor. *Bioresour Technol*, Sep; 192: 68–74.

Zhang P, Liu Y, Fan M, and Jiang P (2016a). Catalytic performance of a novel amphiphilic alkaline ionic liquid for biodiesel production: influence of basicity and conductivity. *Renew Energy*, 86: 99–105.

Zhang Q, Chang J, Wang T, and Xu Y (2007). Review of biomass pyrolysis oil properties and upgrading research. *Energy Convers Manag*, 48: 87–92.

Zhou N, Thilakarathna WPDW, He QS, and Rupasinghe HPV (2022). A review: depolymerization of lignin to generate high-value bio-products: opportunities, challenges, and prospects. *Front Energy Res*, 9: 758744.

Zhou XF (2014). Conversion of Kraft lignin under hydrothermal conditions. *Bioresour Technol*, 170: 583–586.

Zhou ZY, Jin HF, Zhao L, Wang YZ, Wen W, Yang JZ, Li YP, and Qi F (2017). A thermal decomposition study of pine wood under ambient pressure using thermogravimetry combined with synchrotron vacuum ultraviolet photoionization mass spectrometry. *Proc Combust Inst*, 36(2): 2217–2224.

Zhu G, Qiu X, Zhao Y, Qian Y, Pang Y, and Ouyang X (2016). Depolymerization of lignin by microwave-assisted methylation of benzylic alcohols. *Bioresour Technol*, 218: 718–722.

Zhu JL, Yang H, Hu HQ, Zhou Y, Li JG, and Jin LJ (2020). Novel insight into pyrolysis behaviors of lignin using in-situ pyrolysis-double ionization time-of-flight mass spectrometry combined with electron paramagnetic resonance spectroscopy. *Bioresour Technol*, 312: 123555.

Zhu L, Cui C, Liu H, Zhou Z, and Qi F (2022). Thermochemical depolymerization of lignin: process analysis with state-of-the-art soft ionization mass spectrometry. *Front Chem Eng*, 4: 982126. https://doi.org/10.3389/fceng.2022.982126.

Zhuo K, Du Q, Bai G, Wang C, Chen Y, and Wang J (2015). Hydrolysis of cellulose catalyzed by novel acidic ionic liquids. *Carbohydr Polym*, 115: 49–53.

3.2 Biological Depolymerization

The term "biological depolymerization of lignin" refers to the relatively simple process by which fungi, bacteria, or isolated enzymes degrade lignin under relatively milder conditions. According to Chen and Wan (2017), in vitro lignin depolymerization uses a variety of fungi and bacteria-derived lignin-degrading enzymes, so depolymerization of lignin with enzymes overlaps with bacteria and fungi-assisted depolymerization. As a result, biological depolymerization of lignin can fundamentally be regarded as enzymatic depolymerization in both in vivo and in vitro settings.

Based on the discovery of multiple classes of potent oxidative enzymes produced by white rot fungi, several studies have been conducted on biological depolymerization of lignin over the past few decades. The name "white-rot" comes from the white appearance of the wood when these fungi attack it. When lignin is removed from the wood, it takes on a bleached appearance. (Barr and Aust, 1994; Eaton and Hale, 1993; Heinzkill et al., 1998; Kaal et al., 1995; Kirk and Farrell, 1987; Leonowicz et al., 1999; Martinez et al., 2004; Pointing, 2001; Robinson et al., 2001; Ruttimann-Johnson et al., 1993; Tien and Kirk, 1983; Wan and Li, 2012; Wesenberg et al., 2003; Wong, 2009).

Lignin peroxidase (LiP), manganese peroxidase (MnP), laccase, glyoxal oxidase, versatile peroxidase (VP), aryl-alcohol oxidase (AAO), and dye-decolorizing peroxidase (DyPs) are some of the enzymes that belong to this group (Martínez et al., 2005; Salvachúa et al., 2013a, 2013b, 2020b, 2020a).

The majority of these enzymes use radical intermediates for the depolymerization of lignin and have broad substrate specificity. Because enzymes that are specific to particular stereochemistry and linkages were discovered, the range of enzymatic systems that microbes use to depolymerize lignin has expanded (Picart et al., 2015; Bugg et al., 2011a, 2011b; Sonoki et al., 2002).

The discovery that some bacteria can produce many of the same kinds of enzymes has also shown that fungi are not the only organisms capable of biological lignin depolymerization (Ahmad et al., 2011; Salvachúa et al., 2015). Despite these efforts, it is still not entirely clear how ligninolytic enzymes interact with one another to depolymerize lignin, which restricts our ability to utilize these potent enzymes for industrial depolymerization of lignin in an efficient manner (Beckham et al., 2016). Several studies that combine in-depth lignin characterization and proteomic analysis of fungal secretomes are providing deeper insights into the nature of these interactions (Fernandez Fueyo et al., 2016; Salvachúa et al., 2013a, 2013b). The production of aromatic compounds that could be isolated and turned into bioproducts, which is a developing goal in biorefineries to increase overall process economics, as well as the removal of lignin from biomass, which is a common goal to overcome resistance, would ideally be made possible by the ability to depolymerize lignin with enzymes (Ragauskas et al., 2014).

Biological depolymerization, is cost-efficient and specific so it is deemed superior to other lignin depolymerization methods, which need higher energy requirement and strict conditions (Xu et al., 2018a, 2018b). Numerous studies have shown that biological depolymerization of lignin is possible (Giri and Sharma, 2020; Hemati et al., 2022; Hermosilla et al., 2018; Kumar and Chandra, 2020; Lee et al., 2019; Numata and Morisaki, 2015; Nurika et al., 2022; Salvachua et al., 2020a, 2020b; Srivastava et al., 2022; Weiss et al., 2020; Xu et al., 2019; Xu et al., 2018a, 2018b, 2020; Zhang et al., 2021). However, additional research is required to address the typical obstacles, like low yield and low productivity, before looking into potential industrial applications (Chen and Wan, 2017).

3.2.1 Lignin Depolymerization by Fungi

Increasing numbers of microorganisms have shown that they can convert lignin into fungible fuels and products due to the fast advancement of multi-omics technologies (Floudas et al., 2020). For a long time, fungi, most of which are white rot and some of which are brown rot, have been extensively utilized in the decomposition of lignin and the elimination of pollutants with similar structures (Bai et al., 2017; Goodell, 2020). Vanillic acid, syringyl alcohol, protocatechuic acid, and ferulic acid, are among the valuable phenolic precursors that have been found to be released from lignocellulose biomass by some fungi, according to many studies (Andlar et al., 2018; Dashtban et al., 2010). However, hardly any of these valuable compounds are captured as intermediates and are therefore incapable of being fermented on a large scale. Tables 3.3 and 3.4 list a few significant fungi that are effective at degrading lignin (Alhassan et al. 2020; Weng et al., 2021).

Table 3.3 Lignin depolymerization through fungi.

Fungal Strains	Substrate	Strategy/Pathway	Enzymes	References
Phanerochaete chrysosporium	Synthetic lignin and free-hydroxyl phenolic groups	Multi enzyme approach, ortho-cleavage pathway, and phenanthrene metabolism	Lignin peroxidase, manganese peroxidase, dehydrogenase, engineered 4-O-methyltransferase	(Hong et al., 2017; Pham and Kim, 2016)
Phanerochaete chrysosporium	*Irpex lacteus* CD2	Alkali lignin	The synergistic approach of the fungal co-culture Non-specific lignin-degrading enzymes	(Ruhong et al., 2020)
Physisporinus vitreus	Monomeric and dimeric, phenolic and non-phenolic lignin model compounds	Enzymatic hydrolysis of corn stover in vitro	Versatile peroxidase	(Kong et al., 2017; Weng et al., 2021)
Ceriporiopsis subvermispora	Non-phenolic lignin model monomers and dimers	One-oxidation-electron mechanism	Unidentified	(Nguyen et al., 2022; Wan and Li, 2010)
Ceriporiopsis subvermispora, Pleurotus eryngii Lentinula edodes	Wheat straw lignin and structural motifs	Selective delignification	Unidentified	(van Erven et al., 2019)
Ceriporiopsis subvermispora	Specificities for typical LiP/VP substrates	Lignocellulose de lignification through high selectivity	Newly discovered peroxidases	(Fernández-Fueyo et al., 2012)
Pleurotus ostreatus	Lignin model dimer and synthetic lignin corn stover lignin	Heterologous expression in *E.coli*, and heterologous expression in *Pichia pastoris*	Versatile peroxidases, manganese peroxidases, and aryl-alcohol dehydrogenase laccase	Fernández-Fueyo et al. (2014); Wan and Li (2010).
Trametes versicolor	Phenol, p-creosol	Eukaryotic β-ketoadipate pathway	Laccases	(Alexieva et al., 2010; Xu et al., 2020)
Anthracophyllum discolor	Pyrene, phenanthrene, fluoranthene, anthracene, and benzo pyrene	Polycyclic aromatic hydrocarbon degradation in Kirk medium	Mn peroxidase; laccase and lignin peroxidase	(Acevedo et al., 2011)

Alhassan et al. (2020) / MDPI / CC BY 4.0.

Table 3.4 Fungi degradation of lignin in various biomass sources.

Fungi	Strains	Biomass materials	Lignin degradation	References
WHITE-ROT	*Phanerochaete chrysosporium*	Wheat straw and cornstalk	30%, 34.3%	(Singh et al., 2011; Zhao et al., 2012)
	Pleurotus ostreatus	Rice straw	41%	Taniguchi, 2005
	Lentinula edode LE16	Sugarcane bagasse	87.6%	Dong et al., 2013
	Phlebia sp. MG-60	Oak wood	40.7%	Kamei et al., 2012
	Ceriporiopsis subvermispora	Pinus taeda wood chips	22%	Guerra et al., 2004
	Trametes versicolor	Radiata pine wood chips	22%	Shirkavand, 2017
	Dichomytus squalens	Wheat straw	34.1%	Knežević et al., 2013
	Gloeophyllum trabeum	Wafers of spruce wood	16%	Yelle et al., 2008
	Fomitopsis pinicola	Wheat straw	32.4%	Knežević et al., 2013
	Polyporus ostreiformis	Rice straw	18.6%	Dey, 1994

Weng et al. 2021 / Springer Nature / CC BY 4.0.

Different lignin-degrading enzymes and microorganism metabolic systems have evolved to depolymerize and convert lignin, which is necessary for its utilization (Brown and Chang, 2014). Fungi are able to produce numerous enzymes that break down lignin, making them the most efficient lignin-degrading microorganisms. There are three main categories of lignin-degrading fungi, according to the lignin degradation mechanism: fungi that cause soft, brown, and white rot (Andlar et al., 2018). White-rot fungi are the only of the three fungi that can completely degrade lignin into carbon dioxide and water (Blanchette, 1995). These fungi are superior to that of soft-rot and brown-rot fungi (Andlar et al., 2018).

White-rot fungi are ideal for delignification because they produce extracellular oxidative enzymes which degrade lignin. The ability of white-rot fungi, including *Phanerochaete chrysosporium*, *Ceriporiopsis subvermispora*, *Pleurotus ostreatus*, *Trametes versicolor* and *Cyathus stercoreus*, to degrade lignin has been studied (Martínez et al., 2004; Ruttimann-Johnson et al., 1993; Wan and Li, 2012). White-rot fungi degrade lignin, resulting in fibrous, whitish-colored decayed wood. *Phlebia* spp., *C. subvermispora*, *Phellinus pini*, and other white-rot fungi and the genus *Pleurotus* delignify wood by attacking lignin in a more aggressive manner than cellulose and hemicellulose resulting in enhanced cellulose. However, other white-rot fungi, including *Heterobasidion annosum*, *T. versicolor* and *Irpex lacteus*, simultaneously degrade the components of the cell wall (Wong, 2009).

The majority of basidiomycetes and a small number of ascomycetes are among the white-rot fungi that degrade lignin (Abdel-Hamid et al., 2013; Sigoillot et al., 2012; Wong, 2009)

White-rot fungi's main extracellular enzymes for breaking down lignin were oxidases and peroxidases. For lignin decomposition, oxidoreductase is responsible for catalyzing the breakage of carbon–carbon bonds, ether links, side chains, and aromatic rings (Zabel and Morrell, 2020).

P. chrysosporium has been utilized for the biological pre-treatment of lignocelluloses (Singh et al., 2011; Zhao et al., 2012). It is a model white-rot fungus that breaks down lignin. *P. chrysosporium's* manganese and lignin peroxidase enzymes do not oxidize lignin specifically (Zeng et al., 2013a).

It was also discovered that sugarcane bagasse lignin was degraded by the fungi *Pleurotus ostreatus* PO45 and *Lentinula edode* LE16 which produced MnP and polyphenol oxidase (PPO) (Dong et al., 2013). Laccases and peroxidases for lignin oxidation and decomposition can be produced by some white-rot fungal species. Through fermentation with *Phlebia* sp., lignin from oak wood was converted into ethanol directly. The culture contained the identifiers MG-60, manganese peroxidase, and laccase (Kamei et al., 2012).

In solid-state fermentation, the lignin of *Pinus taeda* wood chips was primarily degraded by the manganese peroxidase and laccase's β-O-aryl ether cleavage by *C. subvermispora* (Guerra et al., 2004). The laccase (Lcc1) that was discovered in *Ganoderma tsugae* has the ability to speed up stipe elongation, mycelium growth, the formation of pigment, and the breakdown of lignin (Jin et al., 2018).

White-rot fungi are helpful in the biopulping process used in the paper industry because they can break down lignin. These fungi have also been used for other applications in industry such as biorefinery and bioremediation (Asgher et al., 2008; Reddy, 1995).

The brown-rot fungi, such as *Laetiporus portentosus*, *Gloeophyllum trabeum* and *Fomitopsis lilacinogilva* account for 7% of wood-rotting basidiomycetes. These fungi grow primarily on conifers, and are another group of wood-rotting fungi. Brown-rot fungi, in contrast to white-rot fungi, partially alter lignin while degrading wood polysaccharides. The wood shrinks, turns brown from oxidized lignin, and cracks into roughly cubic pieces as a result of this kind of decay (Gilbertson, 1980; Monrroy et al., 2011).

Brown-rot fungi were found to degrade lignin using Fenton oxidation chemistry-generated hydroxyl radicals (Bugg et al., 2011a, 2011b). Brown-rot fungi produce extracellular hydroquinones which are able to reduce the Fe^{3+} in the Fe–oxalate complex to Fe^{2+}. This Fe^{2+} then reacts with hydrogen peroxide and produce hydroxyl radicals. Redox cycling can be achieved by converting the oxidized quinone into hydroquinone (Jensen et al., 2001). The lignin intermonomer side-chain linkages can be broken non-specifically by *Gloeophyllum trabeum*, and its aging can result in 16% loss of lignin in spruce wood (Yelle et al., 2008). Aspen wood treated with *Postia placenta*, which breaks down lignin with hydroxyl radicals and produces an extracellular Fenton system, had lower levels of arylglycerol-aryl ether (Yelle et al., 2011). *Fomitopsis pinicola* degraded about 32.4% of the lignin after two weeks of fermentation (Knežević et al., 2013).

Soft-rot fungi, like brown-rot fungi, are able to break down lignin by attacking the syringyl units (Zabel and Morrell, 2020). Ascomycetes and deuteromycetes are the most common soft-rot fungi, and they preferentially degrade hardwood (Kuhad, 1997).

Aspergillus niger and *Penicillium chrysogenum*, two soft-rot fungi, rapidly degraded sycamore and pine wood (Hamed, 2013). Additionally, few soft-rot fungi were found to rapidly degrade vanillic acid and phenols (Aarti and Agastian, 2015).

Although the enzymes that soft-rot fungi use to break down lignin are poorly understood, it has been suggested that soft-rot fungi might alter lignin instead of mineralizing it.

3.2.2 Lignin Depolymerization by Bacteria

Bacteria do not degrade lignin to the same extent as fungi or other organisms. Various habitats, including animal gut, rotting wood, a wastewater treatment plant, and soil, have been found to contain lignin-degrading bacteria (Xu et al., 2018a, 2018b). Despite being not much more effective than fungi at degrading lignin, bacteria are more adaptable to their environment. According to a research, proteobacteria, actinobacteria, and firmicutes are significant lignin-degrading bacteria (Bugg et al., 2011a, 2011b). When oxygen is present, bacteria which grow on lignin release oxidative enzymes that break down lignin. In addition, extreme anaerobic conditions can degrade lignin. Under aerobic conditions, bacterial lignin depolymerization occurs primarily (Zimmermann, 1990). The typical bacteria for the degradation of lignin are *Streptomyces* and *Rhodoccocus* of actinobacteria. The filamentous form of *Streptomyces viridosporus* T7A secretes extracellular enzymes that break down lignin (Kamimura et al., 2019). *S. viridosporus* T7A is able to degrade native lignin of the wheat straw lignin, reducing its guaiacyl units (Zeng et al., 2013b). After 12 weeks of fermentation with *S. setonii* 75Vi2 and *S. viridosporus* T7A, 30 to 45% of the lignin from grass, hardwood, and softwood was removed (Antai and Crawford, 1981). *Rhodoccocus* is thought to be a strong microorganism for breaking down lignin because it can handle toxic metabolites well and has hydrolytic activity for them. Kraft lignin and wheat straw can be transformed into aromatic dicarboxylic acids and vanillin by *R. jostii* RHA1 which is soil bacterium degrading polychlorinated biphenyl (Mycroft et al., 2015; Sainsbury et al., 2013) and about 19% of the lignin can be utilized. In the lignin model compound, the dyp-type peroxidase DypB from *R. jostii* RHA1 was found to cleave the β-aryl ether linkage (Ahmad et al., 2011). In addition, *R. erythropolis* isolated from soil and wood demonstrated significant degradation activity against wheat straw nitrated lignin (Taylor et al., 2012). *Pseudomonas*, *Pandoraea*, and *Comamonas*, among other genera of protobacteria were used to depolymerize lignin. *P. putida* KT2440 and *P. putida mt-2* were found to depolymerize and catabolize less than 30% of the lignin in alkaline pretreated liquor (APL) (Salvachua et al., 2015). *P. putida NX-1* secretes extracellular lignin degrading enzymes and uses kraft lignin as the only carbon source for growth (Xu et al., 2018a, 2018b). *P. putida* is a very good chassis bacterium for metabolic engineering of aromatics obtained from lignin into biobased products. *Pandoraea B-6* degraded kraft lignin efficiently and produced aromatic and acidic compounds with low molecular weights (Shi et al. 2013). *Comamonas* sp. B-9 decolorized and depolymerized kraft lignin by 54% and 45% after seven days of treatment (Chai et al., 2014).

High-throughput sequencing was used to identify the *Firmicutes Bacillus* genus that could degrade lignin. Processing of alkaline lignin with *B. ligniniphilus* L1 yielded fifteen distinct phenol ring aromatic compounds (Zhu et al., 2017). Genomic and proteomic research revealed the lignin degradation pathways for benzoic acid, gentisate and β-ketoadipate in *Bacillus*.

B. amyloliquefaciens SL-7 can grow and secrete ligninolytic enzymes on tobacco straw lignin (Mei et al., 2020).

Anaerobic bacteria, like aerobic bacteria, are known to turn lignin and the aromatics it contains into methane and carbon dioxide. Modified lignin has a better anaerobic degradation performance than natural lignin, having a higher degree of methoxylation (Ahring et al., 2015).

Many bacteria target the methoxy group first during the anaerobic degradation of methoxylated aromatics. Aromatic ring cleavage, demethoxylation, and methanogenesis are the most common components of the anaerobic lignin degradation process (Khan and Ahring, 2019).

Methane is produced by methanogenic microorganisms that anaerobically digest aromatics extracted from lignin (Kato et al., 2015). In anaerobic environments, it has been discovered that many bacteria can break down lignin. A tropical forest soil containing only alkali-treated lignin and an isolated facultative anaerobe known as *Enterobacter lignolyticus* SCF1 (Deangelis et al., 2011). Transcriptomic and proteomic analyses showed that glutathione S-transferases and catalase/peroxidase genes for the 4-hydroxyphenylacetate pathway of lignin degradation were up-regulated (Deangelis et al., 2013).

Klebsiella sp. BRL6-2 and *Tolumonas lignolytica* BRL6-1 strain were isolated and described as lignin degraders under anaerobic conditions, and a few putative proteins for lignin degradation were distinguished (Billings et al., 2015; Woo, 2014). *Acetoanaerobium* WJDL-Y2, which could oxidize kraft lignin to low molecular weight aromatic and acidic compounds like hexanoic acid, ferulic acid, and syringic acid, was found in the pulp and paper mill sludge (Duan et al., 2016).

Extremophilic microbes were more aggressive for lignin usage and degradation due to their unique enzyme and metabolic pathways. The thermophilic bacteria demonstrated promising capabilities to transform and degrade lignin. A β-aryl ether lignin model compound can be oxidized by Dyp-type peroxidase from *Thermobifida fusca* and kraft lignin can be degraded (Rahmanpour et al., 2016). *Clostridium thermocellum*, an anaerobic thermophile, decreased the amount of β-O-4 linkage in *Populus trichocarpa* hardwood and increased the ratio of syringyl/guaiacyl (S/G) in lignin (Akinosho et al., 2017).

The *Caldicellulosiruptor kronotskyensis*, exceptionally thermophilic bacterium is able to degrade natural rice straw without requiring any prior treatment, resulting in soluble carbohydrates, organic acids, and aromatics obtained from lignin (Peng et al., 2018). A psychrotrophic bacteria called *Arthrobacter* sp. C2 was also found to degrade lignin at lower temperatures. After the treatment, intermediates of acids, alcohols, phenols, and aldehydes were found (Jiang et al., 2018).

Tables 3.5 and 3.6 list some important bacteria that can break down lignin.

Actinobacteria, alpha *proteobacteria*, beta *proteobacteria*, gamma *proteobacteria* are examples of well-known lignin degraders (Bugg et al., 2011a, 2011b; Lee et al., 2019; Morya et al., 2019). These bacteria collect various compounds like coniferyl-alcohol and vanillate and transform them via peripheral pathways into intermediates like catechol and protocatechuate.

These bacteria can secrete enzymes that turn lignin or compounds obtained from lignin from different sources into precursors for bioproducts. As depicted in Figure 3.11, the central carbon metabolism further converts the β-ketoadipate pathway's end products, like acetyl-CoA and succinyl-CoA into polyhydroxyalkanoate (PHA), lipids, and other chemicals. As a result, these strategies provide a straightforward and adaptable method for directing the wide range of molecules made by the depolymerization of lignin into specific intermediates. Through a process called "biological funneling," the intermediates then produce chemicals, fuels, and other materials. *Pseudomonas* and *Cupriavidus Necator* H16, two well-studied bacterial strains, are examples of this process.

Putida KT2440 has been shown to be an excellent candidate for converting aromatic compounds obtained from lignin into polyhydroxyalkanoates in biorefinery streams enriched with lignin and they can even build up substantial amounts of muconate from aromatic lignin compounds (Kuatsjah et al., 2022; Ravi et al., 2017; Tomizawa et al., 2014). Likewise, *R. jostii* RHA1 and a few other species of *Rhodococcus*, such as *R. opacus* PD630 and *R. opacus* DSM1069 have demonstrated the capacity to accumulate triglyceride lipids through the conversion of various lignin sources (Alvarez et al., 2021; Bugg et al., 2021). A few bacteria like *Pandoraea* sp. B-6, few other *Bacillus* species, and a fresh discovery from a thermophilic environment have also revealed their potential for conversion of lignin into value added products (Levy-Booth et al., 2022; Liu et al., 2019; Salvachúa et al., 2015).

The majority of significant lignin-degrading bacteria are responsible for lignin's bioconversion into useful products such as PHA, lipids, and mucanoic acid (Table 3.6).

The ability of certain strains of *Streptomyces* to break down lignin is well-known. One of the best-studied strains of *Streptomyces viridosporus* T7A is an actinobacterium that produces extracellular lignin-degrading enzymes and resembles fungi in its filamentous form.

Crawford et al. (1983) found that acid precipitable polymeric lignin (APPL) rich in phenolic hydroxyl groups derived from lignin was identified as the intermediate produced by *S. viridosporus* T7A culture on corn stover lignocellulose. More in-depth analysis of chemical composition, such as Fourier-transform infrared spectroscopy (FTIR), pyrolysis gas chromatography–mass spectrometry and 13C cross polarization/magic angle spinning nuclear magnetic resonance spectroscopy has demonstrated the biodecomposition of lignin. The ratio of carbohydrate to lignin increased and the guaiacyl unit was removed from lignin following the growth of *S. viridosporus* T7A on

Table 3.5 Lignin depolymerization through bacteria.

Bacterial Strains	Substrate	Product	Strategy/Pathway	Enzymes	References
Rhodococcus rhodochrous	4-hydroxybenzoic acid, vanillic acid and glucose as the co-substrates	Lipid	Uses lignin model monomer	Enzymes involved in aromatic degradation and lipid accumulation	Shields-Menard et al., 2017
R. opacus PD630 *R. jostii* RHA1 VanA	Kraft lignin	Lipid	β-ketoadipate pathway and phenylacetic acid pathway, the co-culture of *R. jostii* RHA1 and *R. opacus*	Multiple peroxidases with accessory oxidases	Li et al., 2019
R. opacus DSM 1069	O_2 pretreated kraft lignin	Lipid	β-ketoadipate pathway, O_2-based lignin pretreatment	Peroxidases and lipid biosynthetic enzymes	Kosa and Ragauskas, 2012
Pandoraea sp. ISTKB	*Pandoraea* sp. ISTKB	Kraft lignin	PHA CoA-mediated degradation pathways of phenylacetate and benzoate	Peroxidase-accessory enzyme system	Kumar et al., 2017
Engineered *P. putida* KT2440 *Ralstonia Eutropha.* H16	Lignin Alkaline pretreated liquor	PHA	The knocking out of phaZ and the overexpression of alkK, phaG, phaC1, and phaC2 genes Two-step enzymatic hydrolysis	Multiple enzyme systems for PHA biosynthesis PHB biosynthetic pathway	Salvachúa et al., 2020b; Saratale and Oh, 2015
Cupriavidus basilensis B-8	Lignin	PHA	β-ketoadipate pathway and the gentisate pathway	Manganese peroxidase (MnP) and laccase	Shi et al., 2017
Oceanimonas doudoroffii Corynebacterium. glutamicum MA-2	Lignin	PHA	Direct microbial conversion from lignin to biopolyester	Pathway not identified	Numata and Morisaki (2015).

(*Continued*)

Table 3.5 (Continued)

Bacterial Strains	Substrate	Product	Strategy/Pathway	Enzymes	References
Amycolatopsis sp. ATCC 39116	Softwood lignin hydrolysate mainly guaiacol	cis, cis-MA	β-ketoadipate pathway, uses a metabolically engineered cell factory	Catechol 1,2-dioxygenase Enzymes involved in the ketoadipate pathway, catechol dioxygenase and _α-glucuronidase	(Barton et al., 2018; Becker et al., 2018)
Pseudomonas. putida KT2440	Alkaline pretreated liquor, p-coumaric acid	cis, cis- Muconate, miconic acid	Integration of the aroY gene and the deletion of the catB and catC genes, controlling carbon catabolite repression	CatA and CatA2 dioxygenases	Johnson et al., 2017
Sphingobium sp. SYK-6, *Pseudomonas. putida* KT2440	Hardwood lignin hydrolysate	cis, cis-MA	Using G-lignin components for cis, cis-MA production	Multiple enzyme system for cis, cis-MA production	Sonoki et al., 2018

Alhassan et al. (2020) / MDPI / CC BY 4.0.

Table 3.6 Bacterial degradation of lignin in various biomass sources.

Fungi	Strains	Biomass materials	Lignin degradation	References
Aerobic bacteria	*Streptomyces viridosporus* T7A	Softwood spruce, hardwood maple, and grass	30.9%, 32% and 44.2%	Antai and Crawford, 1981
	Rhodoccocus Jostii RHA1	Soluble and lignin rich stream	18.9%	Salvachúa et al., 2015
	R. pyridinivorans CCZU-B16	Alkali lignin	30.2%	Chong et al., 2018
	Pseudomonas putida KT2440	Alkaline pre-treated liquor	~ 30%	Salvachúa et al., 2015
	P. putida NX-1	Kraft lignin	28.5%	Xu et al., 2018a
	Comamonas sp. B-9	Kraft lignin	45%	Chai et al., 2014
	Bacillus ligniniphilus L1	Alkaline lignin	38.9%	Zhu et al., 2017
	B. amyloliquefaciens SL-7	Tobacco straw lignin	28.55%	Mei et al., 2020
Facultative anaerobe bacteria	*Enterobacter lignolyticus* SCF1	Alkali lignin	56%	Deangelis et al., 2013
	Acetoanaerobium sp	Kraft lignin	24.9%	Duan et al., 2016
Extremophile bacteria	*Caldicellulosiruptor kronotskyensis*	Natural rice straw	52.5%	Peng et al., 2018
	Arthrobacter sp. C2	Sodium lignin sulfonate	40.1%	Jiang et al., 2018

Weng et al. 2021 / Springer Nature / CC BY 4.0.

wheat straw (Zeng et al., 2013b). This class of *Streptomyces* also contained strains that were active against lignin decay; *S. coelicolor* A3(2) used a small laccase to generate APPL in a medium containing grass lignocellulose, while *S. badius* ATCC 39117 broke down 14C-labeled plant lignin (Majumdar et al., 2014; McCarthy, 1987).

On hardwood, softwood, and grass lignocellulose, *S. amycolatopsis* 75iv2, and *S. griseus* 75iv2, was grown for two months. After that, it lost 34, 29, and 39% of its weight in lignin, comparable to *S. viridosporus* T7A's weight loss (Antai and Crawford, 1981).

Amycolatopsis sp. 75iv2 was found to produce APPL in the same manner as *Streptomyces* on grass lignocellulose. The *Rhodoccocus* species are also desirable strains for breaking down lignin. *R. jostii* RHA1 in soil was first discovered to degrade PCB (Brown et al., 2011; Seto et al., 1995).

Ahmad et al. (2010) directly used wheat straw lignocellulose and kraft lignin as the only carbon source in the medium to produce bioproducts like vanillin and aromatic dicarboxylic acids in an assay utilizing fluorescently modified lignin to measure its lignin degrading activity (Mycroft et al., 2015; Sainsbury et al., 2013).

R. erythropolis, a different bacterium degrading PCB isolated from termite gut, also displayed lignin metabolizing activity that was discovered by a UV-vis assay using wheat lignin (Chung et al., 1994; Taylor et al., 2012). Kosa and Ragauskas (2013) reported that when organosolv lignin was used as the sole energy and carbon source, two oleaginous *R. opacus* PD630 and DSM 1069 and PD630 increased their colony forming unit by more than 300 times compared to the amount in the inoculum after seven days. This implies that the organisms have the ability to degrade lignin. It was also noted that

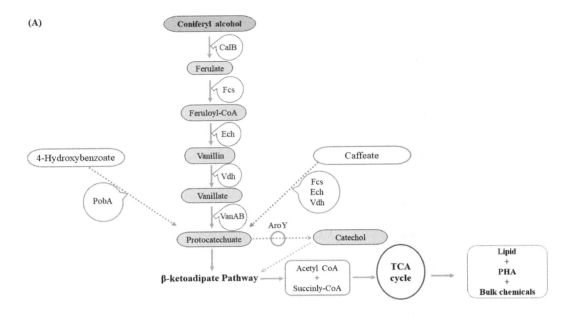

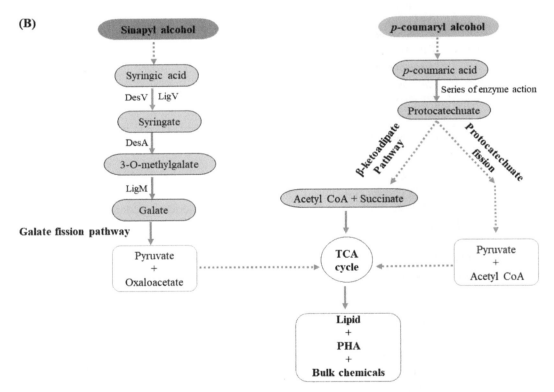

Figure 3.11 Aromatic catabolism from the coniferyl, sinapyl and p-coumaryl branch.
(**A**) Conversion of diverse compounds like the coniferyl-alcohol, 4-hydroxybenzoate and caffeate to aromatics protocatechuate and catechol occurs, which in turn are processed through the different enzyme systems involved in the β-oxidation pathway. The acetyl CoA and succinyl CoA formed then goes through the TCA cycle, leading to the formation of triglycerols lipids, PHA, and other fine chemicals.
(**B**) The sinapyl alcohol forms pyruvate and oxaloacetate through the galate fission pathway, which enters to TCA cycle, leading to the formation of fine chemicals. In contrast, the p-coumaryl alcohol forms succinate + acetyl CoA through the β-ketoadipate pathway and pyruvate + acetyl CoA through the protocatechuate fission pathway, which in turn enter the TCA cycle and help in the formation of lipids, PHA or other bulk chemicals.
Alhassan et al. (2020) / MDPI / CC BY 4.0.

R. opacus PD630 grew exclusively on alkali corn stover lignin, albeit more slowly than *R. jostii* RHA1 (He et al., 2017).

There are various types of ligninolytic bacteria that can solubilize lignin and carbohydrate in lignocellulose at the same time. One illustration is the aerobic thermophile *Thermobifica fusca*, whose genome contains enzymes for both cellulose hydrolysis and lignin modification (Rahmanpour et al., 2016; Wilson, 2004). Deng and Fong (2011) claim that this strain, which expresses heterologous alcohol dehydrogenase, was able to produce 1-propanol from biomass made of corn stover and switch grass. This implies that lignin was disintegrated to allow cellulase sufficient access to produce the carbon source. Anaerobic and thermophilic organism *Clostridium thermocellum* performs similar functions. Akinosho et al. (2017), in their investigation of the plant cell wall following *C. thermocellum* treatment, discovered that this strain increased the syringyl/guaiacyl (S/G) index and decreased the β-O-4 linkage, which is primarily present in lignin. *Caldicellulosiruptor bescii*, which was initially obtained from a geothermally heated pool, has also caught the attention of researchers due to its ligninolytic and cellulolytic activity. Various aromatic compounds obtained from lignin were released into the culture fluid following this bacteria's successful growth at 78 °C on untreated switchgrass biomass (Kataeva et al., 2013). Like *T. fusca*, engineered *C. bescii* also produced ethanol from untreated biomass (Chung et al., 2014).

Despite the fact that they derive their primary energy from sugars extracted from biomass rather than lignin, these strains are notable for their strength as strong lignin decomposers. *Enterobacter lignolyticus* SCF1, a facultative anaerobic organism, was isolated from rain forest soil using an alkali-treated lignin as the only carbon source (Deangelis et al., 2011). When *E. lignolyticus* SCF1 was grown on xylose under anaerobic conditions, its cell density increased more than twofold when lignin was added, indicating that the strain can depolymerize lignin. In their subsequent study, transcriptomic and proteomic analyses showed that in kraft lignin-amended samples, four enzymes involved in depolymerization of lignin were up-regulated, including glutathione S-transferase and DyP-type peroxidase (Deangelis et al., 2013).

E. aerogenes and *E. soil* sp. nov. among others were also isolated from soil using kraft lignin as the only carbon source in a culture medium. *E. aerogenes* can absorb lignin and aromatics obtained from lignin. Utilizing *Pseudomonas* strains in lignin depolymerization is appealing (Deschamps et al., 1980; Manter et al., 2011; Nikel and de Lorenzo, 2018; Poblete-Castro et al., 2012). Initially, carbon dioxide release from 14C-labeled lignin was used to investigate lignin degradation by *Pseudomonas* species using a lake isolate (Haider et al., 1978). The findings of numerous studies on ligninolytic *Pseudomonas* strains were published after many years, with *P. putida* serving as a notable example. The capacity of *P. putida* KT2440 to depolymerize lignin was indirectly investigated through analysis of mcl-PHA created from alkali-pretreated liquor supplemented with a 13C-labeled substance (Linger et al., 2014). Gel permeation chromatography and Klason lignin analysis showed that both *P. putida* KT2440 and its parent strain, *P. putida* mt-2, reduced the quantity of high molecular weight components in alkali pre-treated liquor by approximately 30% after seven days of treatment (Salvachúa et al., 2015).

Another *P. putida* strain A514, chosen for its capacity to utilize carbon as a source of energy, was grown on the minimal medium containing 1% alkali-insoluble lignin. It was also discovered that a novel isolate, *P. putida* NX-1, could modify lignin. Using scanning electron microscopy and FT-IR analysis, the morphology and chemical bonds of the treated lignin were found to change, resulting in a smaller particle size and a smaller guaiacyl unit. Although the ligninolytic potential of *P. fluorescens* was demonstrated by identifying and characterizing extracellular enzyme involved in the decomposition of lignin (Lin et al., 2016; Xu et al., 2018a; Rahmanpour and Bugg, 2015), how lignin actually breaks down has not yet been documented. The lignin polymer can be broken down by many *Bacillus* genus strains. Particularly, a few confines from soil, dregs, and slop have as of late been observed and portrayed for lignin treatment. The kraft lignin-screened soil bacterium *Bacillus* sp. was found using 16S rRNA genotyping to demonstrate 99% resemblance to *B. thuringiensis* and *B. cereus*. After 72 hours of culture with the high molecular weight fraction of kraft lignin as the only carbon source, this bacterium showed an increase in CFU (Bandounas et al., 2011). *B. atrophaeus*, and *B. pumilus*, two of which were selected because of their potent ability to degrade lignin, were also isolated from rainforest soil in Peru (Huang et al., 2013). When these two strains degraded kraft lignin products, the dioxane-extracted large lignin fragments were removed from the GPC by 50 to 70%. Several compounds were found by LCMS analysis in the subsequent study using poplar biomass, indicating that LMW compounds were produced through lignin modification. From a pulp industry sludge sample, one isolate, ITRC-S8, showed good growth on kraft lignin and decolorization of the culture with time, with a strong correlation between the color change and the presence of kraft lignin in the medium (Raj et al., 2007). Formation of multiple compounds by *B. ligniniphilus* L1 with a single phenol ring, in addition to the decolorization of the alkaline lignin-containing culture medium was reported (Zhu et al., 2017). The degradation of polymeric lignin was investigated using *Paenibacillus* strains, which were previously members of the *Bacillus* family (Chandra et al., 2008; Matthews et al., 2016). The molecular weight of BioChoice lignin in the culture fluid was reduced, according to GPC analysis of *Paenibacillus glucanolyticus*, which was studied with a variety of substrates, including BioChoice lignin, under either aerobic or anaerobic conditions (Mathews et al., 2016). *Citrobacter* and *Klebsiella* deconstruction of lignin has been proposed, and it has been reported that a culture containing a mixture of these two species increases efficiency. *C. freundii* did not grow well on APL and lignin conversion as compared to other strains tested, such as *P. putida* KT2440 (Salvachúa et al., 2015). However, when *C. freundii* and another *Citrobacter* sp. (isolated from sludge sample) were co-cultured, up to 62% of kraft lignin was decolourized (Chandra and Bharagava, 2013). Additionally, the formation of new metabolites, such as tri-, tetra-, and penta-chlorophenols, was confirmed by GC-MS and high-performance liquid chromatography analyses.

According to Xu et al. (2018b), *K. pneumoniae* NX-1 can alter lignin by reducing its absorbance at 280 nm (A280) by 23.8%. According to Yadav and Chandra (2015), the sludge-isolated mixed culture of *K. pneumonia* and *B. subtilis* was also more effective in terms of reduction of lignin as well as growth. In addition, *Norcadia*, *Novosphingobium*, *Cupriavidus basilensis* and *Pandoraea* species have all been found to undergo bacterial lignin depolymerization (Chen et al., 2012; Haider et al., 1978; Shi et al., 2013; Si et al., 2018). Additionally, because of their lignolytic capability, *Pandoraea* strains were utilized in the pretreatment of biomass (Kumar et al., 2016; Liu et al., 2018).

3.2.3 Lignin Depolymerization by Enzymes

An increasing number of bacterial and fungal enzymes that break down lignin have been found and used for lignin depolymerization and mineralization. Heme-containing peroxidases (lignin peroxidases, manganese peroxidases, and versatile peroxidases) and laccases (phenol oxidase) are receiving considerable attention as the primary enzymes responsible for degrading lignin (Chan et al., 2019; Zhang, 2015). Also, a number of accessory enzymes, like glyoxal oxidases and alcohol oxidases which contribute hydrogen peroxide to the peroxidase reactions are found to play an important role during degradation of lignin (Guillén et al., 1992; Kersten, 1990; Kersten and Kirk, 1987). Dye decolorizing peroxidases which are a new class of enzymes, have also been found to help in the breakdown of lignin. This suggests that natural lignin degrading enzymes vary widely (Ahmad et al., 2011).

Ligninolytic microorganisms typically produce extracellular oxidases due to the chemical and structural complexity of lignin molecules. However, the features of each enzyme system, like the type of main enzyme component and the physiological conditions for production of enzyme, differ greatly among ligninolytic microorganisms. As a result, it has been discovered that a few white-rot fungi only produce one or two classes of enzymes catalyzing the oxidative, reactions whereas others produce several classes (Floudas et al., 2012; Lundell et al., 2010).

The characteristics and reactions of major ligninolytic enzymes are presented in Table 3.7. The role of the different enzymes involved in lignin degradation is illustrated in Figure 3.12, Figures 3.13–3.16 show catalytic mechanism of laccase, lignin peroxidase, manganese peroxidase, and versatile peroxidase (Datta et al., 2017; Weng et al., 2021). Figure 3.17 shows 3D structures of the ligninolytic enzymes (Furukawa et al., 2014; Perez-Boada et al., 2005; Piontek et al., 2002; Poulos, 1993; Roberts et al., 2011; Sundaramoorthy, 2005).

3.2.3.1 Laccase

Laccases are multi-copper oxidase enzymes. They are found in plants, bacteria, and fungi. In comparison to laccases found in bacteria and plants, the reduction potential of the fungal laccase is generally higher. Laccase was first discovered by Yoshida in 1883, in the sap of Rhus vernicifera (the Japanese lacquer tree). Since this finding, laccases have been obtained from many organisms, including plants, insects, a few microorganisms, and various ascomycetous and basidiomycetous growths (Baldrian, 2006; Claus, 2004; Dwivedi et al., 2011; Kramer et al., 2001; Mayer and Staples, 2002). Among the organisms that produce laccases, white-rot fungi that are members of the class basidiomycetes typically produce the highest level of laccase activity. *Pycnoporus cinnabarinus* secretes more than 1 gram per liter of laccases into the culture fluid. Given the widespread distribution of laccases throughout the living world, it should come

Table 3.7 Major ligninolytic enzymes.

Laccase EC 1.10.3.2
Source: widely distributed in fungi and bacteria (e.g., *Ascomycetes, Basidiomycetes* and *Streptomyces*)
Substrate: phenolic compounds, aromatic amines and dye molecules
General reaction: 4 benzenediol + $O_2 \rightleftharpoons$ 4 benzosemiquinone + $2H_2O$

Lignin peroxidase EC 1.11.1.14
Source: white-rot fungal genera (e.g., *Bjerkandera, Phanerochaete, Phlebia* and *Trametes*)
Substrate: phenols, aromatic amines, aromatic ethers and polycyclic aromatics
General reaction: 1,2-bis(3,4-dimethoxyphenyl)propane- 1,3-diol + H_2O_2
$\rightleftharpoons$ 3,4-dimethoxybenzaldehyde + 1-(3,4-dimethoxyphenyl) ethane-1,2-diol + H_2O

Manganese peroxidase EC 1.11.1.13
Source: wood and litter-decomposing white-rot fungi (e.g., *Dichomitus squalens, Agaricus bisporus* and *Agrocybe praecox*)
Substrate: phenolic compounds
General reaction: $2Mn(II) + 2H^+ + H_2O_2 \rightleftharpoons 2Mn(III) + 2H_2O$

Versatile peroxidase EC 1.11.1.16
Source: white-rot species (e.g., *Pleurotus ostreatus, Bjerkandera adusta*)
Substrate: high-redox-potential aromatic compounds and recalcitrant dyes
General reaction: (1) reactive black 5 + $H_2O_2 \rightleftharpoons$ oxidized reactive black 5 + 2 H_2O
(2) donor + H2O2 = oxidized donor + 2 H_2O

Dye-decolorizing peroxidase EC 1.11.1.19
Source: fungi and bacteria (e.g., *Ascomycetes, Basidiomycetes* and *Bacillus*)
Substrate: dye compounds, carotenoids and phenolics
General reaction: reactive blue 5 + 2 $H_2O_2 \rightleftharpoons$ phthalate + 2,2′-disulfonyl azobenzene + 3-[(4-amino-6-chloro-1,3,5- triazin-2-yl)amino]
benzenesulfonate + 2 H_2O

Basd on (Kumar and Chandra, 2020; Makela et al., 2017; Weng et al. 2021)

66 | *3 Lignin Depolymerization Technologies*

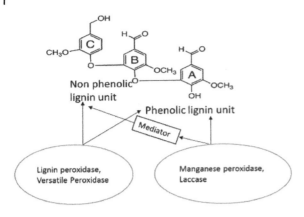

Figure 3.12 Ligninolytic enzymes and their selective action on lignin components. Datta et al. (2017) / MDPI / CC BY 4.0.

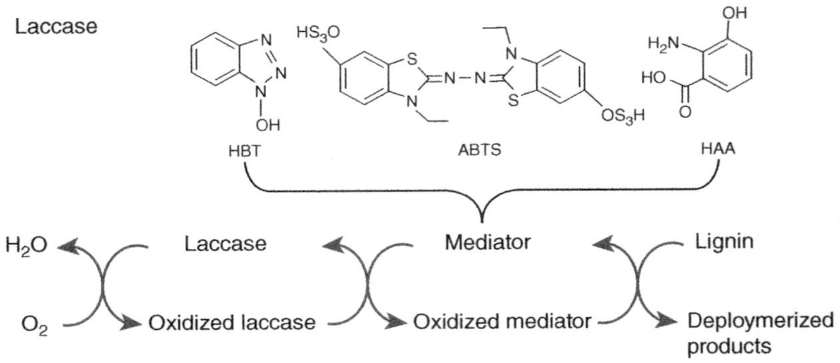

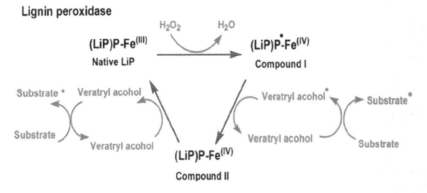

Figure 3.13 Catalytic mechanism of laccase mediated lignin degradation. Weng et al. 2021 / Springer Nature / CC BY 4.0.

Figure 3.14 Catalytic mechanism of lignin peroxidase (LiP) mediated lignin degradation.
LiP indirectly degrades lignin via oxidizing veratryl alcohol to the corresponding diffusible cation radical as a direct oxidant on lignin. Two electrons of the native ferric enzyme are oxidized by H_2O_2 to form compound one, which receives one electron to form compound two. Finally, compound two is returned to the resting native ferric state by gaining one more electron from the reducing substrate. Weng et al. 2021 / Springer Nature / CC BY 4.0.

Manganese peroxidase

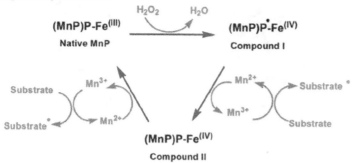

Figure 3.15 Catalytic mechanism of manganese peroxidase (MnP) mediated lignin degradation. MnP oxidizes the one-electron donor Mn^{2+} to Mn^{3+}, which in turn oxidizes a large number of phenolic substrates. The native ferric enzyme initially reacts with H_2O_2 to form compound one, and an Mn^{2+} ion donates one electron to the porphyrin intermediate to form compound two. The native enzyme is similarly produced from compound two by obtaining one electron from Mn^{2+}. Weng et al. 2021 / Springer Nature / CC BY 4.0.

Versatile peroxidase

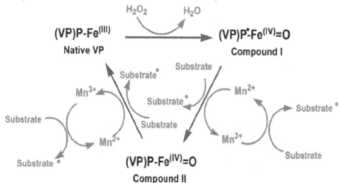

Figure 3.16 Catalytic mechanism of versatile peroxidase (VP) mediated lignin degradation. The basic catalytic cycle of VP is similar to the MnP and LiP with the two intermediary compounds one and two. Weng et al. 2021 / Springer Nature / CC BY 4.0.

as no surprise that laccases are involved in a number of biological processes, which include the biosynthesis of lignin in the plant cell wall, insect cuticle sclerotization in the epidermis, morphogenesis, pigmentation and resistance to stress, copper and iron homeostasis, environmental detoxification, and lignin degradation in bacteria and fungi (Baldrian, 2006; Boerjan et al., 2003; Enguita et al., 2003; Kramer et al., 2001; Levasseur et al., 2014; Sharma et al., 2007; Stoj and Kosman, 2003; Williamson et al., 1998).

Some inorganic ions, aminophenols, substituted phenols, and aromatic thiols, as well as a number of phenolic lignin compounds, have been shown to be oxidized by laccases (Baldrian, 2006; Giardina et al., 2010; Wong, 2009). Phenoxy radicals are formed when phenolic substrates undergo one-electron oxidation by laccases. Phenoxy radicals can either randomly rearrange molecules through

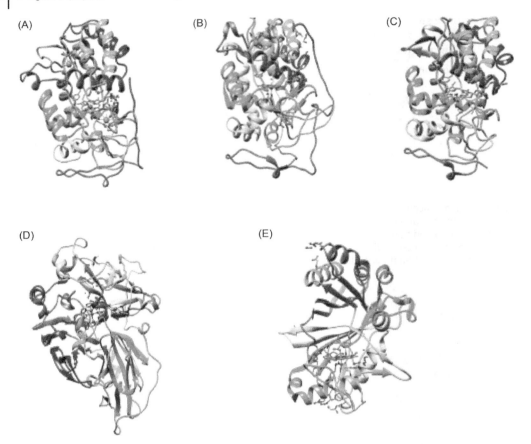

Figure 3.17 Three-dimensional structures of ligninolytic enzymes (A) A lignin peroxidase from *P. chrysosporium*; Protein Data Bank (PDB) ID: 1LGA (B) A manganese peroxidase from *P. chrysosporium*; 3M5Q. (C) A versatile peroxidase from *P. erynjii*; PDB ID: 3FM1 (D) A laccase from *T. versicolor*; PDB ID: 1GYC. (E) A dye-decolorizing type peroxidase from *R. jostii*; PDB ID: 3QNR. Reproduced with permission Furukawa et al. (2014) / Springer Nature.

a variety of cleavage reactions or spontaneously cause radical polymerization, as in lignin biosynthesis. According to research conducted with the β-1 lignin structure model compound, cleavage reactions like alkyl–alkyl cleavage, C oxidation, C–C cleavage, and aromatic ring cleavage is found to take place during the oxidation reaction catalysed by laccases indicating that these enzymes play some part in degradation of lignin. Typically, ligninolytic fungi produce laccases as an extracellular monomeric glycoprotein with a variety of isozymes (Baldrian, 2006; Wong, 2009).

There are four copper atoms in each laccase molecule type-1, type-2, and type-3. Type-1 copper is paramagnetic and oxidizes the substrate. Type-2 copper and two copper atoms of type-3 conforming to a trinuclear cluster have a major role in converting the molecular oxygen to two water molecules (Figure 3.13) (Weng et al. 2021). The laccase catalytic cycle begins in the resting state, where each of the four copper atoms is in the 2+ oxidation state (Claus, 2004; Dooley et al., 1979; Malkin and Malmström 1970; Wong, 2009). The primary electron acceptor, the T1 copper, mediates the generation of free radicals through one-electron oxidations of reducing substrates. The T2 and T3 copper atoms in the trinuclear site receive the captured electrons next following four cycles of one-electron oxidation which results in the enzyme being completely reduced. At the trinuclear center, one oxygen molecule is reduced by the formation of peroxy-intermediates, native intermediates and oxygen-bound intermediates where two successive electron transfers form two water molecules. It is for the most part acknowledged that laccases' electrochemical capabilities are affected by the molecular environment in which the copper atom is located in the T1 copper site which characterize its reactivity towards various reducing substrates (Gianfreda et al., 1999; Giardina et al., 2010; Morozova et al., 2007; Wong, 2009; Yaropolov et al., 1994).

Laccases can be divided into three groups based on the T1 site's redox potential. The redox potential of laccase can be moderate (0.5–0.6 V), high (0.7–0.8 V), or low (0.4–0.5 V) (Xu et al., 1996). It has been found that laccases with a high redox potential are appealing for many industrial applications due to their higher reactivity towards a variety of substrates. Laccases find use in the textile industry, pulp and paper processing, and other industries due to their broader substrate specificity (Cañas and Camarero, 2010; Kunamneni et al., 2008; Osma et al., 2010; Rodríguez Couto and Toca Herrera, 2006; Shraddha et al., 2011). It has been discovered that laccases are unable to attack non-phenolic units in lignin molecules directly due to their moderately lower redox capability (0.8–1.4 V) than the ligninolytic peroxidases. But, as few ligninolytic fungi, like the *Pycnoporus* and *Trametes* species utilize laccases as the major ligninolytic enzymes, laccases can also attack non-phenolic lignin structures when certain compounds, mediators, are present (Eggert et al., 1997). The enzymatic system that utilizes the blend of laccase and mediator is referred to as the laccase-mediator system. The mediators are the compounds that are able to serve as electron carriers between the final substrate and enzyme (Bourbonnais and Paice, 1990; Cañas and Camarero, 2010). *P. cinnabarinus*' secondary metabolite 3-hydroxyanthranilic acid (3-HAA) was found to mediate laccase-mediated oxidation of non-phenolic compounds, supporting the natural role of the laccase mediator system in lignin degradation (Eggert et al., 1996). More than 100 synthetic chemicals are found to act as mediators. The mediators—HBT (1-hydroxybenzotriazole) and ABTS (2,2′-azino-bis-3-ethylbenzothiazoline-6-sulfonic acid) are frequently utilized for the oxidation of non-phenolic compounds (Bourbonnais and Paice, 1990; Call and Mücke, 1997; Cañas and Camarero, 2010). In addition to the synthetic mediators, it has been discovered that some naturally occurring phenolic compounds associated with the lignin polymer also function as effective mediators (Camarero et al., 2005). Sinapic acid, vanillin, acetosyringone, syringaldehyde, and p-coumaric acid are examples of these compounds that significantly enhance laccases' capacity for oxidation. For many applications, including the delignification of lignocelluloses, these compounds have been effectively applied to the laccase mediator system (Andreu and Vidal, 2011; Fillat et al., 2010; Rico et al., 2014). These reports propose that laccases likewise utilize these natural mediators to help break down non-phenolic lignin structures during the natural ligninolysis process. Laccases are typically produced by multiple isozymes in wood-rotting fungi, but their enzymatic properties, such as optimal conditions, redox potentials, and molecular weight, differ significantly from one another. However, it has been discovered that bacteria only produce a small number of laccase isozymes, most of which were thought to be spore-bound or intracellular (Baldrian, 2006; Claus, 2004; Santhanam et al., 2011). Despite the fact that laccase-encoding genes are found to be present in the genomes of virtually all wood-decaying fungi, few basidiomycetes, including the model ligninolytic fungi—*Phanerochaete chrysosporium*, lacks conventional laccase genes, indicating that the biological system that breaks down lignin has a purposeful diversity (Floudas et al., 2012; Martínez et al., 2004).

A genomic analysis (Riley et al., 2014) suggests that in comparison to brown-rot fungi, white-rot fungi possess more laccase-encoding genes, with 16 putative genes predicted for *Ganoderma* sp. (white-rot fungus) and six in *Coniophora puteana* (brown-rot fungus). Unlike fungal laccases, bacterial laccases have received relatively little attention. However, a number of unusual bacteria-derived laccases, such as those with a high thermoresistance, are being acknowledged as a significant potential source of enzymes for upcoming biotechnological applications (Miyazaki, 2005).

The laccase of *Trametes versicolor* has been characterized in terms of its structure (Piontek et al., 2002). There are about 500 amino acid residues in it. A His–Cys–His tripeptide pathway transports four electrons to the tri-nuclear center during the type 1 laccase-driven reaction. With oxygen serving as the final electron acceptor, laccase can degrade phenolic as well as non-phenolic compounds (Figure 3.13). Phenoxyl free radical, an unstable intermediate produced by laccase oxidation of phenolic substrates, facilitates C oxidation, alkyl-aryl cleavage, and C–C cleavage (Kawal, 1988). When degrading non-phenolic substrates, laccase needs to work with mediators such as 3-hydroxyanthranilic acid (HAA), HBT and ABTS. The C–C cleavage, C oxidation, aromatic ring cleavage, and ether cleavage can all be aided by mediators and oxidized non-phenolic compounds (Kawai et al., 2004).

Most people agree that mediators aid in overcoming the steric barrier that exists between the substrate and laccase and enhance the oxidation capabilities of laccase. These mediators allow laccase to be used in the delignification process as well. With 2.5% HBT acting as a mediator, the *T. villosa* laccase was able to remove approximately 48% lignin from *Pennisetum purpureum* and 32% of lignin from *Eucalyptus globules* (Gutierrez et al., 2012; Martínez et al., 2004; Thurston, 1994). Innovative fermentation techniques, genetic modification, the addition of inducers and cofactors, and other strategies have all been used to increase microbes' laccase yield.

3.2.3.2 Lignin Peroxidase

In 1983, Tien and Kirk discovered lignin peroxidase for the first time in *P. chrysosporium*. It was further identified as a genuine lignase due to its higher redox potential. Lignin peroxidases stand out from other peroxidases due to their higher redox potential and lower pH optimum closer to pH 3.0. The ligin peroxidases that have a heme glycoprotein are members of the plant peroxidase family's class II secreted fungal heme peroxidase group. This is divided into three subfamilies—plant, bacterial, and fungal peroxidases. By using one-electron oxidation mechanisms in the presence of hydrogen peroxide, lignin peroxidases can oxidize a broad range of aromatic compounds that are not susceptible to the action of other peroxidases, such as lignin and compounds similar to lignin. One of the most significant features of lignin peroxidases in the degradation of lignin is their capacity to oxidize both phenolic and non-phenolic lignin units, which can share as much as 90% of lignin's structure. Consequently, it has been hypothesized that lignin peroxidases play a crucial role in the natural degradation of lignin. The cleavage of C–C propyl side chains in lignin and lignin model compounds is one reaction that lignin peroxidases catalyze; the hydroxylation of the groups in benzylic methylene; the transformation of benzyl alcohols into ketones or aldehydes by oxidation; non-phenolic lignin model compounds undergo phenol oxidation and aromatic cleavage (Eriksson and Ander, 1990; Hammel et al., 1993; Kirk and Farrell, 1987; Paliwal et al., 2012; Welinder et al., 1992; Wong, 2009).

Lignin peroxidase is a glycoprotein. It has a molecular weight between 38 and 43 kilodaltons and an isoelectric point between 3.3 and 4.7 (Chan et al., 2019). The majority of the lignin peroxidase crystal structure in *P. chrysosporium* is made up of α-helices (Poulos, 1993) and the three-dimensional structure is stabilized by two calcium ions and four disulfide bonds. An iron atom containing heme makes up the active site of lignin peroxidase. The trp171 residue preserved in lignin peroxidase sequences is crucial for lignin peroxidase's catalytic activity. With hydrogen peroxide and veratryl alcohol acting as an electron donor and cofactor, lignin peroxidase oxidizes both phenolic and non-phenolic compounds.

For degradation of lignin by lignin peroxidase, one oxidation and two reduction steps are typically required. The oxidation of ferric [Fe(III)] lignin peroxidase and the reduction of hydrogen peroxide to water results in the formation of the oxoferryl iron porphyrin radical cation [Fe(IV)=O+]. Then, through a full catalytic cycle and two steps of one-electron reduction in succession, [Fe(IV)=O+] was transformed into two [Fe(IV)=O] (Datta et al., 2017; Francesca et al., 2001). Because lignin peroxidase can break down a wider variety of phenolic and non-phenolic compounds, it is a potential candidate for depolymerization of lignin. Lignin peroxidase is the primary enzyme that is responsible for the degradation of lignin because of its higher reduction potential as compared to other peroxidases. Submerged fermentation of *Aspegillus oryzae* CGMCC 5992 yielded lignin peroxidase with high activity on hydrogen peroxide-treated corn stover lignin (Zhang et al., 2015) and genetic modification and the addition of minerals were used to boost lignin peroxide yield.

3.2.3.3 Manganese Peroxidase

Manganese peroxidase is the primary ligninolytic peroxidase found in basidiomycetes. It is a hemeprotein that has been glycosylated and has a molecular weight of 45 to 60 kDa (Chan et al., 2019). In the mid-1980s, this enzyme was also found in *P. chrysosporium* culture fluid. Using H_2O_2 as an oxidant, the oxidation of Mn^{2+} to Mn^{3+} is mediated by manganese peroxidases, a heme-containing glycoprotein of the class II peroxidase family. The Mn^{3+} that is produced is a powerful, easy-to-diffuse oxidizer, but it is very unstable in water. As a result, Mn^{3+} can be stabilized by chelating it with organic acids like oxalate and malonate to form a Mn^{3+}- chelator complex that serves as a small diffusible oxidizer for the oxidation of lignin. In point of fact, many white-rot fungi that produce manganese peroxidase secrete organic acids as a secondary metabolite, primarily oxalate. It has been demonstrated that *P. chrysosporium* culture filtrate contains a physiologic amount of oxalate to stimulate manganese peroxidase activity. According to Wong (2009), through one-electron oxidation of the substrates, the generated Mn^{3+}-chelator complex is able to oxidize a wide variety of phenolic substrates, such as monomeric phenols and phenolic lignin structures, to produce a phenoxy-radical intermediate that ultimately results in the decomposition of the compounds. Nevertheless, the Mn^{3+} complex's electron potential decreases in comparison to that of the unchelated Mn^{3+} cation after getting stabilized by the chelator molecule, making it a mild oxidant. In this manner it can't straightforwardly go after the prevailing non-phenolic structures in the lignin polymer, in contrast to lignin peroxidases.

Manganese peroxidase has been found to contribute to the oxidation of non-phenolic lignin structures by forming extremely reactive radical species in the presence of a second mediator (Glenn et al., 1986; Glenn and Gold, 1985; Gold et al., 1984; Kishi et al., 1994; Kuan et al., 1993; Kuwahara et al., 1984; Paszczynski et al., 1985; Reddy et al., 2003; Wariishi et al., 1989). Both thiols (like glutathione) and unsaturated lipids (like linoleic acid) can be oxidized to thiyl and lipid peroxyl radicals, respectively, by the enzymatically generated Mn^{3+}. Through hydrogen abstraction mechanisms, the produced radicals can oxidize a broader range of non-phenolic compounds to produce a benzylic radical. The degradation compounds are produced through non-enzymatic reactions of the resulting radical.

It has been reported that non-phenolic β-O-4 lignin models are decomposed by microbial enzymes for lignin degradation. These systems catalyze the breakdown of Cα-Cβ and β-aryl ether bonds. A number of peroxidizable unsaturated fatty acids were found in *C. subvermispora* wood decaying culture, supporting the manganese peroxidase-lipid oxidation system's role in lignin degradation. As this fungus grew in aspen wood, significant up-regulation of manganese peroxidase genes and putative fatty acid desaturase-encoding genes, which might play a role in the formation of these unsaturated lipids, was also seen (Bao et al., 1994; Dunford, 1999; Fernandez-Fueyo et al., 2012; Gutiérrez et al., 2002; Reddy et al., 2003). Under physiological conditions, these findings propose that the oxidation reaction of non-phenolic lignin structures is probably mediated by unsaturated fatty acids.

The published crystal structure of manganese peroxidase from *P. chrysosporium* shows similarities to lignin peroxidase (Sundaramoorthy, 2005) and it consists of one heme propionate, one Mn^{2+} ion, and the side chains of Asp179, Glu39, and Glu35. The lignin degradation catalyzed by manganese peroxidase incorporates both reduction and oxidation steps (Figure 3.14) (Weng et al. 2021). By binding hydrogen peroxide to the native ferric enzyme, manganese peroxidase initiates the catalytic cycle. In the absence of chelators, manganese peroxidase oxidizes Mn^{2+} to Mn^{3+}, which then transforms lignin phenolic compounds into phenoxy-radicals. Oxalate and malonate, two organic acid chelators, can both increase enzyme activity and stabilize Mn3+ (Hilden et al., 2014; Hofrichter, 2002). Manganese peroxidase plays a crucial role in the initial depolymerization of lignin, just like lignin peroxidase does. In addition, it was discovered that lignin depolymerization can be sped up when manganese peroxidase is added to the culture medium. Manganese peroxidase was found to be able to encourage lignin degradation and methane yield, and it was found that it removed 68.4% of the lignin from municipal solid waste (Jayasinghe et al., 2011).

3.2.3.4 Versatile Peroxidase

The white-rot fungi *Pleurotus* and *Bjerkandera* contain the distinctive enzyme versatile peroxidase (VP), which breaks down lignin. (Hofrichter, 2010). *P. eryngii's* versatile peroxidase has a crystal structure that is comparable with that of *P. chrysosporium's* lignin peroxidase and manganese peroxidase (Pérez-Boada et al., 2005). The protein structure contained a Mn2+-binding site that enabled the direct transfer of electrons to the heme. Additionally, a tryptophan residue demonstrated the possibility of oxidizing aromatic compounds at the protein surface via long-distance electron transfer. The versatile peroxidase prefers a wide variety of substrates because it has a catalytic tryptophan, a Mn oxidation site, and an access channel for heme. The term "hybrid peroxidase" refers to a versatile peroxidase that shares similar catalytic mechanisms with both manganese peroxidase and lignin peroxidase (Figure 3.15) (Weng et al., 2021). In contrast to manganese peroxidase and lignin peroxidase, versatile peroxidase can oxidize Mn2+ independently and directly degrade high reduction potential substrates (Camarero, 1999). Due to its unique bifunctionality, versatile peroxidase has attracted research interest in biotechnological applications and genetic manipulation. Reduced saccharification reluctance and enhanced enzymatic hydrolysis of corn stover were achieved through the utilization of the adaptable peroxidase from *Physisporinus vitreus* (Kong et al., 2017). For large-scale production, the versatile peroxidase from *Bjerkandera adusta* was cloned and overexpressed in *Escherichia coli* (Mohorcic et al., 2009).

3.2.3.5 β-etherase

Other than peroxidases, β-etherases are found to degrade lignin fragment in vivo, which could cleave β-aryl ether and biphenyl linkages inside lignin atoms. A reliable method for depolymerizing and converting lignin can be found in the β-etherases that are involved in the β-O-4 ether and biphenyl catabolic pathways. Because the β-O-4 ether bond is the most common linkage in lignin and makes up more than half of all ether links (Reiter et al., 2013), its degradation is necessary for depolymerizing lignin. Enzymatic cleavage of the β-O-4 ether bond (Figure 3.18) (Weng et al.,

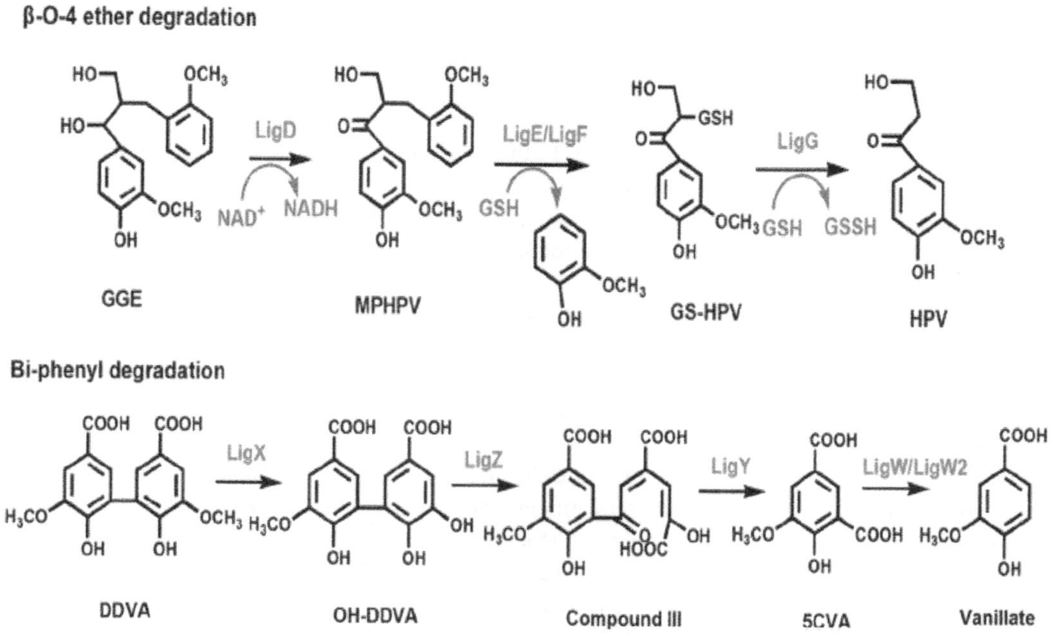

β-O-4 ether degradation

Figure 3.18 Mechanisms of β-O-4 ether and biphenyl linkage degradation. Weng et al. 2021 / Springer Nature / CC BY 4.0.

2021) was examined in *Sphingobium* sp. bacteria. *Novosphingobium* sp., SYK-6, *Dichomitus squalens* and PP1Y (Gall et al., 2014; Picart et al., 2014; Sato et al., 2009). C-dehydrogenase LigD, which uses NAD+ to oxidize the hydroxyl group at the C position, initiates the degradation of the β-O-4 ether bond. The intermediate is then broken down by β-etherase LigE or LigF into GS-HPV, which contains glutathione at the C position. While the glutathione is oxidized to glutathione by glutathione lyase LigG and releases the final product of β-hydroxyproppiovanillone (Chio et al., 2019). To break the β-O-4 aryl ether bond, LigD, LigG, and LigF were all expressed heterologously in *Arabidopsis thaliana*, enhancing lignin digestibility (Mnich et al., 2017). The breakdown of lignin was only possible with the help of the enzymes LigD, E, F, and G.

3.2.3.6 Biphenyl Bond Cleavage Enzyme

Another important bond, the biphenyl linkage, accounts for about 10% of softwood lignin (Pandey and Kim, 2011). Biphenyl linkages are found to be present in polychlorinated biphenyls (PCB), which are significant carcinogens and environmental pollutants (Pieper, 2005). Numerous studies have been conducted on PCB degradation. The non-heme iron-dependent demethylase enzyme LigX removes one methoxy group and forms a hydroxyl group prior to the catalytic process of 5, 5′-dehydrodivanillate (Pieper, 2005). LigX's product serves as the extradiol dioxygenase LigZ's substrate for oxidative meta-cleavage (Peng, 1998).

The product of ring fission is converted into 5-carboxyvanillic acid and 4-carboxy-2-hydroxypentadienoic acid by the C–C hydrolase LigY. Finally, the decarboxylases LigW and LigW2 transform 5CVA into the metabolic central intermediate vanillate or vanillic acid for bioproduct synthesis (Masai et al., 2007). It has been demonstrated that lignin degradation can be enhanced by the cleavage of the biphenyl linkage.

Bibliography

Aarti MVA and Agastian P (2015). Lignin degradation: a microbial approach. *South Indian J Biol Sci*, 1: 119–127.

Abdel-Hamid AM, Solbiati JO, and Cann IK (2013). Insights into lignin degradation and its potential industrial applications. *Adv Appl Microbiol*, 82: 1–28.

Acevedo F, Pizzul L, del Pilar Castillo M, Cuevas R, and Diez MC (2011). Degradation of polycyclic aromatic hydrocarbons by the Chilean white-rot fungus *Anthracophyllum discolor*. *J Hazard Mater*, 185: 212–219.

Ahmad M, Roberts JN, Hardiman EM, Singh R, Eltis LD, and Bugg TD (2011). Identification of DypB from *Rhodococcus jostii* RHA1 as a lignin peroxidase. *Biochemistry*, 50(23): 5096–5107.

Ahmad M, Taylor CR, Pink D, Burton K, Eastwood D, Bending GD, and Bugg TD (2010). Development of novel assays for lignin degradation: comparative analysis of bacterial and fungal lignin degraders. *Mol Biosyst* 2010 May, 6(5): 815–821.

Ahring BK, Biswas R, Ahamed A, Teller PJ, and Uellendahl H (2015). Making lignin accessible for anaerobic digestion by wet-explosion pretreatment. *Bioresour Technol*, 175: 182–188.

Akinosho HO, Yoo CG, Dumitrache A, Natzke J, Muchero W, Brown SD, and Ragauskas AJ (2017). Elucidating the structural changes to Populus lignin during consolidated bioprocessing with *Clostridium thermocellum*. *ACS Sustain Chem Eng*, 5: 7486–7491.

Alexieva Z, Yemendzhiev H, and Zlateva P (2010). Cresols utilization by *Trametes versicolor* and substrate interactions in the mixture with phenol. *Biodegradation*, 21: 625–635.

Alvarez HM, Hernández MA, Lanfranconi MP, Silva RA, and Villalba MS (2021). *Rhodococcus* as biofactories for microbial oil production. *Molecules*, 26: 4871.

Andlar M, Rezic̓ T, Marḑ̄etko N, Kracher D, Ludwig R, and Šantek B (2018). Lignocellulose degradation: an overview of fungi and fungal enzymes involved in lignocellulose degradation. *Eng Life Sci*, 18: 768–778.

Andreu G and Vidal T (2011). Effects of laccase-natural mediator systems on kenaf pulp. *Bioresour Technol*, 102(10): 5932–5937.

Antai SP and Crawford DL (1981). Degradation of softwood, hardwood, and grass lignocelluloses by two *Streptomyces* strains. *Appl Environ Microbiol*, 42: 378–380.

Asgher M, Bhatti HN, Ashraf M, and Legge RL (2008). Recent developments in biodegradation of industrial pollutants by white rot fungi and their enzyme system. *Biodegradation*, 19: 771–783.

Bai Z, Ma Q, Dai Y, Yuan H, Ye J, and Yu W (2017). Spatial heterogeneity of SOM concentrations associated with white-rot versus brown-rotwood decay. *Sci Rep*, 7: 13758.

Baldrian P (2006). Fungal laccases- occurrence and properties. *FEMS Microbiol Rev*, 30(2): 215–242.

Bandounas L, Wierckx NJ, De Winde JH, and Ruijssenaars HJ (2011). Isolation and characterization of novel bacterial strains exhibiting ligninolytic potential. *BMC Biotechnol*, 11: 94.

Bao W, Fukushima Y, Jensen KA Jr, Moen MA, and Hammel KE (1994). Oxidative degradation of non-phenolic lignin during lipid peroxidation by fungal manganese peroxidase. *FEBS Lett*, 354(3): 297–300.

Barr DP and Aust SD (1994). Mechanisms of white fungi use to degrade pollution. *Crit Rev Environ Sci Technol*, 28(2): 79–87.

Barton N, Horbal L, Starck S, Kohlstedt M, Luzhetskyy A, and Wittmann C (2018). Enabling the valorization of guaiacol-based lignin: integrated chemical and biochemical production of Cis,Cis-muconic acid using metabolically engineered *Amycolatopsis* Sp ATCC 39116. *Metab Eng*, 45: 200–210.

Becker J, Kuhl M, Kohlstedt M, Starck S, and Wittmann C (2018). Metabolic engineering of *Corynebacterium glutamicum* for the production of cis, cis-muconic acid from lignin. *Microb Cell Fact*, 17: 115.

Beckham GT, Johnson CW, Karp EM, Salvachúa D, and Vardon DR (2016). Opportunities and challenges in biological lignin valorization. *Curr Opin Biotechnol* Dec, 42: 40–53.

Billings AF, Fortney JL, Hazen TC, Simmons B, Davenport KW, Goodwin L, Ivanova N, Kyrpides NC, Mavromatis K, Woyke T, and DeAngelis KM (2015). Genome sequence and description of the anaerobic lignin degrading bacterium *Tolumonas lignolytica* sp. nov. *Stand Genomic Sci*, 10: 106.

Blanchette RA (1995). Degradation of the lignocellulose complex in wood. *Can J Bot*, 73: 999.

Boerjan W, Ralph J, and Baucher M (2003). Lignin biosynthesis. *Annu Rev Plant Biol*, 54(1): 519–546.

Bourbonnais R and Paice MG (1990). Oxidation of non-phenolic substrates. An expanded role for laccase in lignin biodegradation. *FEBS Lett*, 267(1): 99–102.

Brown ME and Chang MC (2014). Exploring bacterial lignin degradation. *Curr Opin Chem Biol*, 19: 1–7.

Brown ME, Walker MC, Nakashige TG, Iavarone AT, and Chang MC (2011). Discovery and characterization of heme enzymes from unsequenced bacteria: application to microbial lignin degradation. *J Am Chem Soc*, 133: 18006–18009.

Bugg TD, Ahmad M, Hardiman EM, and Rahmanpour R (2011b). Pathways for degradation of lignin in bacteria and fungi. *Nat Prod Rep*, 28: 1883–1896.

Bugg TD, Ahmad M, Hardiman EM, and Singh R (2011a). The emerging role for bacteria in lignin degradation and bio-product formation. *Curr Opin Biotechnol*, 22: 394–400.

Bugg TDH, Williamson JJ, and Alberti F (2021). Microbial hosts for metabolic engineering of lignin bioconversion to renewable chemicals. *Renew Sustain Energy Rev*, 152: 111674.

Call HP and Mücke I (1997). History, overview and applications of mediated lignolytic systems, especially laccase-mediator-systems (Lignozym®-process). *J Biotechnol*, 53(2–3): 163–202.

Camarero S (1999). Description of a versatile peroxidase involved in the natural degradation of lignin that has both manganese peroxidase and lignin peroxidase substrate interaction sites. *J Biol Chem*, 274: 10324–10330.

Camarero S, Ibarra D, Martínez MJ, and Martínez AT (2005). Lignin derived compounds as efficient laccase mediators for decolorization of different types of recalcitrant dyes. *Appl Environ Microbiol*, 71(4): 1775–1784.

Cañas AI and Camarero S (2010). Laccases and their natural mediators: biotechnological tools for sustainable eco-friendly processes. *Biotechnol Adv*, 28(6): 694–705.

Chai LY, Chen YH, Tang CJ, Yang ZH, Zheng Y, and Shi Y (2014). Depolymerization and decolorization of kraft lignin by bacterium *Comamonas* sp. B-9. *Appl Microbiol Biotechnol*, 98: 1907–1912.

Chan JC, Paice M, and Zhang X (2019). Enzymatic oxidation of lignin: challenges and barriers toward practical applications. *ChemCatChem*, 12: 401–425.

Chandra R and Bharagava RN (2013). Bacterial degradation of synthetic and kraft lignin by axenic and mixed culture and their metabolic products. *J Environ Biol*, 34: 991–999.

Chandra R, Singh S, Purohit HJ, and Kapley A (2008). Isolation and characterization of bacterial strains *Paenibacillus* sp. and *Bacillus* sp. for kraft lignin decolorization from pulp paper mill waste. *J Gen Appl Microbiol Biol*, 54: 399–407.

Chen Y, Chai L, Tang C, Yang Z, Zheng Y, Shi Y, and Zhang H (2012). Kraft lignin biodegradation by *Novosphingobium* sp. B-7 and analysis of the degradation process. *Bioresour Technol*, 123: 682–685.

Chen Z and Wan C (2017). Biological valorization strategies for converting lignin into fuels and chemicals. *Renew Sust Energ Rev*, 73: 610–621.

Chio C, Sain M, and Qin W (2019). Lignin utilization: a review of lignin depolymerization from various aspects. *Renew Sust Energ Rev*, 107: 232–249.

Chong GG, Huang XJ, Di JH, Xu DZ, He YC, Pei YN, Tang YJ, and Ma CL (2018). Biodegradation of alkali lignin by a newly isolated *Rhodococcus pyridinivorans* CCZU B16. *Bioprocess Biosyst Eng*, 41: 501–510.

Chung D, Cha M, Guss AM, and Westpheling J (2014). Direct conversion of plant biomass to ethanol by engineered *Caldicellulosiruptor bescii*. *Proc Natl Acad Sci U.S.A.*, 111: 8931–8936.

Chung S-Y, Maeda M, Song E, Horikoshij K, and Kudo T (1994). A Gram-positive polychlorinated biphenyl-degrading bacterium, *Rhodococcus erythropolis* strain TA421, isolated from a termite ecosystem. *Biosci Biotechnol Biochem*, 58: 2111–2113.

Claus H (2004). Laccases: structure, reactions, distribution. *Micron*, 35(1–2): 93–96.

Crawford DL, Pometto AL, and Crawford RL (1983). Lignin degradation by *Streptomyces viridosporus*: isolation and characterization of a new polymeric lignin degradation intermediate. *Appl Environ Microbiol*, 45: 898–904.

Dashtban M, Schraft H, Syed TA, and Qin W (2010). Fungal biodegradation and enzymatic modification of lignin. *Int J Biochem Mol Biol*, 1: 36–50.

Datta R, Kelkar A, Baraniya D, Molaei A, Moulick A, Meena R, and Formaneck P (2017). Enzymatic degradation of lignin in soil: a review. *Sustainability*, 9: 1163.

Deangelis KM, D'Haeseleer P, Chivian D, Fortney JL, Khudyakov J, Simmons B, Woo H, Arkin AP, Davenport KW, Goodwin L, Chen A, Ivanova N, Kyrpides NC, Mavromatis K, Woyke T, and Hazen TC (2011). Complete genome sequence of "*Enterobacter lignolyticus*" SCF1. *Stand Genomic Sci*, 5: 69.

Deangelis KM, Sharma D, Varney R, Simmons B, Isern NG, Markilllie LM, Nicora C, Norbeck AD, Taylor RC, Aldrich JT, and Robinson EW (2013). Evidence supporting dissimilatory and assimilatory lignin degradation in *Enterobacter lignolyticus* SCF1. *Front Microbiol*, 4: 280.

Deng Y and Fong SS (2011). Metabolic engineering of *Thermobifida fusca* for direct aerobic bioconversion of untreated lignocellulosic biomass to 1-propanol. *Met Eng*, 13: 570–577.

Deschamps A, Mahoudeau G, and Lebeault J (1980). Fast degradation of kraft lignin by bacteria. *Eur J Appl Microbiol Biotechnol*, 9: 45–51.

Dey S (1994). Production of some extracellular enzymes by a lignin peroxidase-producing brown rot fungus, *Polyporus ostreiformis*, and its comparative abilities for lignin degradation and dye decolorization. *Appl Environ Microbiol*, 60: 4216–4218.

Dong XQ, Yang JS, Zhu N, Wang ET, and Yuan HL (2013). Sugarcane bagasse degradation and characterization of three white-rot fungi. *Bioresour Technol*, 131: 443–451.

Dooley DM, Rawlings J, Dawson JH, Stephens PJ, Andreasson LE, Malmstrom BG, Gray HB (1979). Spectroscopic studies of *Rhus vernicifera* and *Polyporus versicolor* laccase. Electronic structures of the copper sites. *J Am Chem Soc*, 101(17): 5038–5046.

Duan J, Huo X, Du WJ, Liang JD, Wang DQ, and Yang SC (2016). Biodegradation of kraft lignin by a newly isolated anaerobic bacterial strain, *Acetoanaerobium* sp. WJDL-Y2. *Lett Appl Microbiol*, 62: 55–62.

Dunford HB (1999). *Heme Peroxidases*. Wiley, New York.

Dwivedi UN, Singh P, Pandey VP, and Kumar A (2011). Structure–function relationship among bacterial, fungal and plant laccases. *J Mol Catal, B Enzym*, 68: 117–128.

Eaton RA and Hale MDC (1993). *Wood: Decay, Pests, Protection*. Chapman and Hall, London.

Eggert C, Temp U, Dean JFD, and Eriksson KEL (1996). A fungal metabolite mediates degradation of non-phenolic lignin structures and synthetic lignin by laccase. *FEBS Lett*, 391(1–2): 144–148.

Eggert C, Temp U, and Eriksson KE (1997). Laccase is essential for lignin degradation by the white-rot fungus *Pycnoporus cinnabarinus*. *FEBS Lett*, 407(1): 89–92.

Enguita FJ, Martins LO, Henriques AO, and Carrondo MA (2003). Crystal structure of a bacterial endospore coat component. A laccase with enhanced thermostability properties. *J Biol Chem*, 278(21): 19416–19425.

Eriksson KELBRA and Ander P (1990). *Microbial and Enzymatic Degradation of Wood and Wood Components*. Springer, Berlin. pp. 1–72.

Fernandez Fueyo E, Ruiz-Duenas FJ, Lopez-Lucendo MF, Perez-Boada M, Rencorat J, Gutierrez A, Pisabarro AG, Ramirez L, and Martinez AT (2016). A secretomic view of woody and nonwoody lignocellulose degradation by *Pleurotus ostreatus*. *Biotechnol Biofuels*, 9(49): 1–18.

Fernandez-Fueyo E, Ruiz-Dueñas FJ, Ferreira P, Floudas D, Hibbett DS, Canessa P, Larrondo LF, James TY, Seelenfreund D, Lobos S, Polanco R, Tello M, Honda Y, Watanabe T, Watanabe T, Ryu JS, Kubicek CP, Schmoll M, Gaskell J, Hammel KE, St John FJ, Vanden Wymelenberg A, Sabat G, Splinter BonDurant S, Syed K, Yadav JS, Doddapaneni H, Subramanian V, Lavín JL, Oguiza JA, Perez G, Pisabarro AG, Ramirez L, Santoyo F, Master E, Coutinho PM, Henrissat B, Lombard V,

Magnuson JK, Kües U, Hori C, Igarashi K, Samejima M, Held BW, Barry KW, LaButti KM, Lapidus A, Lindquist EA, Lucas SM, Riley R, Salamov AA, Hoffmeister D, Schwenk D, Hadar Y, Yarden O, de Vries RP, Wiebenga A, Stenlid J, Eastwood D, Grigoriev IV, Berka RM, Blanchette RA, Kersten P, Martinez AT, Vicuna R, and Cullen D (2012). Comparative genomics of *Ceriporiopsis subvermispora* and *Phanerochaete chrysosporium* provide insight into selective ligninolysis. *Proc Natl Acad Sci U S A*, 109(14): 5458–5463.

Fernández-Fueyo E, Ruiz-Dueñas FJ, Martínez MJ, Romero A, Hammel KE, Medrano FJ, and Martínez AT (2014). Ligninolytic peroxidase genes in the oyster mushroom genome: heterologous expression, molecular structure, catalytic and stability properties, and lignin-degrading ability. *Biotechnol Biofuels*, 7: 2.

Fernández-Fueyo E, Ruiz-Dueñas FJ, Miki Y, Martínez MJ, Hammel KE, and Martínez AT (2012). Lignin-degrading peroxidases from genome of selective ligninolytic fungus *Ceriporiopsis subvermispora*. *J Biol Chem*, 287: 16903–16916.

Fillat A, Colom JF, and Vidal T (2010). A new approach to the biobleaching of flax pulp with laccase using natural mediators. *Bioresour Technol*, 101(11): 4104–4110.

Floudas D, Bentzer J, Ahrén D, Johansson T, Persson P, and Tunlid A (2020). Uncovering the hidden diversity of litter-decomposition mechanisms in mushroom-forming fungi. *ISME J*, 14: 2046–2059.

Floudas D, Binder M, Riley R, Barry K, Blanchette RA, Henrissat B, Martínez AT, Otillar R, Spatafora JW, Yadav JS, Aerts A, Benoit I, Boyd A, Carlson A, Copeland A, Coutinho PM, de Vries RP, Ferreira P, Findley K, Foster B, Gaskell J, Glotzer D, Górecki P, Heitman J, Hesse C, Hori C, Igarashi K, Jurgens JA, Kallen N, Kersten P, Kohler A, Kües U, Kumar TK, Kuo A, LaButti K, Larrondo LF, Lindquist E, Ling A, Lombard V, Lucas S, Lundell T, Martin R, McLaughlin DJ, Morgenstern I, Morin E, Murat C, Nagy LG, Nolan M, Ohm RA, Patyshakuliyeva A, Rokas A, Ruiz-Dueñas FJ, Sabat G, Salamov A, Samejima M, Schmutz J, Slot JC, St John F, Stenlid J, Sun H, Sun S, Syed K, Tsang A, Wiebenga A, Young D, Pisabarro A, Eastwood DC, Martin F, Cullen D, Grigoriev IV, and Hibbett DS (2012). The Paleozoic origin of enzymatic lignin decomposition reconstructed from 31 fungal genomes. *Science*, 336(6089): 1715–1719.

Francesca GM, Lanzalunga O, Lapi A, Piparo MGL, and Mancinelli S (2001). Isotope-effect profiles in the oxidative N-Demethylation of N,N Dimethylanilines catalysed by lignin peroxidase and a chemical model. *Eur J Org Chem*, 2001: 2305–2310.

Furukawa T, Bello FO, and Horsfall L (2014). Microbial enzyme systems for lignin degradation and their transcriptional regulation. *Front Biol*, 9: 448–471.

Gall DL, Ralph J, Donohue TJ, and Noguera DR (2014). A group of sequence related sphingomonad enzymes catalyzes cleavage of beta-aryl ether linkages in lignin β-guaiacyl and β-syringyl ether dimers. *Environ Sci Technol*, 48: 12454–12463.

Gianfreda L, Xu F, and Bollag JM (1999). Laccases: a useful group of oxidoreductive enzymes. *Bioremediat J*, 3(1): 1–25.

Giardina P, Faraco V, Pezzella C, Piscitelli A, Vanhulle S, and Sannia G (2010). Laccases: a never-ending story. *Cell Mol Life Sci*, 67(3): 369–385.

Gilbertson RL (1980). Wood-rotting fungi of North America. *Mycologia*, 72(1): 1–49.

Giri R and Sharma RK (2020). Fungal pretreatment of lignocellulosic biomass for the production of plant hormone by *Pichia Fermentans* under submerged conditions. *Bioresour Bioprocess*, 7: 30.

Glenn JK, Akileswaran L, and Gold MH (1986). Mn(II) oxidation is the principal function of the extracellular Mn-peroxidase from *Phanerochaete chrysosporium*. *Arch Biochem Biophys*, 251(2): 688–696.

Glenn JK and Gold MH (1985). Purification and characterization of an extracellular Mn(II)-dependent peroxidase from the lignin-degrading basidiomycete, *Phanerochaete chrysosporium*. *Arch Biochem Biophys Biophys*, 242(2): 329–341.

Gold MH, Kuwahara M, Chiu AA, and Glenn JK (1984). Purification and characterization of an extracellular H2O2-requiring diarylpropane oxygenase from the white rot basidiomycete, *Phanerochaete chrysosporium*. *Arch Biochem Biophys*, 234(2): 353–362.

Goodell B (2020). Fungi involved in the biodeterioration and bioconversion of lignocellulose substrates. *Genetics and Biotechnology*, Benz JP and Schipper K (eds.). Springer International Publishing, Cham, Switzerland, pp. 369–397. ISBN 978-3-030-49924-2.

Guerra A, Mendonca R, Ferraz A, Lu F, and Ralph J (2004). Structural characterization of lignin during *Pinus taeda* wood treatment with *Ceriporiopsis subvermispora*. *Appl Environ Microbiol*, 70: 4073–4078.

Guillén F, Martínez AT, and Martínez MJ (1992). Substrate specificity and properties of the aryl-alcohol oxidase from the ligninolytic fungus *Pleurotus eryngii*. *Eur J Biochem*, 209(2): 603–611.

Gutiérrez A, del Río JC, Martínez-Iñigo MJ, Martínez MJ, and Martínez AT (2002). Production of new unsaturated lipids during wood decay by ligninolytic basidiomycetes. *Appl Environ Microbiol*, 68(3): 1344–1350.

Gutierrez A, Rencoret J, Cadena EM, Rico A, Barth D, del Rio JC et al. (2012). Demonstration of laccase-based removal of lignin from wood and nonwood plant feedstocks. *Bioresour Technol*, 119: 114–122.

Haider K, Trojanowski J, and Sundman V (1978). Screening for lignin degrading bacteria by means of 14C-labelled lignins. *Arch Microbiol*, 119: 103–106.

Hamed SAM (2013). In-vitro studies on wood degradation in soil by soft-rot fungi: *Aspergillus niger* and *Penicillium chrysogenum*. *Int Biodeterior Biodegrad*, 78: 98–102.

Hammel KE, Jensen KA, Jr, Mozuch MD, Landucci LL, Tien M, and Pease EA (1993). Ligninolysis by a purified lignin peroxidase. *J Biol Chem*, 268(17): 12274–12281.

He Y, Li X, Ben H, Xue X, and Yang B (2017). Lipid production from dilute alkali corn stover lignin by *Rhodococcus* strains. *ACS Sustain Chem Eng*, 5: 2302–2311.

Heinzkill M, Bech L, Halkier T, Schneider P, and Anke T (1998). Characterization of laccases and peroxidases from woodrotting fungi (family Coprinaceae). *Appl Environ Microbiol*, 64: 1601.

Hemati A, Aliasgharzad N, Khakvar R, Delangiz N, Asgari Lajayer B, and van Hullebusch ED (2022). Bioaugmentation of thermophilic lignocellulose degrading bacteria accelerate the composting process of lignocellulosic materials. *Biomass Convers Biorefinery*, 1–15. https://doi.org/10.1007/s13399-021-02238-7.

Hermosilla E, Rubilar O, Schalchli H, da Silva AS, Ferreira-Leitao V, and Diez MC (2018). Sequential white-rot and brown-rot fungal pretreatment of wheat straw as a promising alternative for complementary mild treatments. *Waste Manag*, 79: 240–250.

Hildén K, Mäkelä MR, Steffen KT, Hofrichter M, Hatakka A, Archer DB, and Lundell TK (2014). Biochemical and molecular characterization of an atypical manganese peroxidase of the litter-decomposing fungus *Agrocybe praecox*. *Fungal Genet Biol*, 72: 131–136.

Hofrichter M (2002). Review: lignin conversion by manganese peroxidase (MnP). *Enzyme Microb Technol*, 30: 454–466.

Hofrichter M (2010). New and classic families of secreted fungal heme peroxidases. *Appl Microbiol Biotechnol*, 87: 871–897.

Hong C-Y, Ryu S-H, Jeong H, Lee -S-S, Kim M, and Choi I-G (2017). *Phanerochaete chrysosporium* Multienzyme catabolic system for in vivo modification of synthetic lignin to succinic acid. *ACS Chem Biol*, 12: 1749–1759.

Huang XF, Santhanam N, Badri DV, Hunter WJ, Manter DK, Decker SR, Vivanco JM, and Reardon KF (2013). Isolation and characterization of lignin-degrading bacteria from rain forest soils. *Biotechnol Bioeng*, 110: 1616–1626.

Jayasinghe PA, Hettiaratchi JP, Mehrotra AK, and Kumar S (2011). Effect of enzyme additions on methane production and lignin degradation of landfilled sample of municipal solid waste. *Bioresour Technol*, 102: 4633–4637.

Jensen KA, Jr, Houtman CJ, Ryan ZC, and Hammel KE (2001). Pathways for extracellular Fenton chemistry in the brown rot basidiomycete *Gloeophyllum trabeum*. *Appl Environ Microbiol*, 67: 2705–2711.

Jiang C, Cheng Y, Zang H, Chen X, Wang Y, Zhang Y, Wang J, Shen X, and Li C (2018). Biodegradation of lignin and the associated degradation pathway by psychrotrophic *Arthrobacter* sp. C2 from the cold region of China. *Cellulose* 2019;27: 1423–1440.

Jin W, Li J, Feng H, You S, Zhang L, Norvienyeku J, Hu K, Sun S, and Wang Z (2018). Importance of a laccase gene (*Lcc1*) in the development of *Ganoderma tsugae*. *Int J Mol Sci* Feb 6, 19(2): 471.

Johnson CW, Abraham PE, Linger JG, Khanna P, Hettich RL, and Beckham GT (2017). Eliminating a global regulator of carbon catabolite repression enhances the conversion of aromatic lignin monomers to muconate in *Pseudomonas Putida* KT2440. *Metab Eng Commun*, 5: 19–25.

Kaal EEJ, Field JA, and Joyce TW (1995). Increasing ligninolytic enzyme activities in several white-rot basidiomycetes by nitrogen-sufficient media. *Bioresour Technol*, 53: 133–139.

Kamei I, Hirota Y, and Meguro S (2012). Integrated delignification and simultaneous saccharification and fermentation of hard wood by a white-rot fungus, *Phlebia* sp. MG-60. *Bioresour Technol*, 126: 137–141.

Kamimura N, Sakamoto S, Mitsuda N, Masai E, and Kajita S (2019). Advances in microbial lignin degradation and its applications. *Curr Opin Biotechnol*, 56: 179–186.

Kataeva IA, Foston MB, Yang S, Pattathil S, Biswal AK, Poole FL, Basen M, Rhaesa AM, Thomas TP, Azadi P, Olman V, Saffold TD, Mohler KE, Lewis DL, Doeppke C, Zeng Y, Tschaplinski TJ, York WS, Davis MF, Mohnen D, Xu Y, Ragauskas AJ, Ding S, Kelly RM, Hahn MG, and Adams MW (2013). Carbohydrate and lignin are simultaneously solubilized from unpretreated switchgrass by microbial action at high temperature. *Energy Environm Sci*, 6: 2186–2195.

Kato S, Chino K, Kamimura N, Masai E, Yumoto I, and Kamagata Y (2015). Methanogenic degradation of lignin-derived monoaromatic compounds by microbial enrichments from rice paddy field soil. *Sci Rep*, 5: 14295.

Kawai S, Iwatsuki M, Nakagawa M, Inagaki M, Hamabe A, and Ohashi H (2004). An alternative β-ether cleavage pathway for a non-phenolic β-O-4 lignin model dimer catalyzed by a laccase-mediator system. *Enzyme Microb Technol*, 35: 154–160.

Kawal S (1988). Aromatic ring cleavage of 4,6-di(*tert*-butyl)guaiacol, a phenolic lignin model compound, by lactase of *Coriolus versicolor*. *FEBS Lett*, 236: 309–311.

Kersten PJ (1990). Glyoxal oxidase of *Phanerochaete chrysosporium*: its characterization and activation by lignin peroxidase. *Proc Natl Acad Sci USA*, 87(8): 2936–2940.

Kersten PJ and Kirk TK (1987). Involvement of a new enzyme, glyoxal oxidase, in extracellular H2O2 production by *Phanerochaete chrysosporium*. *J Bacteriol*, 169(5): 2195–2201.

Khan MU and Ahring BK (2019). Lignin degradation under anaerobic digestion: influence of lignin modifications—a review. *Biomass Bioenergy*, 128: 105325.

Kirk TK and Farrell RL (1987). Enzymatic "combustion": the microbial degradation of lignin. *Annu Rev Microbiol*, 41(1): 465–505.

Kishi K, Wariishi H, Marquez L, Dunford HB, and Gold MH (1994). Mechanism of manganese peroxidase compound II reduction. Effect of organic acid chelators and pH. *Biochemistry*, 33(29): 8694–8701.

Knežević A, Milovanović I, Stajić M, Lončar N, Brčeski I, Vukojević J, and Cilerdžić J (2013). Lignin degradation by selected fungal species. *Bioresour Technol*, 138: 117–123.

Kong W, Fu X, Wang L, Alhujaily A, Zhang J, Ma F, Zhang X, and Yu H (2017). A novel and efficient fungal delignification strategy based on versatile peroxidase for lignocellulose bioconversion. *Biotechnol Biofuels*, 10: 218.

Kosa M and Ragauskas AJ (2012). Bioconversion of lignin model compounds with Oleaginous *Rhodococci*. *Appl Microbiol Biotechnol*, 93: 891–900.

Kosa M and Ragauskas AJ (2013). Lignin to lipid bioconversion by oleaginous *Rhodococci*. *Green Chem*, 15: 2070–2074.

Kramer KJ, Kanost MR, Hopkins TL, Jiang H, Zhu YC, Xu R, Kerwin JL, and Turecek F (2001). Oxidative conjugation of catechols with proteins in insect skeletal systems. *Tetrahedron*, 57(2): 385–392.

Kuan IC, Johnson KA, and Tien M (1993). Kinetic analysis of manganese peroxidase. The reaction with manganese complexes. *J Biol Chem*, 268(27): 20064–20070.

Kuatsjah E, Johnson CW, Salvachúa D, Werner AZ, Zahn M, Szostkiewicz CJ, Singer CA, Dominick G, Okekeogbu I, Haugen SJ, Woodworth SP, Ramirez KJ, Giannone RJ, Hettich RL, McGeehan JE, and Beckham GT (2022). Debottlenecking 4-Hydroxybenzoate hydroxylation in *Pseudomonas putida* KT2440 improves muconate productivity from p-Coumarate. *Metab Eng*, 70: 31–42.

Kuhad RC (1997). Microorganisms and enzymes involved in the degradation of plant fiber cell walls. *Adv Biochem Eng Biotechnol*, 57: 47–125.

Kumar A and Chandra R (2020). Ligninolytic enzymes and its mechanisms for degradation of lignocellulosic waste in environment. *Heliyon*, 6: e03170.

Kumar M, Singhal A, and Thakur IS (2016). Comparison of submerged and solid state pretreatment of sugarcane bagasse by *Pandoraea* sp. ISTKB: enzymatic and structural analysis. *Bioresour Technol*, 203: 18–25.

Kumar M, Singhal A, Verma PK, and Thakur IS (2017). Production and characterization of polyhydroxyalkanoate from lignin derivatives by *Pandoraea* sp. ISTKB. *ACS Omega*, 2: 9156–9163.

Kunamneni A, Camarero S, García-Burgos C, Plou FJ, Ballesteros A, and Alcalde M (2008). Engineering and applications of fungal laccases for organic synthesis. *Microb Cell Fact*, 7(1): 32.

Kuwahara M, Glenn JK, Morgan MA, and Gold MH (1984). Separation and characterization of two extracelluar H2O2-dependent oxidases from ligninolytic cultures of *Phanerochaete chrysosporium*. *FEBS Lett*, 169(2): 247–250.

Lee S, Kang M, Bae J-H, Sohn J-H, and Sung BH (2019). Bacterial valorization of lignin: strains, enzymes, conversion pathways, biosensors, and perspectives. *Front Bioeng Biotechnol*, 7: 209.

Leonowicz A, Matuszewska A, Luterek J, Ziegenhagen D, Wojtaś-Wasilewska M, Cho NS, Hofrichter M, and Rogalski J (1999). Biodegradation of lignin by white rot fungi. *Fungal Genetics and Biology*, 27(2–3): 175–185.

Levasseur A, Lomascolo A, Chabrol O, Ruiz-Dueñas FJ, Boukhris-Uzan E, Piumi F, Kües U, Ram AF, Murat C, Haon M, Benoit I, Arfi Y, Chevret D, Drula E, Kwon MJ, Gouret P, Lesage-Meessen L, Lombard V, Mariette J, Noirot C, Park J, Patyshakuliyeva A, Sigoillot JC, Wiebenga A, Wösten HA, Martin F, Coutinho PM, deVries RP, Martinez AT, Klopp C, Pontarotti P, Henrissat B, and Record E (2014). The genome of the white-rot fungus *Pycnoporus cinnabarinus*: a basidiomycete model with a versatile arsenal for lignocellulosic biomass breakdown. *BMC Genomics*, 15(1): 486.

Levy-Booth DJ, Navas LE, Fetherolf MM, Liu L-Y, Dalhuisen T, Renneckar S, Eltis LD, and Mohn WW (2022). Discovery of lignin-transforming bacteria and enzymes in thermophilic environments using stable isotope probing. *ISME J*, 16: 1944–1956.

Li X, He Y, Zhang L, Xu Z, Ben H, Gaffrey MJ, Yang Y, Yang S, Yuan JS, Qian W, and Yang B (2019). Discovery of potential pathways for biological conversion of poplar wood into lipids by co-fermentation of *Rhodococci* strains. *Biotechnol Biofuels*, 12: 60.

Lin L, Cheng Y, Pu Y, Sun S, Li X, Jin M, Pierson EA, Gross DC, Dale BE, Dai SY, Ragauskas AJ, and Yuan JS (2016). Systems biology-guided biodesign of consolidated lignin conversion. *Green Chem*, 18: 5536–5547.

Linger JG, Vardon DR, Guarnieri MT, Karp EM, Hunsinger GB, Franden MA, Johnson CW, Chupka G, Strathmann TJ, Pienkos PT, and Beckham GT (2014). Lignin valorization through integrated biological funneling and chemical catalysis. *Proc Natl Acad Sci USA*, Aug 19, 111(33): 12013–12018.

Liu D, Yan X, Si M, Deng X, Min X, Shi Y, and Chai L (2019). Bioconversion of lignin into bioplastics by *Pandoraea* Sp. B-6: molecular mechanism. *Environ Sci Pollut Res*, 26: 2761–2770.

Liu D, Yan X, Zhuo S, Si M, Liu M, Wang S, Ren L, Chai L, and Shi Y (2018). *Pandoraea* sp. B-6 assists the deep eutectic solvent pretreatment of rice straw via promoting lignin depolymerization. *Bioresour Technol* 2018 Jun, 257: 62–68.

Lundell TK, Mäkelä MR, and Hildén K (2010). Lignin-modifying enzymes in filamentous basidiomycetes—ecological, functional and phylogenetic review. *J Basic Microbiol*, 50(1): 5–20.

Majumdar S, Lukk T, Solbiati JO, Bauer S, Nair SK, Cronan JE, and Gerlt JA (2014). Roles of small laccases from *Streptomyces* in lignin degradation. *Biochemistry* Jun 24, 53(24): 4047–4058.

Makela MR, Bredeweg EL, Magnuson JK, Baker SE, Vries RP, and Hilden K (2017). Fungal ligninolytic enzymes and their applications. *Microbiol Spectr*, 4(6): 1049–1061. http://doi.org/10.1128/microbiolspec.FUNK-0017-2016.

Malkin R and Malmström BG (1970). The state and function of copper in biological systems. *Adv Enzymol Relat Areas Mol Biol*, 33: 177–244.

Manter DK, Hunter WJ, and Vivanco JM (2011). *Enterobacter* soli sp. nov.: a lignin-degrading γ-proteobacteria isolated from soil. *Curr Microbiol*, 62: 1044–1049.

Martínez AT, Speranza M, Ruiz-Dueñas FJ, Ferreira P, Camarero S, Guillén F, Martínez MJ, Gutiérrez A, and del Río JC (2005). Biodegradation of lignocellulosics: microbial, chemical, and enzymatic aspects of the fungal attack of lignin. *Int Microbiol* Sep, 8(3): 195–204.

Martínez D, Larrondo LF, Putnam N, Gelpke MD, Huang KH, Chapman J, Helfenbein KG, Ramaiya P, Detter J, Larimer FW, Coutinho PM, Henrissat B, Berka RM, Cullen D, and Rokhsar DS (2004). Genome sequence of the lignocellulose degrading fungus *Phanerochaete chrysosporium* strain RP78. *Nature Biotechnol*, 22: 695–700.

Masai E, Katayama Y, and Fukuda M (2007). Genetic and biochemical investigations on bacterial catabolic pathways for lignin-derived aromatic compounds. *Biosci Biotechnol Biochem*, 71: 1–15.

Mathews SL, Grunden AM, and Pawlak J (2016). Degradation of lignocellulose and lignin by *Paenibacillus glucanolyticus*. *Int Biodeterior Biodegrad*, 110: 79–86.

Mayer AM and Staples RC (2002). Laccase: new functions for an old enzyme. *Phytochemistry*, 60(6): 551–565.

McCarthy A (1987). Lignocellulose-degrading actinomycetes. *FEMS Microbiol Lett*, 46: 145–163.

Mei J, Shen X, Gang L, Xu H, Wu F, and Sheng L (2020). A novel lignin degradation bacteria-*Bacillus amyloliquefaciens* SL-7 used to degrade straw lignin efficiently. *Bioresour Technol*, 310: 123445.

Miyazaki K (2005). A hyperthermophilic laccase from *Thermus thermophilus* HB27. *Extremophiles*, 9(6): 415–425.

Mnich E, Vanholme R, Oyarce P, Liu S, Lu F, Goeminne G, Jørgensen B, Motawie MS, Boerjan W, Ralph J, Ulvskov P, Møller BL, Bjarnholt N, and Harholt J (2017). Degradation of lignin β-aryl ether units in *Arabidopsis thaliana* expressing LigD, LigF and LigG from *Sphingomonas paucimobilis* SYK-6. *Plant Biotechnol J*, 15: 581–593.

Mohorcic M, Bencina M, Friedrich J, and Jerala R (2009). Expression of soluble versatile peroxidase of *Bjerkandera adusta* in *Escherichia coli*. *Bioresour Technol*, 100: 851–858.

Monrroy M, Ortega I, Ramírez M, Baeza J, and Freer J (2011). Structural change in wood by brown rot fungi and effect on enzymatic hydrolysis. *Enzyme Microb Technol* Oct 10, 49(5): 472–477.

Morozova OV, Shumakovich GP, Gorbacheva MA, Shleev SV, and Yaropolov AI (2007). "Blue" laccases. *Biochemistry (Mosc)*, 72(10): 1136–1150.

Morya R, Kumar M, Singh SS, and Thakur IS (2019). Genomic analysis of burkholderia Sp. ISTR5 for biofunneling of lignin-derived compounds. *Biotechnol Biofuels*, 12: 277.

Mycroft Z, Gomis M, Mines P, Law P, and Bugg TD (2015). Biocatalytic conversion of lignin to aromatic dicarboxylic acids in *Rhodococcus jostii* RHA1 by re-routing aromatic degradation pathways. *Green Chem*, 17: 4974–4979.

Nguyen H, Kondo K, Yagi Y, Iseki Y, Okuoka N, Watanabe T, Mikami B, Nagata T, and Katahira M (2022). Functional and structural characterizations of lytic polysaccharide monooxygenase, which cooperates synergistically with cellulases, from *Ceriporiopsis subvermispora*. *ACS Sustain Chem Eng*, 10: 923–934.

Nikel PI and de Lorenzo V (2018). *Pseudomonas putida* as a functional chassis for industrial biocatalysis: from native biochemistry to trans-metabolism. *Met Eng*, 50: 142–155.

Numata K and Morisaki K (2015). Screening of marine bacteria to synthesize polyhydroxyalkanoate from lignin: contribution of lignin derivatives to biosynthesis by *Oceanimonas doudoroffii*. *ACS Sustain Chem Eng*, 3: 569–573.

Nurika I, Shabrina EN, Azizah N, Suhartini S, Bugg TDH, and Barker GC (2022). Application of ligninolytic bacteria to the enhancement of lignocellulose breakdown and methane production from oil palm empty fruit bunches (OPEFB). *Bioresour Technol Rep*, 17: 100951.

Osma JF, Toca-Herrera JL, and Rodríguez-Couto S (2010). Uses of laccases in the food industry. *Enzyme Res*, 2010: 918761.

Paliwal R, Rawat AP, Rawat M, and Rai JP (2012). Bioligninolysis: recent updates for biotechnological solution. *Appl Biochem Biotechnol*, 167(7): 1865–1889.

Pandey MPK and Kim CS (2011). Lignin depolymerization and conversion: a review of thermochemical methods. *Chem Eng Technol*, 34: 29–41.

Paszczynski A, Huynh VB, and Crawford R (1985). Enzymatic activities of an extracellular, manganese-dependent peroxidase from *Phanerochaete chrysosporium*. *FEMS Microbiol Lett*, 29: 37–41.

Peng X (1998). Cloning of a *Sphingomonas paucimobilis* SYK-6 gene encoding a novel oxygenase that cleaves lignin-related biphenyl and characterization of the enzyme. *Appl Environ Microbiol*, 64: 2520–2527.

Peng X, Kelly RM, and Han Y (2018). Sequential processing with fermentative *Caldicellulosiruptor kronotskyensis* and chemolithoautotrophic *Cupriavidus necator* for converting rice straw and CO2 to polyhydroxybutyrate. *Biotechnol Bioeng*, 115: 1624–1629.

Pérez-Boada M, Ruiz-Dueñas FJ, Pogni R, Basosi R, Choinowski T, Martínez MJ, Piontek K, and Martínez AT (2005). Versatile peroxidase oxidation of high redox potential aromatic compounds: site-directed mutagenesis, spectroscopic and crystallographic investigation of three long-range electron transfer pathways. *J Mol Biol*, 354: 385–402.

Pham LTM and Kim YH (2016). Discovery and characterization of new O-Methyltransferase from the genome of the lignin-degrading fungus *Phanerochaete chrysosporium* for enhanced lignin degradation. *Enzym Microb Technol*, 82: 66–73.

Picart P, de María PD, and Schallmey A (2015). From gene to biorefinery: microbial β-etherases as promising biocatalysts for lignin valorization. *Front Microbiol* Sep 4, 6: 916.

Picart P, Müller C, Mottweiler J, Wiermans L, Bolm C, Domínguez de María P, and Schallmey A (2014). From gene towards selective biomass valorization: bacterial β-etherases with catalytic activity on lignin-like polymers. *Chemsuschem*, 7: 3164–3171.

Pieper DH (2005). Aerobic degradation of polychlorinated biphenyls. *Appl Microbiol Biotechnol*, 67: 170–191.

Piontek K, Antorini M, and Choinowski T (2002). Crystal structure of a laccase from the fungus *Trametes versicolor* at 1.90-A resolution containing a full complement of coppers. *J Biol Chem*, 277: 37663–37669.

Poblete-Castro I, Becker J, Dohnt K, Dos Santos VM, and Wittmann C (2012). Industrial biotechnology of *Pseudomonas putida* and related species. *Appl Microbiol Biotechnol*, 93: 2279–2290.

Pointing SB (2001). Feasibility of bioremediation by white-rot fungi. *Appl Microbiol Biotechnol*, 57: 20–33.

Poulos TL (1993). Crystallographic refinement of lignin peroxidase at 2 A. *J Biol Chem*, 268: 4429–4440.

Ragauskas AJ, Beckham GT, Biddy MJ, Chandra R, Chen F, Davis MF, Davison BH, Dixon RA, Gilna P, Keller M, Langan P, Naskar AK, Saddler JN, Tschaplinski TJ, Tuskan GA, and Wyman CE (2014). Lignin valorization: improving lignin processing in the biorefinery. *Science*, 344: 1246843.

Rahmanpour R and Bugg TD (2015). Characterisation of Dyp-type peroxidases from *Pseudomonas fluorescens* Pf-5: oxidation of Mn (II) and polymeric lignin by Dyp1B. *Arch Biochem Biophys*, 574: 93–98.

Rahmanpour R, Rea D, Jamshidi S, Fülöp V, and Bugg TD (2016). Structure of *Thermobifida fusca* DyP-type peroxidase and activity towards Kraft lignin and lignin model compounds. *Arch Biochem Biophys*, 594: 54–60.

Raj A, Reddy MK, Chandra R, Purohit HJ, and Kapley A (2007). Biodegradation of kraft-lignin by *Bacillus* sp. isolated from sludge of pulp and paper mill. *Biodegradation*, 18: 783–792.

Ravi K, García-Hidalgo J, Gorwa-Grauslund MF, and Lidén G (2017). Conversion of lignin model compounds by *Pseudomonas putida* KT2440 and isolates from compost. *Appl Microbiol Biotechnol*, 101: 5059–5070.

Reddy C (1995). The potential for white-rot fungi in the treatment of pollutants. *Curr Opin Biotechnol*, 6: 320–328.

Reddy GVB, Sridhar M, and Gold MH (2003). Cleavage of nonphenolic β-1 diarylpropane lignin model dimers by manganese peroxidase from *Phanerochaete chrysosporium*. *Eur J Biochem*, 270(2): 284–292.

Reiter J, Strittmatter H, Wiemann LO, Schieder D, and Sieber V (2013). Enzymatic cleavage of lignin β-O-4 aryl ether bonds via net internal hydrogen transfer. *Green Chem*, 15: 1373.

Rico A, Rencoret J, Del Río JC, Martínez AT, and Gutiérrez A (2014). Pretreatment with laccase and a phenolic mediator degrades lignin and enhances saccharification of Eucalyptus feedstock. *Biotechnol Biofuels*, 7(1): 6.

Riley R, Salamov AA, Brown DW, Nagy LG, Floudas D, Held BW, Levasseur A, Lombard V, Morin E, Otillar R, Lindquist EA, Sun H, LaButti KM, Schmutz J, Jabbour D, Luo H, Baker SE, Pisabarro AG, Walton JD, Blanchette RA, Henrissat B, Martin F, Cullen D, Hibbett DS, and Grigoriev IV (2014). Extensive sampling of basidiomycete genomes demonstrates inadequacy of the white-rot/ brown-rot paradigm for wood decay fungi. *Proc Natl Acad Sci U S A*, 111(27): 9923–9928.

Roberts JN, Singh R, Grigg JC, Murphy ME, Bugg TD, and Eltis LD (2011). Characterization of dye-decolorizing peroxidases from *Rhodococcus jostii* RHA1. *Biochemistry*, 50(23): 5108–5119.

Robinson T, Chandran B, and Nigam P (2001). Studies on the production of enzymes by white-rot fungi for the decolourisation of textile dyes. *Enzym Microb Technol*, 29: 575–579.

Rodríguez Couto S and Toca Herrera JL (2006). Industrial and biotechnological applications of laccases: a review. *Biotechnol Adv*, 24(5): 500–513.

Ruhong L, Liao Q, Xia A, Deng Z, Huang Y, Zhu X, and Zhu X (2020). Synergistic treatment of alkali lignin via fungal coculture for biofuel production: comparison of physicochemical properties and adsorption of enzymes used as catalysts. *Front Energy Res*, 8: 231.

Ruttimann-Johnson C, Salas L, Vicuna R, and Kirk TK (1993). Extracellular enzyme production and synthetic lignin mineralization by *Ceriporiopsis subvermispora*. *Appl Environm Microbiol*, 59(6): 1792–1797.

Sainsbury PD, Hardiman EM, Ahmad M, Otani H, Seghezzi N, Eltis LD, and Bugg TD (2013). Breaking down lignin to high-value chemicals: the conversion of lignocellulose to vanillin in a gene deletion mutant of *Rhodococcus jostii* RHA1. *ACS Chem Biol*, 8: 2151–2156.

Salvachúa D, Karp EM, Nimlos CT, Vardon DR, and Beckham GT (2015). Towards lignin consolidated bioprocessing: simultaneous lignin depolymerization and product generation by bacteria. *Green Chem*, 17: 4951–4967.

Salvachúa D, Martínez AT, Tien M, López-Lucendo MF, García F, de Los Ríos V, Martínez MJ, and Prieto A (2013a). Differential proteomic analysis of the secretome of *Irpex lacteus* and other white-rot fungi during wheat straw pretreatment. *Biotechnol Biofuels* Aug 10, 6(1): 115.

Salvachúa D, Prieto A, Martínez ÁT, and Martínez MJ (2013b). Characterization of a novel dye-decolorizing peroxidase (DyP)-type enzyme from *Irpex lacteus* and its application in enzymatic hydrolysis of wheat straw. *Appl Environ Microbiol* Jul, 79(14): 4316–4324.

Salvachúa D, Rydzak T, Auwae R, De Capite A, Black BA, Bouvier JT, Cleveland NS, Elmore JR, Furches A, Huenemann JD, Katahira R, Michener WE, Peterson DJ, Rohrer H, Vardon DR, Beckham GT, and Guss AM (2020b). Metabolic engineering of *Pseudomonas putida* for increased polyhydroxyalkanoate production from lignin. *Microb Biotechnol* May, 13(3): 813.

Salvachúa D, Werner AZ, Pardo I, Michalska M, Black BA, Donohoe BS, Haugen SJ, Katahira R, Notonier S, Ramirez KJ, Amore A, Purvine SO, Zink EM, Abraham PE, Giannone RJ, Poudel S, Laible PD, Hettich RL, and Beckham GT (2020a). Outer membrane vesicles catabolize lignin-derived aromatic compounds in *Pseudomonas putida* KT2440. *Proc Natl Acad Sci USA*, 2020 Apr 28, 117(17): 9302–9310.

Santhanam N, Vivanco JM, Decker SR, and Reardon KF (2011). Expression of industrially relevant laccases: prokaryotic style. *Trends Biotechnol*, 29(10): 480–489.

Saratale GD and Oh M-K (2015). Characterization of Poly-3-Hydroxybutyrate (PHB) produced from *Ralstonia eutropha* using an alkali-pretreated biomass feedstock. *Int J Biol Macromol*, 80: 627–635.

Sato Y, Moriuchi H, Hishiyama S, Otsuka Y, Oshima K, Kasai D, Nakamura M, Ohara S, Katayama Y, Fukuda M, and Masai E (2009). Identification of three alcohol dehydrogenase genes involved in the stereospecific catabolism of arylglycerol-β-aryl ether by *Sphingobium* sp. strain SYK-6. *Appl Environ Microbiol*, 75: 5195–5201.

Seto M, Kimbara K, Shimura M, Hatta T, Fukuda M, and Yano K (1995). A novel transformation of polychlorinated biphenyls by *Rhodococcus* sp. strain RHA1. *Appl Environ Microbiol*, 61: 3353–3358.

Sharma P, Goel R, and Capalash N (2007). Bacterial laccases. *World J Microbiol Biotechnol*, 23(6): 823–832.

Shi Y, Chai L, Tang C, Yang Z, Zheng Y, Chen Y, and Jing Q (2013). Biochemical investigation of kraft lignin degradation by *Pandoraea* sp. B-6 isolated from bamboo slips. *Bioprocess Biosyst Eng* Dec, 36(12): 1957–1965.

Shi Y, Yan X, Li Q, Wang X, Xie S, Chai L, and Yuan J (2017). Directed bioconversion of Kraft lignin to polyhydroxyalkanoate by *Cupriavidus basilensis* B-8 without any pretreatment. *Process Biochem*, 52: 238–242.

Shields-Menard SA, AmirSadeghi M, Green M, Womack E, Sparks DL, Blake J, Edelmann M, Ding X, Sukhbaatar B, Hernandez R, Donaldson JR, and Todd F (2017). The effects of model aromatic lignin compounds on growth and lipid accumulation of *Rhodococcus rhodochrous*. *Int Biodeterior Biodegrad*, 121: 79–90.

Shirkavand E (2017). Pretreatment of radiata pine using two white rot fungal strains *Stereum hirsutum* and *Trametes versicolor*. *Energy Convers Manag*, 142: 13–19.

Shraddha SR, Shekher R, Sehgal S, Kamthania M, and Kumar A (2011). Laccase: microbial sources, production, purification, and potential biotechnological applications. *Enzyme Res*, 2011: 217861.

Si M, Yan X, Liu M, Shi M, Wang Z, Wang S, Zhang J, Gao C, Chai L, and Shi Y (2018). In situ lignin bioconversion promotes complete carbohydrate conversion of rice straw by *Cupriavidus basilensis* B-8. *ACS Sustain Chem Eng*, 6(6): 7969–7978.

Sigoillot J-C, Berrin J-G, Bey M, Lesage-Meessen L, Levasseur A, Lomascolo A, Lomascolo A, Record E, and Boukhris-Uzan E (2012). Fungal strategies for lignin degradation. *Adv Bot Res*, 61: 263–308.

Singh D, Zeng J, Laskar DD, Deobald L, Hiscox WC, and Chen S (2011). Investigation of wheat straw biodegradation by *Phanerochaete chrysosporium*. *Biomass Bioenergy*, 35: 1030–1040.

Sonoki T, Iimura Y, Masai E, Kajita S, and Katayama Y (2002). Specific degradation of β-aryl ether linkage in synthetic lignin (dehydrogenative polymerizate) by bacterial enzymes of *Sphingomonas paucimobilis* SYK-6 produced in recombinant Escherichia coli. *J Wood Sci*, 48: 429–433.

Sonoki T, Takahashi K, Sugita H, Hatamura M, Azuma Y, Sato T, Suzuki S, Kamimura N, and Masai E (2018). Glucose-free Cis,Cis-muconic acid production via new metabolic designs corresponding to the heterogeneity of lignin. *ACS Sustain Chem Eng*, 6: 1256–1264.

Srivastava N, Singh R, Srivastava M, Syed A, Bahadur Pal D, Bahkali AH, Mishra PK, and Gupta VK (2022). Impact of mixed lignocellulosic substrate and fungal consortia to enhance cellulase production and its application in NiFe2O4 nanoparticles mediated enzymatic hydrolysis of wheat straw. *Bioresour Technol*, 345: 126560.

Stoj C and Kosman DJ (2003). Cuprous oxidase activity of yeast Fet3p and human ceruloplasmin: implication for function. *FEBS Lett*, 554(3): 422–426.

Sundaramoorthy M (2005). High-resolution crystal structure of manganese peroxidase: substrate and inhibitor complexes. *Biochemistry*, 44: 6463–6470.

Taniguchi M (2005). Evaluation of pretreatment with *Pleurotus ostreatus* for enzymatic hydrolysis of rice straw. *J Biosci Bioeng*, 100: 637–643.

Taylor C, Hardiman E, Ahmad M, Sainsbury P, Norris P, and Bugg T (2012). Isolation of bacterial strains able to metabolize lignin from screening of environmental samples. *J Appl Microbiol*, 113: 521–530.

Thurston CF (1994). The structure and function of fungal laccases. *Microbiology*, 140: 19–26.

Tien M and Kirk TK (1983). Lignin-degrading enzyme from the hymenomycete *Phanerochaete chrysosporium* Burds. *Science*, 221(4611): 661–663.

Tomizawa S, Chuah JA, Matsumoto K, Doi Y, and Numata K (2014). Understanding the limitations in the biosynthesis of polyhydroxyalkanoate (PHA) from lignin derivatives. *ACS Sustain Chem Eng*, 2: 1106–1113.

Ullah M, Liu P, Xie S, and Sun S (2022). Recent advancements and challenges in lignin valorization: green routes towards sustainable bioproducts. *Molecules*, 27: 6055.

van Erven G, Wang J, Sun P, de Waard P, van der Putten J, Frissen GE, Gosselink RJA, Zinovyev G, Potthast A, van Berkel WJH, and Kabel MA (2019). Structural motifs of wheat straw lignin differ in susceptibility to degradation by the white-rot fungus *Ceriporiopsis subvermispora*. *ACS Sustain Chem Eng* 2019 Dec 16, 7(24): 20032–20042.

Wan C and Li Y (2010). Microbial pretreatment of corn stover with *Ceriporiopsis subvermispora* for enzymatic hydrolysis and ethanol production. *Bioresour Technol*, 101: 6398–6403.

Wan C and Li Y (2012). Fungal pretreatment of lignocellulosic biomass. *Biotechnol Adv*, 30(6): 1447–1457.

Wariishi H, Valli K, Renganathan V, and Gold MH (1989). Thiol-mediated oxidation of nonphenolic lignin model compounds by manganese peroxidase of *Phanerochaete chrysosporium*. *J Biol Chem*, 264(24): 470. Microbial enzyme systems for lignin degradation 14185–14191.

Weiss R, Guebitz GM, Pellis A, and Nyanhongo GS (2020). Harnessing the power of enzymes for tailoring and valorizing lignin. *Trends Biotechnol*, 38: 1215–1231.

Welinder KG, Mauro JM, and Nørskov-Lauritsen L (1992). Structure of plant and fungal peroxidases. *Biochem Soc Trans*, 20(2): 337–340.

Weng C, Peng X, and Han Y (2021). Depolymerization and conversion of lignin to value-added bioproducts by microbial and enzymatic catalysis. *Biotechnol Biofuels*, 14: 84 https://doi.org/10.1186/s13068-021-01934-w.

Wesenberg D, Kyriakides I, and Agathos SN (2003). White-rot fungi and their enzymes for the treatment of industrial dye effluents. *Biotechnol Adv*, 22(1–2): 161–187.

Williamson PR, Wakamatsu K, and Ito S (1998). Melanin biosynthesis in *Cryptococcus neoformans*. *J Bacteriol*, 180(6): 1570–1572.

Wilson DB (2004). Studies of *Thermobifida fusca* plant cell wall degrading enzymes. *Chem Rec*, 4: 72–82.

Wong DW (2009). Structure and action mechanism of ligninolytic enzymes. *Appl Biochem Biotechnol*, 157: 174–209.

Woo HL (2014). Complete genome sequence of the lignin degrading bacterium *Klebsiella* sp. strain BRL6-2. *Stand Genomic Sci*, 9: 19.

Xu F, Shin W, Brown SH, Wahleithner JA, Sundaram UM, and Solomon EI (1996). A study of a series of recombinant fungal laccases and bilirubin oxidase that exhibit significant differences in redox potential, substrate specificity, and stability. *Biochim Biophys Acta*, 1292: 303–311.

Xu L, Sun K, Wang F, Zhao L, Hu J, Ma H, and Ding Z (2020). Laccase production by *Trametes versicolor* in solid-state fermentation using tea residues as substrate and its application in dye decolorization. *J Environ Manag*, 270: 110904.

Xu R, Zhang K, Liu P, Han H, Zhao S, Kakade A, Khan A, Du D, and Li X (2018b). Lignin depolymerization and utilization by bacteria. *Bioresour Technol*, 269: 557–566.

Xu Z, Lei P, Zhai R, Wen Z, and Jin M (2019). Recent advances in lignin valorization with bacterial cultures: microorganisms, metabolic pathways, and bio-products. *Biotechnol Biofuels*, 12: 32.

Xu Z, Qin L, Cai M, Hua W, and Jin M (2018a). Biodegradation of kraft lignin by newly isolated *Klebsiella pneumoniae, Pseudomonas putida*, and *Ochrobactrum tritici* strains. *Environ Sci Pollut Res Int*, 25: 14171–14181.

Yadav S and Chandra R (2015). Syntrophic co-culture of *Bacillus subtilis* and *Klebsiella pneumonia* for degradation of kraft lignin discharged from rayon grade pulp industry. *J Environ Sci*, 33: 229–238.

Yaropolov AI, Skorobogat'ko OV, Vartanov SS, and Varfolomeyev SD (1994). Laccase. *Appl Biochem Biotechnol*, 49(3): 257–280.

Yelle DJ, Ralph J, Lu F, and Hammel KE (2008). Evidence for cleavage of lignin by a brown rot basidiomycete. *Environ Microbiol*, 10: 1844–1849.

Yelle DJ, Wei D, Ralph J, and Hammel KE (2011). Multidimensional NMR analysis reveals truncated lignin structures in wood decayed by the brown rot basidiomycete *Postia placenta*. *Environ Microbiol*, 13: 1091–10100.

Yoshida H (1883). Chemistry of lacquer (Urushi). Part I. Communication from the chemical society of Tokio. *J Chem Soc Trans*, 43: 472–486.

Zabel RA and Morrell JJ (2020). *Wood Microbiology Decay and Its Prevention*, 2nd ed., Elsevier Inc.

Zeng G, Zhao M, Huang D, Lai C, Huang C, Wei Z, Xu P, Li N, Zhang C, Li F, and Cheng M (2013a). Purification and biochemical characterization of two extracellular peroxidases from *Phanerochaete chrysosporium* responsible for lignin biodegradation. *Int Biodeterior Biodegrad*, 85: 166–172.

Zeng J, Singh D, Laskar D, and Chen S (2013b). Degradation of native wheat straw lignin by *Streptomyces viridosporus* T7A. *Int J Environ Sci Technol*, 10: 165–174.

Zhang H, Han L, and Dong H (2021). An insight to pretreatment, enzyme adsorption and enzymatic hydrolysis of lignocellulosic biomass: experimental and modeling studies. *Renew Sustain Energy Rev*, 140: 110758.

Zhang YH (2015). Production of biofuels and biochemicals by in vitro synthetic biosystems: opportunities and challenges. *Biotechnol Adv*, 33: 1467–1483.

Zhang Z, Xia L, Wang F, Lv P, Zhu M, Li J, and Chen K (2015). Lignin degradation in corn stalk by combined method of H2O2 hydrolysis and *Aspergillus oryzae* CGMCC5992 liquid-state fermentation. *Biotechnol Biofuels*, 8: 183.

Zhao L, Cao GL, Wang AJ, Ren HY, Dong D, Liu ZN, Guan XY, Xu CJ, and Ren NQ (2012). Fungal pretreatment of cornstalk with *Phanerochaete chrysosporium* for enhancing enzymatic saccharification and hydrogen production. *Bioresour Technol* 2012 Jun, 114: 365–369.

Zhu D, Zhang P, Xie C, Zhang W, Sun J, Qian WJ, and Yang B (2017). Biodegradation of alkaline lignin by *Bacillus ligniniphilus* L1. *Biotechnol Biofuels*, 10: 44.

Zimmermann W (1990). Degradation of lignin by bacteria. *J Biotechnol*, 13: 119–130.

3.3 Chemical Depolymerization

3.3.1 Acid-catalyzed Depolymerization

Since the middle of the 20th century, a well-known method for depolymerizing lignin has been based on acid catalysis. This process uses acid for breaking down the lignin's β-O-4 linkages (Lundquist, 1976).

The mechanism for acid-catalyzed depolymerization of lignin is comparable to base-catalyzed depolymerization, in which hydrolytic cleavage of the α- and β-aryl ether linkage predominates over other cleavages (Meshgini and Sarkanen Kyosti, 1989). This is because for acid-catalyzed hydrolysis, the α-aryl ether linkage has a lower activation energy compared to β-aryl ether linkage (Agarwal et al., 2018).

Various acids, including sulfuric, hydrochloric, peracetic, and formic acid have been used as catalysts for dissolving the ether linkages (Mahmood et al., 2015). When the ether bonds are broken during hydrolysis by the acids, phenolic compounds are created.

Figure 3.19 depicts mechanism for the cleavage of β-O-4 using acid as a catalyst. Such types of depolymerizations are typically carried out at a high temperature and pressure. In a

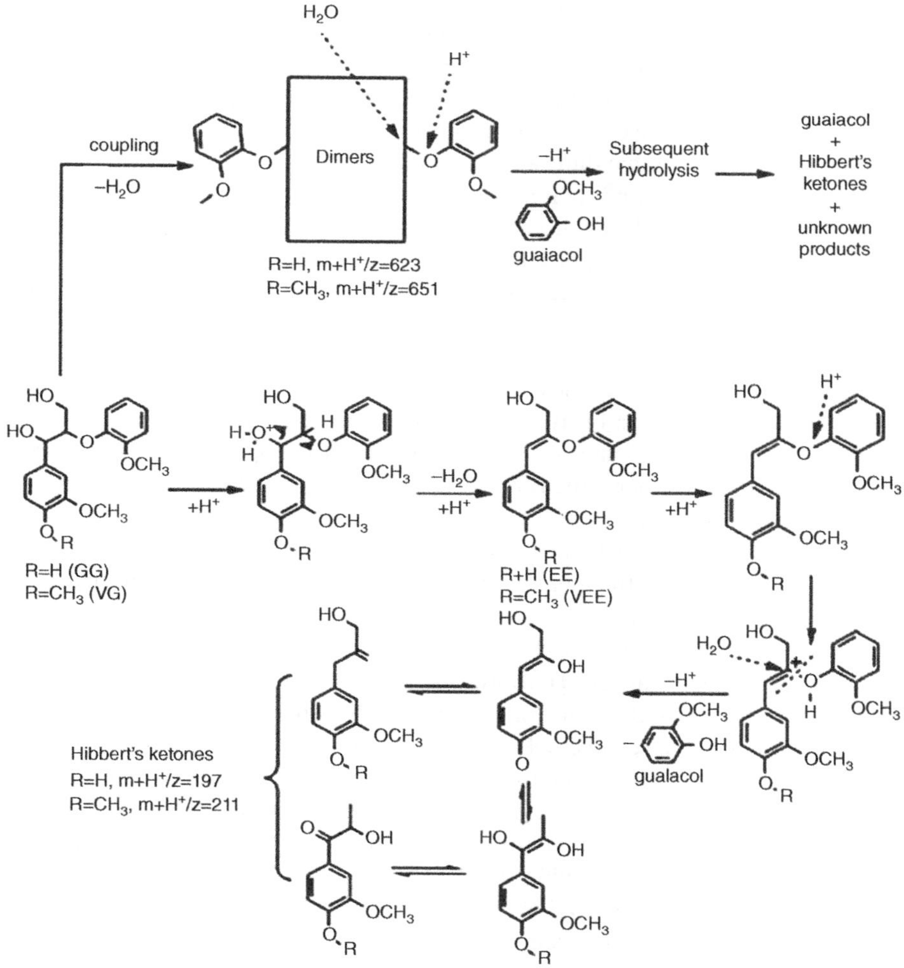

Figure 3.19 Cleavage of β-O-4 linkages by acid catalysts. Reproduced with permission Jia et al. (2010a) / John Wiley & Sons.

depolymerization study, ethanol and acid are typically used for separating the products into water-soluble and water-insoluble lignin by altering the ratio of solvent-acid (such as ethanol/ hydrochloric acid and ethylene glycol/ formic acid). Mahmood et al. (2015) demonstrated that the lignin structure cannot be broken down into monomeric units at temperatures between 78 and 200 °C.

Several research groups have investigated depolymerization using acid in various proportions at higher pressures and temperatures (Forchheim et al., 2012; Gasson et al., 2012). Gasson et al. (2012) reported that formic acid and ethanol in a ratio of 10:77 (weight %) effectively depolymerized the lignin in wheat straw. According to a different group, the best ratio for depolymerizing lignin from wheat straw is 10:81 (weight %) formic acid/ethanol (Forchheim et al., 2012). Utilizing sulfuric acid as a catalyst, the depolymerization rate was approximately 70 weight % in a solvent system containing 1:1 water/ethanol (Mahmood et al., 2015). For the depolymerization reaction, ethanol and water were combined with diluted sulfuric acid. Under 2 MPa and 250 °C for an hour, the reaction produced a higher yield.

The main outcomes and conditions of the acid-catalyzed depolymerization of lignin are presented in Table 3.8 (Roy et al., 2022). One of the most common methods is acid-catalyzed depolymerization, but it has some drawbacks. This method necessitates the use of toxic chemicals, repolymerization, and extreme reaction conditions. Repolymerization takes place primarily between the phenols' reactive sites and the α-carbon of phenol propanol (Forsythe et al., 2013; Yuan et al., 2010). The procedure calls for a higher temperature, a longer response time, and a higher tension. Additionally, it generates waste that is harmful to the environment and significantly raises the cost of reaction due to the need for handling and disposal.

Rahimi et al. (2014) used formic acid and sodium formate to depolymerize oxidized lignin under milder conditions (100 °C), which produced aromatic compounds with a low molecular mass of more than 60 weight %. The greatest proportion of aromatic monomers ever discovered was obtained using this simple C–O cleavage technique. In lignin-based model compounds, the cleavage of the β-O-4 linkage catalyzed by acid at temperatures up to 150 °C revealed two main pathways viz. path A, which results in Hibbert's ketones (C3-fragments), and path B, which involves repolymerization reactions and the loss of the formaldehyde molecule (Sturgeon et al., 2014; Yokoyama, 2015a). Both of these pathways form ethylbenzene derivatives (C2-fragments).

In the rate-determining step of acid catalyzed dehydration, β-O-4 linkages get cleaved to produce an intermediate called enol aryl ether, which quickly gets hydrolyzed to produce α-ketocarbinol and guaiacol. These compounds ultimately get converted into Hibbert's ketones through allylic rearrangement (Jia et al., 2011). The primary monomeric products produced by acidolysis of lignin are Hibbert's ketones (phenolic-C3 fragments). In contrast, the involvement of unstable C2-aldehyde in the repolymerization reaction, which frequently results in the formation of higher molecular weight polymers results in the production of C2-fragments in significantly smaller quantities (Roberts et al., 2011; Xu et al., 2014). Diols were used to capture reactive C2-aldehyde fragments in another study by Deuss et al. (2015). This produced aromatic compounds with low molecular weights like ethanol, acetals, and ethyl and methyl aromatics.

The overall pathways of the reaction are determined by the type of the substrate and the acid utilized for hydrolysis. Utilizing hydrochloric acid and sulfuric acid, for instance, paths A and B were followed in the hydrolysis of lignin-based model compounds in aqueous dioxane solutions under acid catalysis. This is caused by the conjugate base's direct involvement in the elimination reaction, which controls the reaction kinetics (Yokoyama, 2015a, 2015b).

It has been demonstrated that β-O-4 bonds can be effectively broken down by acid catalysts like triflic acid. The depolymerization of lignin model compounds has also been accomplished using the solid acid heterogeneous catalyst Nafion SAC-13 (Fraile et al., 2018). Because it is a highly

Table 3.8 Acid-catalyzed depolymerization of lignin.

Black liquor lignin

Formic acid, 160°C for 30 min

Vanillin, ethanone, phenol, guaiacol, and syringol type compounds

Dong et al., 2014

Aspen wood lignin

350 °C, 13.2 MPa for 1 h

Methoxyphenol and methylated and ethylated derivatives

Kuznetsov et al., 2015

Bagasse lignin

Zeolite HHSZ 100/Ethanol, 250 °C /30 min

Vanillin, methyl vanillate, syringol, homovanillic acid, and vinyl guaiacol

Deepa and Dhepe, 2015

Organosolv lignin

H_2SO_4/methanol, 300 °C /1 h

Phenol, ethyl guaiacol, and syringol

Wanmolee et al., 2016

Soda lignin

Aluminum/copper/nickel triflates, 400 °C for 4 h, 375–400 bars

Phenols and aromatic hydrocarbons

Güvenatam et al., 2016

Diluted acid corn stover lignin

Peracetic acid/Nb_2O_5, 333 K for 60 min

Vanillic acid, 4-hydroxy-2-methoxyphenol, and syringic acid

Ma et al., 2016

Black liquor lignin

Formic acid, 160 °C for 30 min

Ethanone and 1-(4-hydroxy-3-methoxyphenyl)

Wang et al., 2017

Organosolv corn stover lignin

Lewis acid Zr-KIT-5, 250 °C for 60 min

Guaiacol, ethyl guaiacol, ethyl phenol, and 2,4-demethxoyphenol

Nandiwale et al., 2018

Bulrush lignin

Phosphotungstic acid/ethanol/water, 250 °C for 6 h

4-ethylphenol and 4-ethyl-2-methoxyphenol

Du et al., 2019

Bagasse lignin

H_2SO_4, H_3PO_4, and HCl, 350 °C for 3 h

Guaiacol, 4-ethylphenol, vanillin, and ethyl guaiacol

Asawaworarit et al., 2019

Roy et al. (2022) / MDPI / CC BY 4.0.

fluorinated polymer of sulfonic acid, Nafion has a high acidity that is comparable to concentrated sulfuric acid. The acidolysis of α-aryl ether linkage employs an SN1-type mechanism, in contrast to the cleavage of β-aryl ether linkage. Phenols and alcohols typically undergo -aryl ether linkage acidolysis more quickly than water due to the solvent effect (Yuan et al., 2010). Many Lewis acids, like the chlorides of aluminum, boron, copper, iron, zinc, and nickel as well as mineral acid catalysts, have been used to break down lignin (Guvenatam et al., 2015).

Zinc chloride in ethanol was found to be the most effective of the various Lewis acids examined for depolymerizing lignin, yielding liquid products of up to 56 weight % (Zhang et al., 2014a). Combining Lewis acids with water or alcohol encourages depolymerization of lignin by transforming Lewis acids into their corresponding Bronsted acids (Hepditch and Thring, 2000). The performance of reaction was highly dependent upon temperature, despite the fact that it was found to be thermodynamically advantageous (Guvenatam et al., 2015). Lewis acidity that is too high encouraged condensation reactions, which resulted in products with a high molecular weight. Due to the higher solubility of intermediates in ethanol than in water, which prevents recombination, the rate of condensation product formation was higher in water as compared to ethanol.

Lewis acids successfully cut β-O-4 lignin linkages in the corresponding model mixtures when ionic fluids were used as response media (Janesko, 2014). The higher solubility of lignin in ionic liquids, which encouraged the formation of carbocation intermediates, was cited as the reason for this. The hydrolysis reaction is catalyzed by the presence of water, which reacts with metal chlorides to produce hydrochloric acid. A phenol-based model compound, namely guaiacylglycerol-β-guaiacyl ether (GG), could be efficiently broken down by a catalytic system consisting of Lewis acids like $AlCl_3$, $CuCl_3$ or $FeCl_3$, water, and ionic liquid [BMIM]Cl, yielding up to 80 weight % of guaiacol (Jia et al., 2011).

Bronsted acidic ionic liquids, like [HMIM]Cl, were able to hydrolyze GG to guaiacol with more than 70% yield at 150 °C (Cox and Ekerdt, 2012; Jia et al., 2010a), whereas non-acidic ionic liquids were unable to cleave the β-O-4 bonds of GG without using an external acid catalyst. In addition, the non-phenolic lignin model compound veratrylglycerol-β -guaiacyl ether yielded less guaiacol from GG. This was attributed to GG's capacity to generate in situ hydrochloric acid upon interaction of its phenolic group with metal chlorides, resulting in the release of proton. The type of ionic liquid anions and the acidic environment control the pathways of the reaction as well as the quantity of acidolysis product produced (Cox et al., 2011). Anions, for instance, stabilize the intermediates and form hydrogen bonds with the hydroxyl group of the lignin model compound, facilitating the cleavage of ether bonds and the production of guaiacol. Vinyl ether instead of enol ether is formed when an ionic liquid contains anions with fewer coordinating sites, as in the case of BF4. In the meantime, the production of enol ether is necessary for the production of higher guaiacol yield.

3.3.2 Base-catalyzed Depolymerization

Another method that is frequently used for extracting phenolic monomers from lignin is base-catalyzed depolymerization. Numerous affordable inorganic bases, like the hydroxides of alkali metals, have been widely used as catalysts for the base-catalyzed depolymerization of lignin. Under mild conditions, the reaction of hydroxides of alkali metal with lignin frequently results in a mixture of simple aromatic compounds. The depolymerization process is primarily responsible for the breakdown of ether linkages (α- and β-aryl ether), which are the most fragile bonds in the structure of lignin (Dabral et al., 2018a; Lavoie et al., 2011).

Different bases like sodium hydroxide, potassium hydroxide, and calcium hydroxide are used as catalysts in this method, which is carried out at a higher temperature (Evans et al., 1999; Thring 1994). Due to its excellent catalytic performance, this depolymerization method is considered one of the most promising (Gasson et al., 2012; Lavoie et al., 2011). Typically, it is carried out under high pressure and temperatures greater than 300 °C.

The β-O-4 bond is lignin's most common linkage. Cations from the base contribute to the formation of cation adducts during the reaction, which catalyze a six-membered transition on the β-O-4 bond that forms phenolic monomers (Roberts et al., 2011). This bond begins to break at 270 °C. Figure 3.20 depicts mechanism for base catalyzed depolymerization. Another well-known property is the regulated hydrolytic cleavage of lignin's ether bonds by base-catalyzed depolymerization (Beauchet et al., 2012). According to a study, weak bases like calcium hydroxide and lithium hydroxide can depolymerize at lower rates and produce less product overall than strong bases like potassium hydroxide and sodium hydroxide (Evans et al., 1999).

Strong bases depolymerized lignin effectively, but due to char deposition, bases like lithium hydroxide, potassium hydroxide, and calcium hydroxide were ineffective (Lavoie et al., 2011). For selective monomeric products, high yields can be achieved with BCD. Therefore, selectivity and yield are influenced by the type of solvent, temperature, pressure, reaction time, and base concentration (Miller et al., 1999; Yuan et al., 2010).

A homogeneous catalyst with an alkaline base produces intricate post separation.

In place of the homogeneous catalyst, magnesium hydroxide is a less expensive solid base catalyst. It is also considered robust with excellent catalytic activity (McFarland and Metiu, 2013).

The majority of base-catalyzed depolymerization of lignin are carried out at 300 °C and more than 10 MPa, where monomeric products vary depending on the reaction conditions.

Table 3.9 provides a summary of the reaction conditions and promising results of using depolymerization of lignin catalyzed by base (Roy et al., 2022).

Roberts et al. (2011) reported that simple aromatic products were obtained using 2% (w/w) sodium hydroxide and 5% (w/w) starting lignin material at a temperature of 300 °C, 250 bar pressure, and 4 minute optimum residence time (Table 3.10).

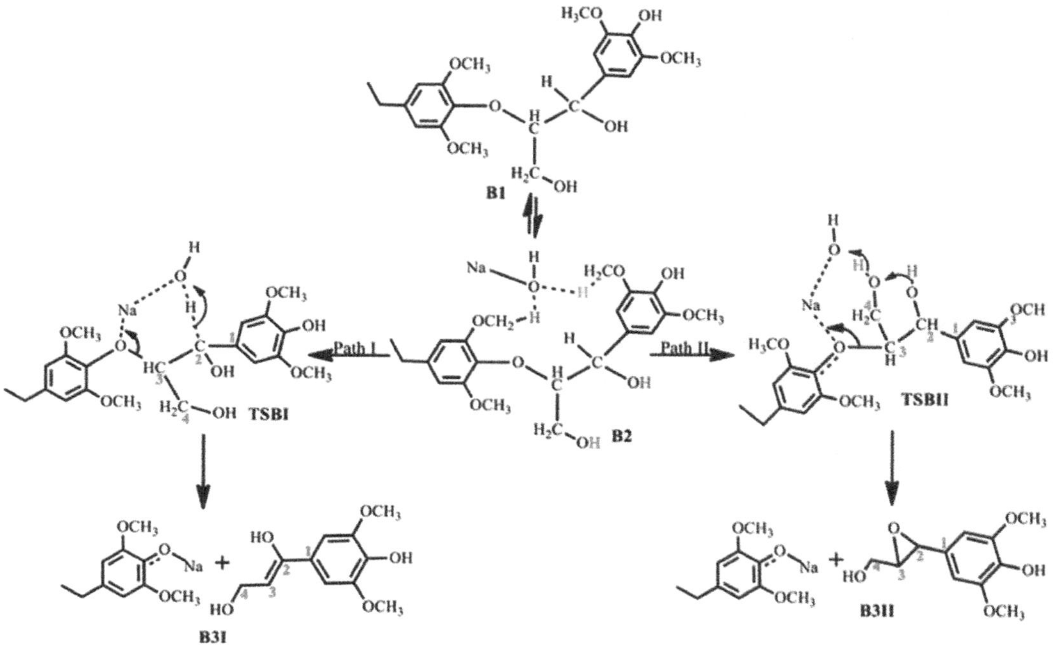

Figure 3.20 Mechanism for base catalyzed depolymerization of lignin Mensah et al. (2022) / Frontiers Media / CC BY 4.0.

Figure 3.21 shows low-molecular-weight products resulting from the depolymerization of lignin with the use of sodium hydroxide.

Beauchet et al. (2012) depolymerized kraft lignin at two different sodium hydroxide concentrations at 130 bar and 270–315 °C. Their findings demonstrated that aromatic monomer production was boosted by sodium hydroxide acting as a catalyst effectively. At 315 °C, pyrocatechol phenolic

Table 3.9 Reaction conditions and the products of base-catalyzed depolymerization of lignin.

Pine lignin

NaOH/Ru/C, 260 °C, 4 MPa
Guaiacol and syringol
Long et al., 2014

Corn stover lignin

2–4% NaOH, 270, 300, and 330 °C for 40 min
Phenol, guaiacol, syringol, vanillin, and vanillic acid
Katahira et al., 2016

Oak hardwood lignin

12.9–13.7 MPa, 30 min, 100 °C with NaOH and subH$_2$O
1,2-dimethoxylbenzene and 3,4-dimethoxytoluene
Hidajat et al., 2017

Alkaline lignin

250 °C, 1 h, NaOH, CsOH
Guaiacol, 2-methoxy-4-methylphenol, and 4-hydroxy benzyl alcohol
Chaudhary and Dhepe, 2017

Softwood kraft lignin

320 °C, 10 min, 250 bar, NaOH
Phenol formaldehyde resols
Solt et al., 2018

Klason lignin

200–250 °C, 1 h, 1000 rpm,
NaOH aromatic compounds
Chaudhary and Dhepe, 2019

Softwood kraft lignin

100 °C, 2–6 h, 10 bar, NaOH with O$_2$
Aromatic compounds
Paananen et al., 2020,

Pine wood-based kraft lignin

513–573 K, 1 h, 250 bar, NaOH,
Phenolic oligomers
Bernhardt et al., 2021,

Alkali lignin

300 °C, 4–9 min, 11 MPa,
NaOH phenol, guaiacol, and syringol
Kudo et al., 2019

Roy et al. (2022) / MDPI / CC BY 4.0.

Table 3.10 Various phenolic products of sodium hydroxide catalyzed hydrolysis of lignin at 300 °C and 250 bar.

Monomer		Structure	Monomer distribution [wt %]
syringol			40.9
syringyl aldehyde	$C_9H_{10}O_4$		18.0
3,5-dimethoxy-4-hydroxyacetophenone	$C_{10}H_{12}O_4$		16.3
4-methyl syringol	$C_9H_{12}O_3$		6.7
guaiacol	$C_7H_8O_2$		10.5
vanillin	$C_8H_8O_3$		3.5
4-hydroxy-3-methoxyphenylacetone	$C_{10}H_{12}O_3$		2.3
o-methoxy catechol	$C_7H_8O_3$		1.6

Reproduced with permission from Roberts et al. (2011) / John Wiley & Sons.

monomers were produced in abundance (up to 25.8%) and highly selective, leading to a yield of 8.4 weight % at this temperature. Temperature significantly increased the production of monomers.

Quinone methide intermediate is created from phenolate units during the base-catalyzed depolymerization process by cleaving α-aryl ether bonds. Guaiacol and coniferyl alcohol are the final products of this intermediate transformation. In addition, reactive formaldehyde and alkyl-stable vinyl

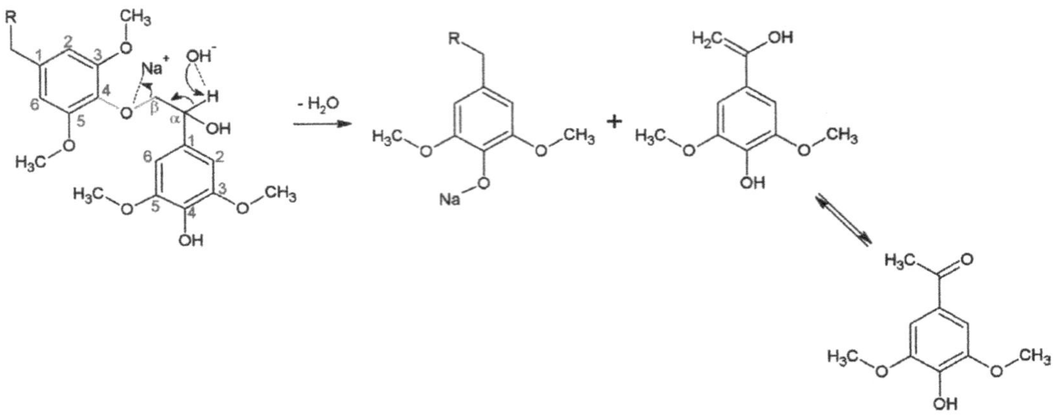

Figure 3.21 Low-molecular-weight products resulting from the depolymerization of lignin with the use of sodium hydroxide. Reproduced with permission from Roberts et al. (2011) / John Wiley & Sons.

ether can be produced by the base-catalyzed de-alkylation reaction of the quinone methide intermediate (Jia et al., 2010b). In contrast, non-phenolic units undergo β-aryl ether (β-O-4) bond cleavage. An oxirane ring is formed, and when a hydroxide ion is added, it eventually opens up to produce a glycol group (Li et al., 2015). Alkali metal cation forms adducts that are a catalyst for the formation of a six-membered transition state over β-O-4 bonds. For base-catalyzed lignin degradation, the concentration of the base and the ratio of lignin to solvent are crucial (Roberts et al., 2011). Also, parameters such as reaction time, pressure and temperature influence the yield of monomeric aromatic compounds (Mahmood et al., 2013) during base-catalyzed lignin depolymerization. The nature of the base and solvent determines the overall yield of bio-oil. Strong bases and organic solvents like phenols and alcohols are expected to yield more (Limarta et al., 2018; Yuan et al., 2010).

Strong bases cause a lot of polarization, but organic solvents encourage the solvolysis of ether links, which improves the kinetics and overall mechanism of the depolymerization reaction (Toledano et al., 2012). One of the main issues that leads to the formation of a solid residue is the repolymerization of degraded intermediates during base-catalyzed lignin depolymerization. The key to controlling repolymerization is to use a suitable radical scavenger to capture reactive radical species. Repolymerization has been avoided by employing a number of radical scavengers and proton-donating solvents, including formaldehyde, formic acid, p-cresol, phenol, and 2-naphthol (Gosselink et al., 2012; Huang et al., 2015; Toledano et al., 2014b).

Radical scavengers trap reactive fragments and cover active sites in lignin, encouraging the complete conversion of lignin with no solid residue. Another issue with base-catalyzed depolymerization is the formation of acidic molecules and subsequent acid/base neutralization, which results in the deactivation of catalysts. During base-catalyzed lignin depolymerization, problems with product isolation arise due to the significant amount of aqueous waste produced and the large amount of base consumed (Zakzeski et al., 2010). Lastly, the higher reaction temperature and uncontrolled selectivity of the products make this method unsuitable for large-scale applications. In addition to inorganic bases, organic N-bases are used to depolymerize guaiacylglycerol-guaiacyl ether (GG), a phenolic lignin model compound dissolved in an imidazolium-based ionic liquid (Jia et al., 2010b). 1,5,7-triazabicyclo[4,4,0]dec-5-ene (TBD) demonstrated >40% β-O-4 ether bond cleavage out of the various organic N-bases investigated. TBD breaks up GG bonds by acting as a dibasic nucleophile and attacking the quinone methide's α- and β-carbons.

A mechanochemical approach that uses solvent-free ball milling and ambient conditions as a pre-treatment for the base-catalyzed chemical depolymerization process has gained popularity. The majority of the β-O-4 bonds are degraded and the particle size is effectively reduced by this method (Kleine et al., 2013). High-temperature pockets caused by frictional energy in the mechanochemical process break through the solid–solid diffusion barrier and mechanically deform the molecule. Being a solvent-free, speedy, productive, versatile, and ambient-based alternative; the mechanochemical methodology offers a promising option for depolymerization of lignin (Dabral et al., 2018a, 2018b; Yao et al., 2018).

The ambient-condition mechanochemical strategy outperforms other high-temperature strategies despite the fact that repolymerization is the only disadvantage, that can be easily overcome by adding radical scavengers in controlled amounts to get the desired result (Brittain et al., 2018). This is because of its lower product selectivity and over-reduction of aromatic rings, both of which frequently form substantial amount of biochar (Azadi et al., 2013). The majority of the β-O-4 bonds are degraded and the particle size is effectively reduced by this method (Kleine et al., 2013).

3.3.3 Ionic Liquid-assisted Depolymerization

When it comes to depolymerizing lignin into aromatic compounds, the approaches that were previously discussed are extremely efficient. However, the use of corrosive chemicals, extreme reactions, and the production of harmful waste raise processing costs. To reduce the production waste, processing costs and protect the environment, depolymerizing lignin into aromatic monomers requires a sustainable, environmentally friendly method.

Numerous types of ionic liquids (ILs) have been produced through extensive research. They can be created using a huge range of different cation and anion options (MacFarlane et al., 2017; Pham et al., 2010; Plechkova and Seddon, 2008).

Figure 3.22 shows the three ionic liquid generations and Figure 3.23 shows all the anions and cations that are normally used for the application of ILs in lignin chemistry (Dai et al., 2020).

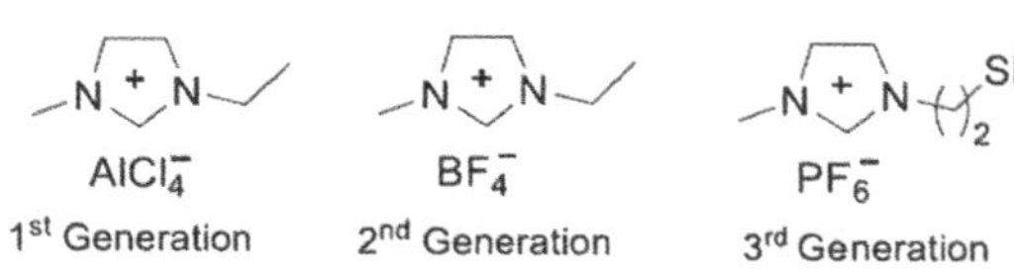

Figure 3.22 The three ionic liquid generations. Reproduced with permission from Dai et al. (2020) / Springer Nature.

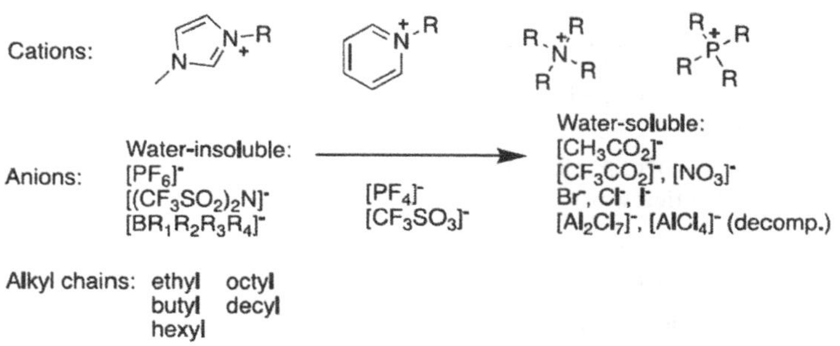

Figure 3.23 Cations and anions of ionic liquids in lignin chemistry. Reproduced with permission from Dai et al. (2020) / Springer Nature.

The extraction of lignin has been studied mainly with imidazolium salts with different alkane chains coupled with different types of common anions, such as chloride, bromide, tetrafluoroborate, and acetate (Zhang, 2013).

Tetrachloroaluminates, ionic liquids at room temperature based on 1-alkyl-3-methylimidazolium salts, were first described by Carlin and Wilkes (1994). These were regarded as the first generation of ionic liquids. The tetrachloroaluminate anion, on the other hand, was sensitive to moisture. By substituting the air- and moisture-stable tetrafluoroborate anion for the tetrachloroaluminates, the second generation of ionic liquids were produced. The wide selection of cations and anions has prompted the improvement of ionic liquids with properties intended for explicit undertakings and dissolvable properties. Ionic liquids with "tunable properties" or "task-specific ionic liquids" have been other names for this. The third generation of ionic liquids are these particular kinds (Figure 3.22).

Currently, ionic liquids are regarded as an environmentally friendly method for cleaving β-O-4 bonds to depolymerize lignin. Low formulation costs, non-volatility, excellent miscibility chemical inertness, lower viscosity and acidity, are just a few of the qualities of ionic liquids that make them suitable for use as a green medium (Vekariya, 2017; Welton, 2011; Zhang, Zhang et al., 2014b).

Because they are liquid over a wider temperature range and exhibit chemical, electrochemical, and thermal stability, ionic liquids are excellent environmentally friendly alternatives to solvents for depolymerizing lignin (Marszałł and Kaliszan, 2007). As they are long-range, non-covalent, interaction-based liquid formulations, it has been reported in recent studies that ionic liquids are excellent solvents for the solubilization of lignin and the fractionation of lignocellulose into its basic components.

Their cations and anions interact with one another to break lignin bonds and maintain a liquid state at temperatures typically below 100 °C (Sun et al., 2014; Underkofler et al., 2015; Zhang et al., 2014a). In conjunction with a variety of catalysts, ionic liquids may have the potential to regulate the degree of oxidative depolymerization (Wang et al., 2014).

Ionic liquids are composed of organic cations and inorganic/organic anions and are in the liquid phase at or below 100 °C. Due to their unique properties, ionic liquids have received a lot of attention and been considered for lignin depolymerization. According to Hossain and Aldous (2012), for instance, in comparison to other solvents, ionic liquids have a high potential to dissolve a wide range

Table 3.11 Ionic liquids used for depolymerization of lignin.

Substrate	Ionic liquids	Condition	Yield of lignin (%)	References
Kraft lignin	[PyFor][c]	75 °C, 1 h	74	Rashid et al. (2016)
Soda lignin	[Bmim][MeSO$_4$][b]	150 °C, 6 h	48	Prado et al. (2013_
Oil palm	[Bmim][cl]	110 °C, 8 h	54	Mohtar et al. (2015)
Corn Stover	[Pyrr][ac][d]	90 °C, 24 h	70	Achinivu et al. (2014)
Loblolly pine wood	[Emim][OAc]	160 °C, 1.5 h	48	Sathitsuksanoh et al. (2014)
Apple tree pruning	[Bmim][MeSO4]	200 °C, 3 min Microwave	92	Prado et al. (2016)
Populus tomentosa	[Emim][OAc]	130 °C, 3 h	90	Wen et al. (2014)
Bagasse	[Emim][ABS][a]	190 °C, 1 h	96	Tan et al. (2009)
Yellow pine	[Hmim][cl]	130 °C, 5 h	90	Cox and Ekerdt (2013)
Wood flour (Fagus crenata)	[Emim][cl]	120 °C, 8 h	95	Miyafuji et al. (2009)

Based on Dai et al. (2020).

of biomass, thereby enhancing the utilization of bioresources. Using these various kinds of ionic liquids, lignin extraction from biomass has been the subject of extensive research. Work on isolating lignin from biomass using ionic liquids under various conditions is summarized in Table 3.11.

Despite the fact that the dissolution mechanism of lignin in solvents remains unknown, Akiba et al. (2017) reported that dissolution of lignin by ionic liquids is a function of a proton donating and accepting ability of ionic liquids. Ionic liquids also have a very low vapor pressure because they are ionic, which reduces the amount of volatile organic compounds released (Dai et al., 2016; Zhu et al., 2018). Alternately, a particular mix of cations and anions can be used to modify the characteristics of ionic liquids, including their acidity, solubility as a catalyst, melting point, and miscibility with solvents, transforming them into "designer solvents and/or catalysts" that can be tailored to different processes and reactions (Cox et al., 2011; Zhu et al., 2018). Ionic liquid is likewise notable for its higher thermal stability, recyclability, and non-combustibility (Dai et al., 2016; Gillet et al., 2017). A thorough comprehension of the properties of several ionic liquids and the roles that various anions and cations play in the conversion process ought to serve as the foundation for the development of effective strategies for lignin depolymerization.

According to George et al. (2011), alkylsulfonate anions are more effective than lactates, acetates, chlorides, and phosphates at reducing lignin's polydispersity or molecular weight and increasing its reaction activity. Targeted cleavage of β-O-4 lignin structure linkages for producing guaiacol has been broadly investigated, and the yield of breakage product in many ionic liquids was found in the following order (Cox et al., 2011)

$$[Hmim]Cl > [Bmim][HSO4] > [Hmim]Br > [Hmim][HSO4] > [Hmim][BF4]$$

Moderately basic anions (such as Cl, Br, and [CF_3CO_2]-) inhibited the adverse dealkylation reaction in targeted breakage of ether bonds in lignin structure catalyzed by Brønsted acid, whereas weakly basic anions (such as [BF_4]-, [CF_3SO_3]-, and [PF_6]) facilitated dealkylation (Binder et al., 2009). The industrialized depolymerization of lignin is rigourously limited by the high cost of ionic liquids, so their recycling or reuse is crucial. Product separation from ionic liquids is also essential for the analysis of reaction products and the purification of industrial chemicals (Zakzeski et al., 2010). Ionic liquids and aromatic lignin structure interact strongly, making it difficult to separate them from products derived from lignin without the use of organic solvents (Hossain and Aldous, 2012; Wang et al., 2013). Mass spectrometry, ultraviolet-visible and infrared spectroscopy, nuclear magnetic resonance spectroscopy, and light scattering methods, are some of the technologies currently in use for analyzing dissolved lignin-derived products in ionic liquids (Zakzeski et al., 2010). Scalability, toxicity, and lifecycle analysis should all be taken into account when using ionic liquids in lignin depolymerization in the future (Stark, 2010; Weldemhret et al., 2020).

The scientific community has paid a lot of attention to ionic liquids as a way to depolymerize lignin into products with added value. To depolymerize lignin, various combinations of ionic liquids and metal catalysts, like cobalt, copper, and manganese, have been investigated.

Stärk et al. (2010) have converted lignin using the catalyst manganese(II) nitrate and various combinations of ionic liquids. They have reported that the best reaction medium for lignin depolymerization was manganese(II) nitrate and 1-ethyl-3-methylimidazolium trifluoromethanesulfonate [EMIM][CF_3SO_3]. They also discovered that, at 100 °C for 24 hours, with 11.5 weight % of 2,6-dimethoxy-1,4-benzoquinone, more than 63% of organosolv beach lignin can be selectively converted into phenolic monomers.

Li et al. (2017) transformed organosolv bagasse lignin into phenolic monomers by utilizing water and the cooperative [bmim][CF_3SO_3]/[bSmim][HSO_4]. Under milder reaction conditions (250 °C, 30 min), with 14.5 weight % of phenolic compounds, the conversion of lignin reached 66.7%, and the amount of char formed was minimal. Due to improved H ion dissociation at 250 °C, water is subcritical, converting 36.4% of the lignin into phenolic monomers through self-catalytic activity. The carbocation mechanism causes the majority of lignin to become an undesirable char product at this temperature (Zakzeski et al., 2010). However, [bmim][CF3SO3] is a superb hydrogen bond donor, increasing lignin conversion to 45.2% without char formation (Pu et al., 2007). It does this by promoting the breakage of the hydrogen-bonded linkages. When acidic ionic liquid [bSmim][HSO4] was added, 60.1% lignin was converted, and when the concentration of ionic liquid [bSmim][HSO$_4$] was raised from 2.0 to 3.0 mmol, 66.7% lignin was converted.

Ionic liquids can function as bases, acids, and nucleophiles and have a high degree of structural flexibility. In this manner, ionic liquids can emulate the acid or base-catalyzed depolymerization of lignin without utilizing any destructive synthetic compounds. The ionic liquids' cations and anions have a significant impact on how effectively ionic liquids depolymerize lignin (Xu et al., 2014). Research has demonstrated that lignin structure integrity is primarily affected by anions, whereas cations serve as observers (George et al., 2011).

According to another study (Cox et al., 2011), the strength of the coordination interaction between the hydrogen in the hydroxyl group of the lignin backbone and the lignin-like structures affects this anionic activity. By stabilization of the electronic environment of the compounds, the coordination with hydrogen directs the nucleophilic attack toward the carbon double bond. An ionic liquid based on ammonium for selectively breaking the lignin down into phenolic monomers was used by Tolesa et al. (2017). The solvent for the selective depolymerization of lignin consisted of two distinct ionic liquids: equimolar diisopropylethylamine and acetic acid or octanoic acid. They have demonstrated that ionic liquids can also function as a catalyst and a solvent. How the hydroxyl groups in compounds can be stabilized by the anion of an acidic ionic liquid to accelerate the cleavage of lignin's C–O bonds has been investigated (Casas et al., 2013).

One of the most recent approaches to lignin bond depolymerization is deep eutectic solvent-based depolymerization. Electrochemical catalysis, microwave irradiation, and conventional heating with deep eutectic solvents are currently the most commonly used methods. The effectiveness of these methods compared to the traditional methods has been characterized and monitored using ^{13}C-NMR, two-dimensional heteronuclear single quantum correlation NMR, ^{31}P NMR, ^{1}H-NMR, gel permeation chromatography, cyclic voltammetry, gas chromatography-mass spectrometry and scanning electron microscopy (Di Marino et al., 2016; Muley et al., 2019; Shen et al., 2019; Yu et al., 2020).

Table 3.12 presents reaction conditions and the products of ionic liquid and deep eutectic solvent-based depolymerization (Roy et al., 2022).

3.3.4 Supercritical Fluids-assisted Lignin Depolymerization

It is essential to develop an alternative green technology with reduced environmental impact for producing specific desired products from lignin (Roy et al., 2022). Consequently, efficient lignin conversion, the utilization of less toxic materials, and lower energy consumption are essential requirements (Cansell et al., 2003; Reverchon, 2002). In order to successfully prevent condensation reactions, methods based on supercritical solvent have been deemed an effective and greener method for depolymerizing lignin (Chio et al., 2019; Gosselink et al., 2012; Kim et al., 2015a, 2015b; Klapiszewski et al., 2017; Wahyudiono et al., 2008, 2013; Wang et al., 2013).

When compared to the surrounding environment, subcritical or supercritical fluids exhibit a variety of distinct properties and are created at extremely high temperatures and pressures. Lower viscosity and lower dielectric constant, and higher diffusivity are some of the distinct properties of

Table 3.12 Reaction conditions and the products of ionic liquid and deep eutectic solvent-based depolymerization.

Sugarcane bagasse

[bmim][CF$_3$SO$_3$]

483–523 K for 15–45 min

Guaiacol, 4-ethylphenol

Li et al., 2017

Pine and willow lignin

1-butylimidazolium hydrogen-sulfate [HC4im]-[HSO4]

100 °C, 30 min for pine and 4 h for willow

Vanillin, syringaldehyde

De Gregorio et al., 2016

Milled wood lignin

1-ethyl-3-methylimidazolium chloride ([C2mim] [Cl])

120 °C for 96 h

Vanillin, syringaldehyde

Ogawa and Miyafuji, 2015

Kraft lignin

ChCl:Ethylene glycol (1:2), ChCl: Urea (1:2)

Electrochemical depolymerization, Ni catalyst, oxidation potential were 0.5 and 1.0 V, room temperature, 24 h, 5 g L^{-1} lignin concentration.

Relative yields of *Guaiacol, vanillin, acetovanillone*, and *syringaldehyde* are 30–38%, 34–37%, 9%, and 12% respectively.

Di Marino et al., 2016.

Alkali lignin

N-allylpyridinium chloride, [Apy]Cl

150–210 °C for 3 h

4-vinylphenol

Wang et al., 2018.

Kraft lignin

ChCl: oxalic acid (1:1)ChCl: formic acid (1:2)

Microwave reactor (MWR), oxalic acid (OA) DES (130 °C) and formic acid (FA) DES (150 °C), 15 min, At 2450 MHz dielectric constant (DC) of OA and FA are 15.0 and 18.0, respectively. Selective cleavage of *b*-O-4 and *b*-5 cross-peaks bonds than typical heating method and narrower molecular weight (MW) distribution. DC of the HBA is an important factor in MWR-based methods.

Muley et al., 2019

Double enzymatic lignin (DEL)

ChCl: lactic acid (1:10)

0.5 g DEL in 10.0 g DES, 60-140 °C, 6 h.

Dissociation of aryl ether linkage (β-O-4) is dominant, with higher phenolic hydroxyl and less aliphatic hydroxyl groups, in the formation of lignin nanoparticles. At higher temperatures, 8.0 g kg^{-1} lignin produces 2.5 g kg^{-1} guaiacol, 1.0 g kg^{-1} vanillin, 0.6 g kg^{-1} syringaldehyde, 0.31 g kg^{-1} guaiacylacetone, and 0.08 g kg^{-1} acetovanillone

Shen et al., 2019

(Continued)

Table 3.12 (Continued)

Alkali lignin
ChCl: MeOH (1:4), lignin/Cu(OAc)2/1,10-phenanthroline (8:1:1), lignin: solvent (1:30), 0.4–3.0 MPa O2, 60 °C for 3 h.
Vanillic acid, vanillin, acetovanillone (46.09%), 2-methoxy-4-vinylphenol, 2-methoxyphenol, dihydroconiferyl alcohol, and acetic acid (40.64%).
Yu et al., 2020

Herbaceous lignin
ChCl: FeCl3 (1:2)
A 25 mL steel autoclave, herbaceous lignin (50 mg), M-DES catalyst 0.1–0.5 mmol, methanol (5–15 mL), stirring 500 rpm, oil bath, temperature 120–200 °C.
Selective tailoring of H moiety to produce methyl phydroxycinnamate (MPC), yield 105.8 mg g^{-1} or 74.1% under mild conditions (160 °C, 8 h)
Li et al. 2020

Roy et al. (2022) / MDPI / CC BY 4.0.

supercritical fluids, which penetrate lignin and readily solubilize the depolymerized products (Cocero et al., 2018). Supercritical fluids typically possess both liquid and gas properties. Additionally, water can effectively dissolve a number of organic compounds due to its lower dielectric constant, which is comparable to that of a non-polar organic solvent in supercritical conditions (Cocero et al., 2018). Substances that are above their critical temperature and pressure are referred to as supercritical fluids. Under these circumstances, a liquid does not go through a vapor–liquid phase transition; instead, it only exists in a homogeneous stage condition with characteristics similar to those of gases and fluids in terms of diffusivity, viscosity, and density (Ibáñez et al., 2018). In addition, by varying the pressure or temperature, or by combining them with specific liquid solvents, the effects entrained by the chemical association between both modifiers and solutes can be used to modify the properties of supercritical fluids (Walsh et al., 1987). The most prevalent substances found in supercritical fluids include water, carbon dioxide, ammonia, and hydrocarbons like propane and butane (Guo et al., 2017; Numan-Al-Mobin et al., 2016).

Since the 1990s, supercritical methods have been used, and supercritical methanol was used to treat KL and OL at first (Miller et al., 1999). Condensation can be avoided extremely effectively through supercritical depolymerization reactions despite the fact that condensation occurs frequently during depolymerization. Lignin-derived intermediates can bind to hydrogen radicals produced by supercritical fluids to halt the condensation reaction, according to research by Mahmood et al. (2015) and Güvenatam et al. (2016).

Saisu et al. (2003) have reported that practically all of the organosolv lignin can be depolymerized and changed over into 2-cresol with a yield of 7.15 weight % using supercritical water at 400 °C for one hour when phenol is present. Similar studies had been conducted at pressures ranging from 25 to 40 MPa and temperatures ranging from 350 to 400 °C. Soluble and insoluble methanol are further subdivided into the products of depolymerized lignin. Phenol, catechol, and o, m, and p-cresols are the main components of the soluble fraction. Repolymerization of low-molecular-weight molecules is also seen in supercritical conditions (Saisu et al., 2003; Sasaki and Goto, 2008; Yong and Matsumura, 2013). According to previous research (Aida et al., 2002; Fang et al., 2008; Saisu et al., 2003), phenol can be added to lignin to prevent the formation of char and repolymerization. In order to prevent the crosslinking reaction and repolymerization, the phenol can block the active site of the decomposed lignin fragment by reacting with it. There is greater suppression when the ratio of phenol to lignin is higher (Fang et al., 2008; Okuda et al., 2004). But, the use of phenol to stop repolymerization is expensive, so the process as a whole would cost more (Paterson, 2012). In addition to using supercritical water as the solvent, various supercritical organic solvents have been utilized in the depolymerization of lignin.

Since the 1990s (Evans et al., 1999; Dorrestijn et al., 1999), supercritical methanol and methanol have been extensively studied and tested under a variety of conditions. Cheng et al. (2010) reported that supercritical ethanol is more reactive and effective than methanol for pine sawdust lignin depolymerization. After 20 minutes of treatment at 300 °C, only 12 weight % of the solid remains. In addition, compared to 100% methanol and ethanol which can convert more than 95% of pine sawdust lignin into 65 weight % bio-oil, 50% ethanol or methanol can do so twice as well. When it comes to the solubilization and depolymerization of lignin, supercritical fluids can be an efficient solvent. However, its applications were limited by its higher cost and harsh reaction condition. In addition, the chemical conversion pathway requires additional research to enhance selectivity and product separation due to the increased difficulty of mechanism study and intermediate detection caused by the quick hydrolysis of lignin in supercritical fluid.

Wahyudiono et al. (2008) effectively directed supercritical water depolymerization of
lignin at 350 to 400 °C and 25 to 40 MPa. They demonstrated that reaction time decreased the formation of heavier compounds, while temperature increased the formation of heavier compounds. Water is an excellent solvent for conversion of lignin into monomers, as the main products were identified as catechol, phenol, and o, m, p-cresol.

Numan-Al-Mobin et al. (2016) carried out yet another study on the use of water and subcritical carbon dioxide to depolymerize lignin. Subcritical carbon dioxide is a naturally occurring compound that is renowned for its sustainability, non-flammability, and capacity for catalysis. An environmentally friendly and more effective method for synthesizing phenolic monomers was the use of water and subcritical carbon dioxide.

Lignin has also been depolymerized with subcritical and supercritical ethanol, methanol, and water. According to the findings of this study, ethanol has the potential to substantially improve the depolymerization of lignin and lessen the formation of char (Guo et al., 2017). This may be because the hydrogen radicals that come from the ethanol combine with the lignin intermediates to prevent the repolymerization reaction intermediates from forming char. Supercritical methanol, on the other hand, has interesting properties that encourage the depolymerization of lignin, such as its ability to donate hydrogen, higher heat transfer, higher dispersity, and enhanced catalytic activity (Chen et al., 2019a).

Figure 3.24 shows a schematic of the reaction mechanism for lignin decomposition under supercritical water conditions.

Sub- and supercritical ethanol and water can also be used to depolymerize lignin into phenolic compounds (Cheng et al., 2012). When it comes to liquefying woody biomass into crude products, supercritical ethanol performs better than methanol. Additionally, it is anticipated that supercritical ethanol will make it simpler to dissolve and stabilize the liquid products.

For the purpose of lignin depolymerization, the medium of choice was supercritical fluid. In the 1990s, it was reported that potassium hydroxide or other bases were used to treat kraft- and organosolv-derived lignins in supercritical methanol or ethanol, indicating that the supercritical liquid affected lignin depolymerization (Miller et al., 1999; Dorrestijn et al., 1999). Takami et al. (2012) used supercritical water and p-cresol as the medium for treating organosolv lignin at temperatures between 350 and 420 C. The reaction mixtures yielded 2-(hydroxybenzyl)-4-methyl-phenol with a weight yield of 75%.

Gosselink et al. (2012) made another attempt to prepare syringol and guaiacol from organosolv hardwood and wheat straw lignins by treating them with carbon dioxide/acetone/water supercritical fluid under 10 MPa at temperatures ranging from 300–370 °C. Some depolymerization products

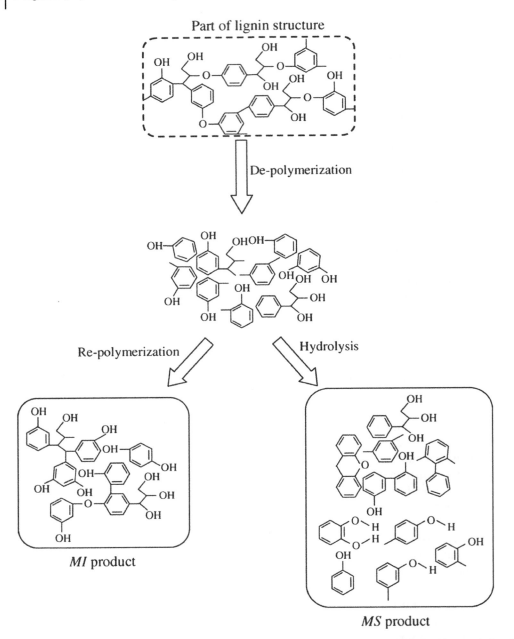

Figure 3.24 Plausible reaction pathway of lignin in supercritical water. Reproduced with permission from Wahyudiono et al. (2008) / ELSEVIER.

have different yields because the linkage content of hardwood and wheat straw lignins is different. Due to its excellent solubility, supercritical fluid, like ionic liquid, was utilized as the solvent in the depolymerization system. Acid and alcohol served as hydrogen sources for the hydrolysis. The high cost of supercritical liquid kept it from becoming a widely used method for lignin conversion, despite its high selectivity and ease of use for product and solvent separation.

Table 3.13 presents reaction conditions and the products of sub- and supercritical depolymerization of lignin (Roy et al., 2022).

Table 3.13 Reaction conditions and the products of sub- and supercritical depolymerization of lignin.

Alkali lignin

50/50 (%v/v) water and ethanol
300 °C for 2 h with 5 MPa
Phenol derivatives
Cheng et al., 2012

Alkali lignin

Subcritical water
330 °C, residence time, and 25 MPa
Eugenol, guaiacyl acetone, guaiacol, and vanillin
Yong and Matsumura, 2013

Organosolv lignin

Ethanol and Ru/Al$_2$O$_3$
260 °C for 8 h and 80 bar
Guaiacol and syringol
Martin et al., 2013.

Alkali lignin

SubcrH2O and Supercritical CO$_2$
350 °C, 10 min, and 22.03 Mpa
Vanillin and guaiacol
Numan-Al-Mobin et al., 2016

Wheat straw lignin

Ethanol and Rh/C
240 °C for 4 h and 7.0 MPa
Phenol derivatives
Guo et al., 2017

Alkali lignin

Ethanol and HZSM-5
360 °C for 120 min and 12.4 Mpa
Guaiacol and syringol
Fan et al., 2018.

Alkali lignin

Subcritical water V/Zeolite, CoO
240 °C, 10–30 min
Guaiacol, vanillin, and eugenol
Jadhav, 2020.

Lignin

Methanol and CoO/m-SEP 260 °C, 4 MPa
Petroleum ether
Chen et al., 2019a.

(Continued)

Table 3.13 (Continued)

Rice husk lignin
Ethanol with CO_2 200 °C, 10 min, 300 rpm, 32 MPa *Phenol, guaiacol, and syringol* Yang et al., 2021
Lignin-rich feedstock
Subcritical water and N_2 300–350 °C, 10–75 min and 5 atm 4-propylguaiacol, *Phenol, and guaiacol* Ciuffi et al., 2021

Roy et al. (2022) / MDPI / CC BY 4.0.

Singh et al. (2016) investigated the thermochemical depolymerization of lignin using various catalysts in supercritical methanol. The process made use of iron filings, sodium hydroxide, and zeolites. Analysis of the results led to possible decomposition reaction mechanisms. The donation of a proton to ether bonds results in the depolymerization and demethoxylation of lignin. The solvent is the proton donor, and as a result, a formaldehyde is formed that can undergo additional reactions, particularly with ring compounds. The addition of sodium hydroxide to the reaction system not only speeds up the process, but it also makes it possible to start a demethylation reaction. A condensation reaction occurs between the intermediate products, resulting in a final product rich in solid carbon residues. Zeolites do not exhibit this phenomenon because they contain acid centers that are able to catalyze the direct methylation of the aromatic ring and thus impede or prevent any potential condensation reactions (Singh et al., 2016).

In another review, Erdocia et al. (2016) used a triple system of methanol, ethanol, and acetone as the supercritical phase to study their effects on the breakdown of lignin. The product that was made in an acetone environment contained the most monomeric phenol derivatives. In-depth analysis confirmed that a reaction involving the detachment of alkyl and methoxy groups resulted in a significant amount of solid carbon residues, catechol, and cresol in the sample.

Depolymerization of the biopolymer in supercritical liquids was additionally explored by Kim et al. (2015a). Systems consisting of various supercritical alcohols—methanol, ethanol, and propan-2-ol—and metallic catalysts deposited on active carbon were used in the process, which was carried out at 350 °C, in the presence of gaseous hydrogen, and at pressures ranging from 13 to 19 MPa. Pd/C/ethanol proved to be the most efficient system, producing the most oil for the least amount of byproducts. This may be due to the large surface area of the catalyst and the fact that among the alcohols used, ethanol is the most effective hydrogen donor.

Gosselink et al. (2012) used acetone, water, and carbon dioxide as the supercritical phase to depolymerize lignin in an organic solvent with formic acid as a catalyst. Under a pressure of 10 MPa, the procedure was conducted at temperatures of 300 and 370 °C. It was discovered that the quantity of product produced was affected by the depolymerization reactions taking place, which led to the formation of byproducts.

3.3.5 Metallic Catalysis

The cleavage of ether bonds is the primary focus of the use of acid or base catalysts. However, specific products cannot be produced using these methods. In addition, acid or base-catalyzed depolymerization typically necessitates relatively harsh reaction conditions, such as high pressure

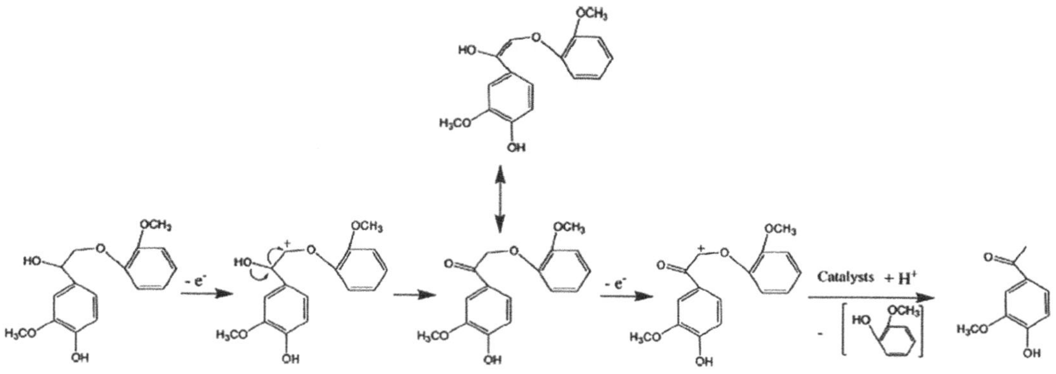

Figure 3.25 The purpose mechanism of cleaving the β-O-4 linkages with the metallic catalyst. Reproduced with permission from Chio et al. (2019) / ELSEVIER.

(ranging from 5 to 10 MPa) and a temperature of 250 °C to more than 300 °C (Forchheim et al., 2012; Lavoie et al., 2011; Xu et al., 2012). Methods or catalysts that, under mild conditions, can catalyze the depolymerization process with high selectivity are the subject of numerous studies. Because these conditions make the depolymerization process expensive and difficult to manage, many metals have been chosen as potential catalysts for further investigation. Nickel is one of the metal catalysts that has received the most research. Figure 3.25 depicts the intended mechanism of β-O-4 linkage metallic catalysis (Chio et al., 2019).

Song et al. (2012) reported that metallic nickel might function as a metal catalyst for the production of phenolic chemicals. Nickel is capable of precisely breaking the ether linkages. Additionally, it can hydrolyze the side chain's carbon–hydroxyl linkage, particularly to alkane.

Nickel can convert more than 50% of the lignin in birch wood into propylguaiacol and propylsyringol in ethyl glycerol in six hours at 200 °C with high selectivity (25 and 72%, respectively) (Song et al., 2013). In addition, these studies demonstrated that nickel can act as a catalyst for the depolymerization process at a low reaction temperature (less than 200 °C). Additionally, a bimetallic alloy known as Ni–M (where M can be Pd, Rh, Ru, Fe, Au, Ti, or Mo) can be formed when nickel and another metal are combined (Grilc et al., 2014; Molinari et al., 2014; Kim et al., 2018; Zhai et al., 2017; Zhang teo et al., 2014a; Zhang, Sun et al. 2014a; Zhang et al., 2016; Zhang et al. 2014b). Due to the synergistic effect, use of a bimetallic catalyst can increase reactivity and selectivity (García-Morales et al., 2015; Shuai and Luterbacher, 2016). Additionally, previous research revealed that these reactions can be carried out at temperatures below 120 °C with higher selectivity (Zhang teo et al., 2014a; Zhang, Sun et al. 2014a; Zhang et al., 2016). But, there are a few drawbacks to using a nickel–bimetallic catalyst. The price of the noble metals used in the bimetallic alloy could make the depolymerization more expensive. Under a variety of reaction conditions, aromatic compounds are over-hydrogenated as a result of the noble metallic catalysts lowering the yield of phenolic products. As a result, numerous efforts have been made to reduce costs, such as substituting inexpensive metal for noble metal in the production of a non-precious alloy and making these catalysts more reusable (Qiu et al., 2018; Song et al., 2013; Zhai et al., 2017).

Numerous metallic catalysts, including inexpensive metals (Al, Fe, Cu, Mo, and Zn), noble metals (Pt, Ru, Pd, and Ti), and their combination and alloys have also been studied (Li et al., 2018; Shu et al., 2018; Yan et al., 2008; Zhai et al., 2017; Zhang et al., 2018a, 2018b). Ye et al. (2012) reported the capability of Ru to particularly convert corn stalk lignin into 4-ethylphenol and

4-ethylguaiacol at 275 °C for 90 minutes under 2 MPa conditions, yielding 3.10 and 1.37%, respectively. With high selectivity, metallic catalysts can catalyze and transform lignin into specific chemicals, but due to the cost of processing them, the use of noble metal, catalyst deactivation, and lower conversion rate (between 50 and 60%) may prevent their use in lignin depolymerization. This is especially the case in comparison to other chemicals such as acid and base, which can almost completely convert lignin during the reaction into other useful chemicals.

Hepditch and Thring (2000) reported $NiCl_2$ or $FeCl_3$ treatment of lignin derived from alcells. At 305 °C, a yield of catechol of only 2.5 weight % was achieved. A number of other attempts demonstrated greater selectivity and efficiency (Bouxin et al., 2010; Liguori and Barth, 2011; Song et al., 2012, 2013; Toledano et al., 2012, 2014a; Xu et al., 2012; Ye et al., 2012; Yoshikawa et al., 2012).

Yoshikawa et al. (2012) used a treatment strategy that consisted of two steps when kraft lignin is treated. To improve the overall recovery of phenols, kraft lignin was first treated with a Si–Al catalyst in water/butanol medium. This was followed by a reaction on a $ZrO2$-Al_2O_3-FeOx catalyst. The yield of phenols ranged from 6.5% to 8.6%, and the lignin conversion into methoxyphenols ranged from 92% to 94%.

Xu et al. (2012) combined formic acid and ethanol with 20% Pt/C catalyst to boost low-molecular-weight fraction yield in the treatment of switchgrass lignin from organosolv. In the presence of a metallic enhancer, the yield of guaiacol derivatives clearly increased.

Liguori and Barth (2011) attempted to improve the formic acid-catalyzed system once more. Both spruce dry lignin as well as lignin model compounds that had been pretreated in a variety of ways in a water medium at 300 °C were treated with a Pd catalyst and Nafion SAC-13. It was possible to isolate guaiacol, resorcinol, and pyrocatechol, but their yields were all below 5 weight %. Guaiacol, pyrocatechol, and resorcinol probably have slightly different yields due to the various lignin pretreatment methods. In comparison to the straightforward acid-catalyzed depolymerization described in Section 3.3.1, Pt and Pd metallic enhancer-mediated depolymerization at a lower temperature did not result in a decrease in activation energy. The metallic enhancer's mechanism and function in acid-catalyzed lignin depolymerization require further investigation. Most of the time, acid or base catalyzed cracking was done at very high temperatures (above 300 °C) (Beauchet et al., 2012; Forchheim et al., 2012; Gasson et al., 2012; Lavoie et al., 2011; Liguori and Barth, 2011; Roberts et al., 2011; Toledano et al., 2012; Xu et al., 2012). But, for carrying out the lignin depolymerization under milder conditions (below 250 °C), attempts were made to reduce the reaction activation energy.

The findings of Song et al. (2012) exhibited a nickel compound-catalyzed method for lignosulfonate depolymerization into guaiacols. The nickel catalyst enabled a higher selectivity of 75% to 95% and a high conversion of more than 60% to guaiacols. Compared to the pyrolysis process only in the presence of base or acid, the reaction temperature was reduced to 200 °C from approximately 380 °C. When birch wood lignin was treated once more using a nickel catalyst, selectivity for propanylguaiacol and propenylsyringol of over 90% and conversion of over 50% were observed (Song et al., 2013)

Ye et al. (2012) presented their findings regarding the gentle enzymatic hydrolysis of corn stalk lignin at temperatures ranging from 200–250 °C. The maximum yields of 4-ethylphenol and 4-ethylguaiacol are 3.1% and 1.4%, respectively, when catalysts are Ru/C, Pd/C or Pt/C.

Toledano et al. (2012) used mesoporous Al-SBA-15 and metal nanoparticles, including nickel, palladium, platinum, and ruthenium, in conjunction with a microwave to treat lignin that had been separated from organosolv olive tree pruning lignin. Diethyl phthalate was the most common product in the presence of nickel or palladium catalyst and tetralin solvent, according to the findings. Even though there was a yield of 1.1 weight %, the reaction temperature was only 140 °C, which was less than the 200–300 °C range found in other published studies. When guaiacyl dehydrogenation oligomers were treated with K10 montmorillonite clay (Al_2O^3-$4SiO^2$-xH_2O) at only 100 °C, 35% of the model compounds were destroyed (Bouxin et al., 2010).

The introduction of metallic catalysts significantly reduced the activation energy of the depolymerization, resulting in a mild reaction state. Even when hydrogen sources such as ethanol or water were present, the C–O and C–C cleavages of the lignin were still the target of metallic catalysis. In the course of the depolymerization process, nickel or other solid catalysts provided accessible metal sites on the external surface. This method had great potential because it reduced the required reaction temperature while also increasing selectivity. Table 3.14 provides a summary of the metallic catalyzed lignin depolymerization (Wang et al., 2013).

Table 3.14 Metallic catalyzed lignin depolymerization.

Lignin	Catalyst	Reaction conditions T (°C)	P (MPa)	Major products	Yield	References
Kraft lignin	Kraft lignin (1) Si-Al catalyst/H2O/butanol (2) ZrO_2-Al_2O_3-FeOx	200–350 300	1.1–23	Phenols	6.5	Yoshikawa et al., 2012
Organosolv switchgrass lignin	16 weight % formic acid 4 weight % Pt/C 80 weight % ethanol	350		4-propylguaiacol 4-methylguaiacol	7.8 5.0	Xu et al., 2012
Acidic hydrolysis spruce lignin	4.4 weight % formic acid 0.15 weight % Pd catalyst 0.94 weight % Nafion SAC-13	300	9.6	Guaiacol Pyrocatechol Resorcinol	2.0 1.8 0.5	Liguori and Barth, 2011.
Enzymatic hydrolysis spruce lignin	4.4 weight % formic acid 0.15 Pd catalyst 0.94 weight % Nafion SAC-13	300	9.6	Guaiacol Pyrocatechol Resorcinol	1.7 1.3 1.0	Liguori and Barth, 2011.
Kraft spruce lignin	4.4 weight % formic acid 0.15 weight % Pd catalyst 0.94 weight % Nafion SAC-13	300	9.6	Guaiacol Pyrocatechol Resorcinol	4.7 4.9 1.2	Liguori and Barth, 2011.
Lignosulfonate	Ni/C, NiLa/C, NiPt/C, NiCu/C, NiPd/C, and NiCe/C	200	5	Guaiacol	10	Song et al., 2012.
Birch sawdust lignin	Ni/C	200		Propenylguaiacol Propenylsyringol	12 36	Song et al., 2013.
Enzymatic hydrolysis corn stalk lignin	Pt/C, Pd/C, Ru/C	200–250	1–6	4-ethylphenol 4-ethylguaicol	0.13–3.1 0.06–1.4	Ye et al., 2012.
Organosolv olive tree pruning lignin	Ni, Pd, Pt, or Ru supported by mesoporous Al-SBA-15	140		Diethyl phthalate	1.1	Toledano et al., 2012
Guaiacyl dehydrogenation oligomers	K10 montmorillonite clay (Al_2O_3-4SiO_2-xH_2O)/HCl	100		Low-molecular weight products	—	Bouxin et al., 2010.

Wang et al. (2013) / Hindawi / CC BY 4.0.

In biomass pyrolysis, the problem of deactivation caused by metallic catalysts is well-known. Their concern on this point was mentioned in some of the research (Song et al., 2013, 2012; Ye et al., 2012). Take, for instance, the work of Song et al. (2012, 2013) on the topic of using a nickel heterogeneous catalyst for lignosulfonate depolymerization, it was mentioned that the nickel catalysts helped produce active H species. Together with nickel sulfides, these active H species formed H2S, which regenerated the Ni(0) site for the subsequent catalytic cycle. Nickel heterogeneous catalysts were recyclable thanks to this mechanism (Song et al., 2012). It was also reported that the Ni/C catalysts were recycled four times without exhibiting any obvious deactivation during the depolymerization of birch wood lignin (Song et al., 2013).

Research conducted by Ye et al. (2012) showed that the Ru/C catalyst had excellent activity when recycled as well. The fact that the recycle of catalysts was not mentioned in the other research cited in this section necessitated further investigation into the deactivation of metallic catalysts.

To study the hydrodeoxygenation of the model compound eugenol, Bjelić et al. (2019) utilized the noble metal compounds Ru, Pd, Pt, and Rh in addition to the non-noble metals Cu and Ni. Ruthenium supported on neutral carbon demonstrated the lowest hydrogenation to deoxygenation ratio.

Kloekhorst and Heeres (2016) used carbon-supported Ru for depolymerization of Alcell lignin through catalytic hydrotreatment using the same lignin as the solvent. The surface area, the small size of the metal particles, and the high dispersion are the primary characteristics of a more effective lignin depolymerization catalyst (Tang et al., 2021). Additionally, it has been demonstrated that repolymerization is prevented by the hydrogenation catalyst by stabilizing the intermediate compounds produced during the depolymerization process, resulting in higher extraction yields of lignin-derived small molecules (Vangeel et al., 2019). The selection of the metal support is an additional important step in maximizing the lignin structure's catalytic depolymerization. The support material's primary objective is to prevent metal agglomeration and improve metal dispersion on its surface. A catalyst that is more stable and makes better use of the metal is the result of this improvement. *Titanium dioxide*, silicon oxide, and aluminum oxide, carbon materials, zeolites, and clays are a few examples of support materials. Depending on its chemical properties, the support has sometimes interacted with the depolymerization reaction (Garcia et al., 2020; Herrero and Ullah, 2020).

According to Zhang et al. (2018b) the metal catalyst is expected to serve as the center of the hydrogenation activity in the reaction, and the support provides the acidic sites necessary for the hydrodeoxygenation reaction. When the active metals and support have the right acidity, the chain scission and ring-opening reactions make it easier to fractionate lignin bonds into smaller molecules.

Carbon and aluminium oxide were used by Hita et al. (2018) for various noble metal catalysts. They discovered that aluminium oxide, which was more acidic, produced more bio-oil than carbon support.

Oregui Bengoechea et al. (2015) investigated three distinct aluminium oxide-supporting noble metal catalysts: ruthenium, rhodium and palladium. The results demonstrated that alumina was involved in depolymerization of lignin. Its presence increased the yield of bio-oil and decreased the oxygen content of compounds, and that it played a part in the depolymerization of lignin. Despite this, Zhang et al. (2018b) demonstrated that hydrodeoxygenation reactions could be more effectively catalyzed by non-acidic supports like SiO_2. A promising alternative to catalytic hydrodeoxygenation is currently available that makes use of hydrogen donor solvents rather than molecular hydrogen. Utilized in conjunction with alcohol or water, formic acid is a hydrogen donor in situ and this reaction may be enhanced by the presence of a catalyst. Lignin depolymerization relies heavily on the addition of a catalyst. For the hydrogenolysis cleavage of C–C or C–O links, this catalytic solvolysis reaction provides a more efficient catalytic route.

Oregui-Bengoechea et al. (2018) studied the effects of ethanol or water on the LtL process of rice straw lignin of Ru supported on activated carbon. The bio-oil yield was increased by ruthenium/Ac, and its active sites, such as hydrodeoxygenation, hydrogenolysis, and alkylation, assisted in the fragmentation of lignin into compounds with low molecular weights.

Yang et al. (2016) reported the utilization of various noble catalysts. However, the highest value achieved in comparison to the other catalysts evaluated was a phenol conversion rate of over 90% achieved by Rh supported on active carbon. Bio-oil compounds with a higher minor oxygen content were produced because rhodium/Ac had a greater capacity for deoxygenation. Rhodium/Ac cleaved primarily the β-O-4, β-β, β-5, and α-O-4 linkages. The O/C ratios obtained from various catalyzed depolymerization reactions are compared in Table 3.15 (Camas and Ullah, 2022).

Wu et al. (2018) reported that Ru/C catalyzed the breakdown of the 4-O-5 linkages in the model compound diphenyl ether.

Hu et al. (2019) demonstrated that during lignin depolymerization, PtRe supported on titanium dioxide produced 18.7% more monophenols than Pd/C. The polarization of the hydroxyl and carbonyl groups made it easier for PtRe/TiO2 to break the C–O bonds. By decreasing the O/C ratio and increasing bio-oil yields, noble metals enhance lignin depolymerization (Barta et al., 2014). But these metals are scarce and expensive, whereas non-noble metals are plentiful and inexpensive (Chen et al., 2019b). Non-noble metals are presently being used in a number of studies. Nickel has gained popularity because it is so effective at dissolving C–O links. However, some supports may have a strong interaction with Ni, reducing its effectiveness (Wang et al., 2019). For a successful catalyzed depolymerization reaction, choosing the right support is essential.

Kong et al. (2019) used isopropanol as a hydrogen donor solvent to investigate porous zeolite support materials for a bimetallic nickel–copper catalyst (Ni–Cu). The H/C ratio went up while the O/C ratio went down and the heating value increased as a result of this bifunctional catalyst's excellent hydrogenation capability and acidic sites. With 98.80%, the support H-Beta attained the highest yield of bio-oil. According to the findings, the fact that zeolites have larger pores makes it easier for lignin to be converted into a low-molecular- weight compound.

The use of nickel nanoparticles supported on nitrogen-doped carbon (NDC) with a hierarchical porosity and prepared through salt-melt synthesis as the catalyst for kraft lignin degradation was successful. It is demonstrated that nickel-NDC performs better than Ni nanoparticles deposited on commercial carbon or on an N-free carbon support produced with comparable porosity characteristics. The effect of the reactor setup on the stability of the recovered catalysts was highlighted by comparing the efficacy of these materials for reactions carried out in flow and batch reactors (Lama et al., 2017). Some of the published papers on lignin depolymerization with nickel as a catalyst are summarized in Table 3.16 (Camas and Ullah, 2022).

In the depolymerization reaction, the nickel catalyst serves as the active center for hydrogenolysis. Consequently, the amount used has a significant impact on the fragmentation of the lignin polymeric structure.

Ma et al. (2019) demonstrated that an increase in Ni loading results in a decrease in biochar and an increase in the yield of monomers. Despite this, the maximum Ni load that yields the most must be determined because it often causes phenolic monomers to hydrogenate too much, which makes more aliphatic compounds.

Klein et al. (2015) reported that regardless of the biomass used, yields were increased by 10% Ni/C loadings.

Compared to noble metal catalysts, which require low hydrogen pressure and operate at temperatures between 200 and 400 °C, molybdenum-based catalysts have a higher potential for hydrodeoxygenation (HDO) reactions (Nolte et al., 2016). In addition, Mo-based catalysts are extremely

Table 3.15 Different lignin sources depolymerized with noble metal catalyst.

Lignin source	Catalyst	Solvents	Reaction conditions	O/C ratio	H/C ratio	Bio-oil yield (%)	Products	References
Rice straw lignin (2 g)	Ru/Ac (0.2 g)	FA (1.5 g) EtOH (2.5 g)	300 °C 10 h	0.20	1.33	75.8	2-ethyl-1-hexanol, ethyl benzoate, diethyl succinate, phenol, guaiacol	Oregui-Bengoechea et al. (2018)
Kraft lignin (0.1 g)	Rh/Ac (0.04 g)	Isopropanol (12 mL)	350 °C 4 h	0.094	1.30	>100	Isophorone, benzene,1,2,3-trinethyl-; cyclohezanone,3,3,5-trimethyl-;cyclohexanol,3,3,5-trimethyl;phenol,3,4-dimethyl-	Yang et al. (2016)
Eucalyptus lignin (335 g)	Ru/Al$_2$O$_3$	FA (427 g) H$_2$O (875 g)	305 °C, 2 h, 1000 rpm	0.272	1.38	78.8	Guaiacol, 2-methoxy-4-methylphenol, 4-ethyl-2-methoxyphenol, 2,6-dimetoxyphenol, 2-methoxy-4-propylphenol	Ghoreishi et al. (2019)
Diphenyl ether (model compound)	Ru/C Isopropanol		120 °C 10 h	NM	NM	99	Benzene, cyclohexane, cyclohexanol	Wu et al. (2018)
Black liquor lignin (0.2 g)	PtRe/TiO$_2$ (0.1g)	Isopropanol (20 mL) Deionized water (10 mL)	240 °C 12 h	NM	NM	NM	4-ethylphenol, 4-propylphenol, guaiacol, 4-ethylguaiacol, 4- methylsyringol, 4-ethylsyringol	Hu et al. (2019)
Kraft lignin (15 g)	Rh/Al$_2$O$_3$ (0.75 g)	No solvent	450 °C 4 h	≈0.1	≈1.2	36.3	Alkylphenolics, aromatics, cyclohexanes, naphthalenes, alkanes	Hita et al. (2018)
Alcell lignin (15 g)	Ru/C (0.75 g)	No solvent	400 °C 4 h H$_2$	≈0.06	≈1.4	≈1.4 64.7	Phenol, cresol, 4-ethylphenol, 4-propylphenol, toluene, ethylbenzene, methylcyclohexane	Kloekhorst and Heeres (2016)
Black locus bark (2 g)	Pd/C (0.2 g)	Methanol (40 mL)	250 °C 2 h H$_2$	NM	NM	35.1	Aliphatic monomers and phenolic monomers	Vangeel et al. (2019)
Acid hydrolysis lignin (2 g)	Ru/Al$_2$O$_3$ (0.2 g Rh/Al$_2$O$_3$ (0.2 g) Pd/Al$_2$O$_3$ (0.2 g)	Water (5 g) FA (3.075 g)	340 °C 6 h	0.18 0.14 0.14	1.17 1.21 1.21	91.5 80.5 82.9	Phenol, cresol, guaiacol, methyl guaiacol, catechol, ethylcatechol, syringol, o-vanillin	Oregui Bengoechea et al. (2015)

NM: not mentioned. FA: formic acid.
Camas and Ullah (2022). Reproduced with permission.

Table 3.16 Nickel catalyst in different support materials to depolymerize lignin.

Lignin source	Catalyst/support	Solvents	Reaction equipment	Reaction conditions	Analysis	Yield (%)	Char (%)	Major products	Catalyst reusability	References
Birch sawdust (2 g)	Ni/AC (0.10 g)	Ethanol/benzene (v/v = 1:2) (total: 40 mL)	Autoclave reactor – sealed with Argon	12 h 200 °C	GC-FID GC-MS Isotopic tests MALDI-TOF	54	–	➢ Propylguaiacol ➢ Propylsyringol	4	Song et al. (2013)
Lignin (0.20 g)	"Inlaid type" Ni/C R (0.10 g)	30 mL of 1,4-dioxane	Stainless batch autoclave reactor. H2	260 °C 10 h	GC-MS	82.4	5	Monomers yield 23.3% ➢ Phenol ➢ Propylguaiacol ➢ Alkenyl-substituted propylguaiacol ➢ Propylsyringol ➢ Alkenyl-substituted propylsyringol	4	Wang et al. (2019)
Organosolv lignin (0.5 g)	15% Ni/ZrP (0.10 g)	Isopropanol (20 mL)	Stainless steel autoclave H2	260 °C 4 h	GC-MS-FID GPC Elemental analyzer	87.3	<5.2	Monomers yield 15.1% ➢ Phenol ➢ Guaiacols ➢ Syringols	4	Ma et al. (2019)
Kraft lignin (0.1 g)	Ni-Cu/H-Beta (0.04 g)	Isopropanol (12 mL)	Autoclave reactor. N2	330 °C 3h	GC-TCD GC-MS FTIR-ATR GPC Elemental analyzer	98.8	0.04	Monomers yield 50.83% ➢ Aromatics ➢ Cyclic ketones ➢ Cycloalkanes ➢ Oxygen-chain compounds ➢ Alkanes	5	Kong et al. (2019)
Alkali lignin (2 g)	5% Ni-rehydrated hydrotalcite (2 g)	Ethanol: water (50 wt %:50 weight %)	Stainless steel autoclave N2	270 °C 600 rpm 30 min	GC-MS FTIR H1-NMR Elemental analyzer	87.9	–	Monomers yield 19.4% ➢ Guaiacol ➢ 4-propylguaiacol ➢ Phenol ➢ 4-ethyl guaiacol ➢ 4-methylguaiacol	3	Kong et al. (2020)

(Continued)

Table 3.16 (Continued)

Lignin source	Catalyst/support	Solvents	Reaction equipment	Reaction conditions	Analysis	Yield (%)	Char (%)	Major products	Catalyst reusability	References
Organosolv lignin (1.25 g)	Ni–Fe–Mo_2C/AC (1.25 g)	Water: methanol (4:1 v/v) (total 50 mL)	Stainless batch autoclave reactor. H_2.	260 °C 4 h	GC-MS HPLC/ MS/ MS GPC HSQC NMR	89.56	6.89	Monomers yield 35.53 % 4-ethylphenol ➤ Phenol ➤ 4-ethyl-2-methoxy phenol	5	Yan et al. (2021)
Kraft lignin (2.5 g)	Ni/NDC (1 g)	Ethanol (500 mL)	Batch and flow	150 °C 24 h	GC-FID GCXGC-MS ICP-OES	NM	NM	➤ Phenols ➤ Guaiacols ➤ Cyclopentene ➤ Benzene	No reuse	Lama et al. (2017)
Birchwood (1 g)	Ni/C (0.1 g)	Methanol (20 mL)	Stainless steel Parr reactor	N_2 200 °C 6 h	GC-FID	32	NM	➤ 2-methoxy-4-propyl phenol ➤ 2,6-dimethoxy-4-propyl phenol ➤ 4-(3-hydroxypropyl)-2-methoxy phenol ➤ Isoeugenol ➤ Methoxyisoeugenol	NM	Klein et al. (2015)

Reproduced with permission from Camas and Ullah (2022) / ELSEVIER.

hydrogen efficient due to their inability to promote hydrogenation in HDO reactions. Over the course of the past few decades, a number of Mo-based catalysts involving oxides, nitrides, carbides, and sulfides have been investigated for the HDO of oxygenated aromatic hydrocarbons obtained from lignin and lignin-based model compounds (Bui et al., 2011; Li et al., 2015; Ranga et al., 2018; Shetty et al., 2015).

The HDO of several oxygenates derived from biomass, as well as other compounds, has been extensively studied for MoO3, the earth's most abundant compound (Shetty et al., 2015). At low hydrogen pressure, MoO3 and Mo2C have high activity and preferentially cleave phenolic Ph-O-Me bonds over aliphatic ones. (Lee et al., 2014; Nolte et al., 2016; Prasomsri et al., 2013, 2014). A reverse Mars–van Krevelen mechanism is used to initiate these reactions, which result in the formation of water by the interaction of hydrogen with oxygen on the surface at a catalyst's active vacant site. The active vacant site on the catalyst is then filled with reactant oxygen, leaving the product unsaturated. The deactivation of the MoO3 catalyst can be reduced by adjusting hydrogen pressure, and the presence of water helps in maintaining the catalytic activity. During the MoO3-catalyzed HDO reaction, a constant stream of hydrogen is required, frequently transforming Mo into the inactive form of Mo4+. However, by forming oxycarbohydrides on the catalyst surface, surface modification of the MoO3 could reduce this reduction. This prolongs the catalyst activity and helps stabilize the active Mo5+ form (Prasomsri et al., 2014). At typical HDO conditions, Mo_2C-based catalysts, in contrast to MoO3, oxidize to inactive MoO_2 and lose their activity when exposed to water (Mortensen et al., 2015).

The partial oxidation of Mo_2C results in the formation of an intermediate state between the oxide and carbide forms of Mo. This intermediate state selectively hydrodeoxygenates organic molecules without going through sequential hydrogenation (Sullivan and Bhan 2016; Sullivan et al., 2016).

Chen et al. (2017a) compared the ethanolysis of kraft lignin using a reduction-modified MoO_3 catalyst to an untreated MoO_3 catalyst under various temperature and reaction conditions. Reductive pretreatment of MoO_3 increased the yield of bio-oil by approximately 15% as compared to untreated MoO_3 catalyst because of the synergistic interaction between Mo^{5+}, Mo^{6+}, and oxygen vacancies which made it easier for molybdenum (V) ethoxide active species to form. In addition, a nitrogen atmosphere helped the small molecules produce 1266 mg g^{-1} of lignin at 593 K after a six hour reaction with a reduction-modified MoO3 catalyst at 623 K. At 553 K, the catalytic ethanolysis of kraft lignin over MoC1x/AC and Mo/Al2O3 catalysts produced 1640 and 1390 mg g-1 lignin yield of bio-oil, respectively, without leaving behind any char or tar (Ma et al., 2015; Rui et al., 2014). Mo_2N/Al_2O_3 and MoOx/CNT showed very good recycle performance following lignin's catalytic depolymerization (Chen et al., 2017b; Xiao et al., 2017).

On the one hand, it was discovered that the support material influenced the Mo-nitride catalyst's product selectivity. However, the HDO process's activity was influenced by Mo's dispersion and nitridation degree (Ghampson et al., 2012b, 2012a).

Unlike the Mo-nitride catalyst, the chemical and physical properties of the support materials had no effect on the active site properties or product selectivity during MoS_2-catalyzed HDO of lignin (Ruiz et al., 2012). Conversely, the presence of cobalt in the MoS_2 phase significantly enhanced the HDO conversion of guaiacol into benzene via the direct deoxygenation pathway (Bui et al., 2011).

Whiffen and Smith (2010) investigated 4-methylphenol's HDO on a low surface area, unsupported MoP, MoO_2 and MoS_2 catalysts. After a 5 hour HDO reaction at 623 K and 4.40 MPa hydrogen pressure, all catalysts, with the exception of MoO_3, remained stable. However, the formation of anionic vacancies and Bronsted acid sites during the reaction led mixed oxides containing Mo, MoO_2, and Mo_4O_{11}, to increase uptake of carbon monoxide by a factor of 100.

Hydrogenolysis and hydrogenation could be catalyzed under HDO conditions by surface defect-associated anionic vacancies produced by reducing metal oxides (Massoth et al., 2006). HDO's catalytic turnover frequency reduced in the following order:

MoP, MoS_2, MoO_2, and MoO_3, with MoP exhibiting the highest selectivity of all.

Ranga et al. (2018) looked into the HDO of anisole catalytic activity of ZrO2-supported Mo oxides. The overall catalytic activity was based on the availability of Mo5+ species, which are active Mo oxide defects. For the majority of Mo-based catalysts, the HDO of simple model compounds with only one or two oxygen atoms has been tested; however, their catalytic activity toward lignin or model compounds based on lignin is still unknown. The efficiency and performance of these catalysts in depolymerizing lignin and model compounds should be investigated urgently.

Bibliography

Achinivu EC, Howard RM, Li G, Gracz H, and Henderson WA (2014). Lignin extraction from biomass with protic ionic liquids. *Green Chem*, 16: 1114–1119.

Agarwal A, Rana M, and Park JH (2018). Advancement in technologies for the depolymerization of lignin. *Fuel Process Technol*, 181: 115–132.

Aida TM, Sato T, Sekiguchi G, Adschiri T, and Arai K (2002). Extraction of Taiheiyo coal with supercritical water–phenol mixtures. *Fuel*, 81: 1453–61.

Akiba T, Tsurumaki A, and Ohno H (2017). Induction of lignin solubility for a series of polar ionic liquids by the addition of a small amount of water. *Green Chem*, 19(9): 2260–2265. https://doi.org/10.1039/c7gc00626h.

Asawaworarit P, Daorattanachai P, Laosiripojana W, Sakdaronnarong C, Shotipruk A, and Laosiripojana N (2019). Catalytic depolymerization of organosolv lignin from bagasse by carbonaceous solid acids derived from hydrothermal of lignocellulosic compounds. *Chem Eng J*, 356: 461–471.

Azadi P, Inderwildi OR, Farnood R, and King DA (2013). Liquid fuels, hydrogen and chemicals from lignin: a critical review. *Renew Sust Energ Rev*, 21: 506–523.

Barta K, Warner GR, Beach ES, and Anastas PT (2014). Depolymerization of organosolv lignin to aromatic compounds over Cu-doped porous metal oxides. *Green Chem*, 16: 191–196.

Beauchet R, Monteil-Rivera F, and Lavoie J (2012). Conversion of lignin to aromatic-based chemicals (L-chems) and biofuels (L-fuels). *Bioresour Technol*, 121: 328–334.

Bernhardt, J.J., Rößiger, B., Hahn, T., & Pufky-Heinrich, D. (2021). Kinetic modeling of the continuous hydrothermal base catalyzed depolymerization of pine wood based kraft lignin in pilot scale. *Industrial Crops and Products*, 159: 113–119.

Binder JB, Gray MJ, White JF, Zhang ZC, and Holladay JE (2009). Reactions of lignin model compounds in ionic liquids. *Biomass and Bioenergy*, 33(9): 1122–1130.

Bjelić A, Grilc M, Huˇs M, and Likozar B (2019). Hydrogenation and hydrodeoxygenation of aromatic lignin monomers over Cu/C, Ni/C, Pd/C, Pt/C, Rh/C and Ru/C catalysts: mechanisms, reaction micro-kinetic modelling and quantitative structure-activity relationships. *Chem Eng J*, 359: 305–320.

Bouxin F, Baumberger S, Pollet B, Haudrechy A, Renault JH, and Dole P (2010). Acidolysis of a lignin model: investigation of heterogeneous catalysis using Montmorillonite clay. *Bioresource Technol*, 101(2): 736–744.

Brittain AD, Chrisandina NJ, Cooper RE, Buchanan M, Cort JR, Olarte MV, and Sievers C (2018). Quenching of reactive intermediates during mechanochemical depolymerization of lignin. *Catal Today*, 302: 180–189.

Bui VN, Laurenti D, Afanasiev P, and Geantet C (2011). Hydrodeoxygenation of guaiacol with CoMo catalysts. Part I: promoting effect of cobalt on HDO selectivity and activity. *Appl Catal B-Environmental*, 101: 239–245.

Camas KL and Ullah A (2022). Depolymerization of Lignin into High-value products. *Biocatal Agric Biotechnol*, 40: 102306. https://doi.org/10.1016/j.bcab.2022.102306.

Cansell F, Aymonier C, and Loppinet-Serani A (2003). Review on materials science and supercritical fluids. *Curr Opin Solid State Mater Sci*, 7: 331–340.

Carlin RT and Wilkes JS (1994). Chemistry and speciation in room-temperature chloroaluminate molten salts. *Chemistry of Nonaqueous Solutions*, Mamantov G and Popov AI (eds.). VCH, New York, p. 277.

Casas A, Omar S, Palomar J, Oliet M, Alonso MV, and Rodriguez F (2013). Relation between differential solubility of cellulose and lignin in ionic liquids and activity coefficients. *RSC Adv*, 3: 3453–3460.

Chaudhary, R., & Dhepe, P.L. (2017). Solid base catalyzed depolymerization of lignin into low molecular weight products. *Green Chemistry*, 19: 778–788.

Chaudhary R and Dhepe PL (2019). Depolymerization of lignin using a solid base catalyst. *Energy Fuels*, 33: 4369–4377

Chen M, Cao Y, Wang Y, Yang Z, Wang Q, Sun Q, and Wang J (2019a). Depolymerization of lignin over CoO/m-SEP catalyst under supercritical methanol. *J Renew Sustain Energy*, 11: 013103.

Chen M, Hao W, Ma R, Ma X, Yang L, Yan F, Cui K, Chen H, and Li Y (2017b). Catalytic ethanolysis of kraft lignin to small-molecular liquid products over an alumina supported molybdenum nitride catalyst. *Catal Today*, 298: 9–15.

Chen MM, Ma XL, Ma R, Wen Z, Yang F, Cui K, Chen H, and Li YD (2017a). Ethanolysis of Kraft lignin over a reduction-modified MoO_3 catalyst. *Ind Eng Chem Res*, 56: 14025–14033.

Chen X, Guan W, Tsang CW, Hu H, and Liang C (2019b). Lignin valorizations with Ni catalysts for renewable chemicals and fuels productions. *Catalysts*, 9: 488.

Cheng S, D'cruz I, Wang M, Leitch M, and Xu C (2010). Highly efficient liquefaction of woody biomass in hot-compressed alcohol– water co-solvents. *Energy Fuels*, 24: 4659–67.

Cheng S, Wilks C, Yuan Z, Leitch M, and Xu CC (2012). Hydrothermal degradation of alkali lignin to bio-phenolic compounds in sub/supercritical ethanol and water–ethanol co-solvent. *Polym Degrad Stab*, 97: 839–848.

Chio C, Sain MM, and Qin W (2019). Lignin utilization: a review of lignin depolymerization from various aspects. *Renewable and Sustainable Energy Rev*, 107(C): 232–249.

Ciuffi B, Loppi M, Rizzo AM, Chiaramonti D, and Rosi L (2021). Towards a better understanding of the HTL process of lignin-rich feedstock. *Sci Rep*, 11: 1–9.

Cocero MJ, Cabeza A, Abad N, Adamovic T, Vaquerizo L, Martinez CM, and Pazo-Cepeda MV (2018). Understanding biomass fractionation in subcritical & supercritical water. *J Supercrit Fluid*, 133(2): 550–565.

Cox BJ and Ekerdt JG (2012). Depolymerization of oak wood lignin under mild conditions using the acidic ionic liquid 1-H-3-methylimidazolium chloride as both solvent and catalyst. *Bioresour Technol*, 118: 584–588.

Cox BJ and Ekerdt JG (2013). Pretreatment of yellow pine in an acidic ionic liquid: extraction of hemicelluloses and lignin to facilitate enzymatic digestion. *Bioresour Technol*, 134: 59–65.

Cox BJ, Jia S, Zhang ZC, and Ekerdt JG (2011). Catalytic degradation of lignin model compounds in acidic imidazolium based ionic liquids: hammett acidity and anion effects. *Polym Degrad Stab*, 96(4): 426–431.

Dabral S, Engel J, Mottweiler J, Spoehrle SSM, Lahive CW, and Bolm C (2018a). Mechanistic studies of base-catalysed lignin depolymerisation in dimethyl carbonate. *Green Chem*, 20: 170–182.

Dabral S, Wotruba H, Hernandez JG, and Bolm C (2018b). Mechanochemical oxidation and cleavage of Lignin β-O-4 model compounds and lignin. *ACS Sustain Chem Eng*, 6: 3242–3254.

Dai J, Patti AF, and Saito K (2016). Recent developments in chemical degradation of lignin: catalytic oxidation and ionic liquids. *Tetrahedron Lett*, 57(45): 4945–4951. https://doi.org/10.1016/j.tetlet.2016.09.084.

Dai J, Patti AF, and Saito K (2020). Depolymerization of lignin by catalytic oxidation in ionic liquids. *Encyclopedia of Ionic Liquids*, Zhang S. (ed.). Springer, Singapore. https://doi.org/10.1007/978-981-10-6739-6_78-1.

De Gregorio GF, Prado R, Vriamont C, Erdocia X, Labidi J, Hallett JP, and Welton T (2016). Oxidative depolymerization of lignin using a novel polyoxometalate-protic ionic liquid system. *ACS Sustain Chem Eng*, 4: 6031–6036.

Deepa AK and Dhepe PL (2015). Lignin depolymerization into aromatic monomers over solid acid catalysts. *ACS Catal*, 5: 365–379.

Deuss PJ, Scott M, Tran F, Westwood NJ, De Vries JG, and Barta K (2015). Aromatic monomers by in situ conversion of reactive intermediates in the acid-catalyzed depolymerization of lignin. *J Am Chem Soc*, 137: 7456–7467.

Di Marino D, Stöckmann D, Kriescher S, Stiefel S, and Wessling M (2016). Electrochemical depolymerisation of lignin in a deep eutectic solvent. *Green Chem*, 18: 6021–6028.

Dong C, Feng C, Liu Q, Shen D, and Xiao R (2014). Mechanism on microwave-assisted acidic solvolysis of black-liquor lignin. *Bioresour Technol*, 162: 136–141.

Dorrestijn E, Kranenburg M, Poinsot D, and Mulder P (1999). Lignin depolymerization in hydrogen-donor solvents. *Holzforschung*, 53(6): 611–616.

Du B, Liu B, Yang Y, Wang X, and Zhou J (2019). A phosphotungstic acid catalyst for depolymerization in bulrush lignin. *Catalysts*, 9: 399.

Erdocia X, Prado R, Fernández-Rodríguez J, and Labidi J (2016). Depolymerization of different organosolv lignins in supercritical methanol, ethanol and acetone to produce phenolic monomers. *ACS Sustain Chem Eng*, 4(3): 1373–1380.

Evans L, Littlewolf A, Lopez M, and Miller J (1999). *Batch Microreactor Studies of Base Catalyzed Ligin Depolymerization in Alcohol Solvents*. Sandia National Laboratories, Albuquerque, NM, USA; Livermore, CA, USA.

Fan D, Xie X-A, Li Y, Li L, and Sun J (2018). Comparative study about catalytic liquefaction of alkali lignin to aromatics by HZSM-5 in sub-and supercritical ethanol. *J Renew Sustain Energy*, 10: 013106.

Fang Z, Sato T, Smith RL Jr, Inomata H, Arai K, and Kozinski JA (2008). Reaction chemistry and phase behavior of lignin in high-temperature and supercritical water. *Bioresour Technol*, 99: 3424–30.

Forchheim D, Gasson JR, Hornung U, Kruse A, and Barth T (2012). Modeling the lignin degradation kinetics in a ethanol/formic acid solvolysis approach. Part 2. validation and transfer to variable conditions. *Ind Eng Chem Res*, 51: 15053–15063.

Forsythe WG, Garrett MD, Hardacre C, Nieuwenhuyzen M, and Sheldrake GN (2013). An efficient and flexible synthesis of model lignin oligomers. *Green Chem*, 15: 3031–3038.

Fraile JM, García JI, Hormigón Z, Mayoral JA, Saavedra CJ, and Salvatella L (2018). Role of substituents in the solid acid-catalyzed cleavage of the β-O-4 linkage in lignin models. *ACS Sustain Chem Eng*, 6: 1837–1847.

Garcia AC, Cheng S, and Cross JS (2020). Solvolysis of kraft lignin to bio-oil: a critical review. *Cleanroom Technol*, 2: 513–528.

García-Morales NG, García-Cerda LA, Puente-Urbina BA, Blanco-Jerez LM, Antaño-López R, and Castañeda-Zaldivar F (2015). Electrochemical glucose oxidation using glassy carbon electrodes modified with Au-Ag nanoparticles: influence of Ag content. *J Nanomater*, 2015: 2.

Gasson JR, Forchheim D, Sutter T, Hornung U, Kruse A, and Barth T (2012). Modeling the lignin degradation kinetics in an ethanol/formic acid solvolysis approach. Part 1. Kinetic model development. *Ind Eng Chem Res*, 51: 10595–10606.

George A, Tran K, Morgan TJ, Benke PI, Berrueco C, Lorente E, Wu BC, Keasling JD, Simmons BA, and Holmes BM (2011). The effect of ionic liquid cation and anion combinations on the macromolecular structure of lignins. *Green Chem*, 13: 3375–3385.

Ghampson IT, Sepúlveda C, Garcia R, García Fierro JL, Escalona N, and Desisto WJ (2012b). Comparison of alumina- and SBA-15-supported molybdenum nitride catalysts for hydrodeoxygenation of guaiacol. *Appl Catal A Gen*, 435–436: 51–60.

Ghampson IT, Sepulveda C, Garcia R, Radovic LR, Fierro JLG, DeSisto WJ, and Escalona N (2012a). Hydrodeoxygenation of guaiacol over carbon-supported molybdenum nitride catalysts: effects of nitriding methods and support properties. *Appl Catal A Gen*, 439–440: 111–124.

Ghoreishi S, Barth T, and Hermundsgård DH (2019). Effect of reaction conditions on catalytic and noncatalytic lignin solvolysis in water media investigated for a 5 L reactor. *ACS Omega*, 4: 19265–19278.

Gillet S, Aguedo M, Petitjean L, Morais ARC, Da Costa Lopes AM, Łukasik RM, and Anastas PL (2017). Lignin transformations for high value applications: towards targeted modifications using green chemistry. *Green Chem*, 19(18): 4200–4233.

Gosselink RJ, Teunissen W, van Dam JE, de Jong E, Gellerstedt G, Scott EL, and Sanders JP (2012). Lignin depolymerisation in supercritical carbon dioxide/acetone/ water fluid for the production of aromatic chemicals. *Bioresour Technol*, 106: 173–177.

Grilc M, Likozar B, and Levec J (2014). Hydrodeoxygenation and hydrocracking of solvolysed lignocellulosic biomass by oxide, reduced and sulphide form of NiMo, Ni, Mo and Pd catalysts. *Appl Catal B*, 150: 275–87.

Guo D, Liu B, Tang Y, Zhang J, Xia X, and Tong S (2017). Catalytic depolymerization of alkali lignin in sub-and super-critical ethanol. *BioResources*, 12: 5001–5016.

Güvenatam B, Heeres EH, Pidko EA, and Hensen EJ (2016). Lewis acid-catalyzed depolymerization of soda lignin in supercritical ethanol/water mixtures. *Catal Today*, 269: 9–20.

Guvenatam B, Heeres EHJ, Pidko EA, and Hensen EJM (2015). Decomposition of lignin model compounds by Lewis acid catalysts in water and ethanol. *J Mol Catal A Chem*, 410: 89–99.

Hepditch MM and Thring RW (2000). Degradation of solvolysis lignin using Lewis acid catalysts. *Can J Chem Eng*, 78: 226–231.

Herrero YR and Ullah A (2020). Metal oxide powder technologies in catalysis. *Met Oxide Powder Technol*, 279–297.

Hidajat, M.J.; Riaz, A.; Park, J.; Insyani, R.; Verma, D.; Kim, J (2017). Depolymerization of concentrated sulfuric acid hydrolysis lignin to high-yield aromatic monomers in basic sub-and supercritical fluids. *Chem. Eng. J.*, 317: 9–19.

Hita I, Deuss PJ, Bonura G, Frusteri F, and Heeres HJ (2018). Biobased chemicals from the catalytic depolymerization of Kraft lignin using supported noble metal-based catalysts. *Fuel Process Technol*, 179: 143–153.

Hossain MM and Aldous L (2012). Ionic liquids for lignin processing: dissolution, isolation, and conversion. *Aust J Chem*, 65(11): 1465–1477.

Hu J, Zhang S, Xiao R, Jiang X, Wang Y, Sun Y, and Lu P (2019). Catalytic transfer hydrogenolysis of lignin into monophenols over platinum-rhenium supported on titanium dioxide using isopropanol as in situ hydrogen source. *Bioresour Technol*, 279: 228–233.

Huang X, Korányi TI, Boot MD, and Hensen EE (2015). Ethanol as capping agent and formaldehyde scavenger for efficient depolymerization of lignin to aromatics. *Green Chem*, 17: 4941–4950.

Ibáñez E, Mendiola JA, and Castro-Puyana M (2018). Supercritical fluid extraction. *Encycl Anal Sci*, 227–233. https://doi.org/10.1016/B978-0-12-384947-2.00675-9.

Jadhav B (2020). *Screening of Catalysts for the Subcritical Water Depolymerization of Lignin*. South Dakota State University, Brookings, SD, USA.

Janesko BG (2014). Acid-catalyzed hydrolysis of lignin β-O-4 linkages in ionic liquid solvents: a computational mechanistic study. *Phys Chem Chem Phys*, 16: 5423–5433.

Jia S, Cox BJ, Guo X, Zhang ZC, and Ekerdt JG (2010a). Cleaving the β-O-4 bonds of lignin model compounds in an acidic ionic liquid, 1-H-3-methylimidazolium chloride: an optional strategy for the degradation of lignin. *ChemSusChem*, 3: 1078–1084.

Jia S, Cox BJ, Guo X, Zhang ZC, and Ekerdt JG (2011). Hydrolytic cleavage of β-O-4 ether bonds of lignin model compounds in an ionic liquid with metal chlorides. *Ind Eng Chem Res*, 50: 849–855.

Jia SY, Cox BJ, Guo XW, Zhang ZC, and Ekerdt JG (2010b). Decomposition of a phenolic lignin model compound over organic N-bases in an ionic liquid. *Holzforschung*, 64: 577–580.

Katahira R, Mittal A, McKinney K, Chen X, Tucker MP, Johnson DK, Beckham GT (2016). Base-catalyzed depolymerization of biorefinery lignins. *ACS Sustain. Chem. Eng.*, 4: 1474–1486.

Kim J-Y, Park J, Hwang H, Kim JK, Song IK, and Choi JW (2015b). Catalytic depolymerization of lignin macromolecule to alkylated phenols over various metal catalysts in supercritical tert-butanol. *J Anal Appl Pyrolysis*, 113: 99–106.

Kim JY, Park J, Kim UJ, and Choi JW (2015a). Conversion of lignin to phenol-rich oil fraction under supercritical alcohols in the presence of metal catalysts. *Energy & Fuels*, 29(8): 5154–5163.

Kim J-Y, Park SY, Choi I-G, and Choi JW (2018). Evaluation of Ru x Ni 1-x/SBA-15 catalysts for depolymerization features of lignin macromolecule into monomeric phenols. *Chem Eng J*, 336: 640–8.

Klapiszewski Ł, Szalaty TJ, and Jesionowski T (2017) Depolymerization and activation of lignin: current state of knowledge and perspectives. *Lignin - Trends and Applications*, Poletto M. (ed.). IntechOpen. https://doi.org/10.5772/intechopen.70376.

Klein I, Saha B, and Abu-Omar MM (2015). Lignin depolymerization over Ni/C catalyst in methanol, a continuation: effect of substrate and catalyst loading. *Catal Sci Technol*, 5: 3242–3245.

Kleine T, Buendia J, and Bolm C (2013). Mechanochemical degradation of lignin and wood by solvent-free grinding in a reactive medium. *Green Chem*, 15: 160–166.

Kloekhorst A and Heeres HJ (2016). Catalytic hydrotreatment of Alcell lignin fractions using a Ru/C catalyst. *Catal Sci Technol*, 6: 7053–7067. https://doi.org/10.1039/c6cy00523c.

Kong L, Liu C, Gao J, Wang Y, and Dai L (2019). Efficient and controllable alcoholysis of Kraft lignin catalyzed by porous zeolite-supported nickel-copper catalyst. *Bioresour Technol*, 276: 310–317.

Kong X, Li W, Li X, Liu L, Geng W, and Liu L (2020). Hydrothermal depolymerization of alkali lignin to high-yield monomers over nickel nitrate modified commercial hydrotalcites catalyst. *J Energy Inst*, 93: 658–665.

Kudo S, Honda E, Nishioka S, Hayashi J-i. (2019). Formation of p-Unsubstituted Phenols in Base-catalyzed Lignin Depolymerization. In: *Proceedings of the MATECWeb of Conferences, 18th Asian Pacific Confederation of Chemical Engineering Congress* (APCChE 2019), Sapporo, Japan, 23–27 September 2019.

Kuznetsov B, Sharypov V, Chesnokov N, Beregovtsova N, Baryshnikov S, Lavrenov A, Vosmerikov A, and Agabekov V (2015). Lignin conversion in supercritical ethanol in the presence of solid acid catalysts. *Kinet Catal*, 56: 434–441.

Lama SMG, Pampel J, Fellinger TP, Bešskoski VP, Slavkovíc-Bešskoski L, Antonietti M, and Molinari V (2017). Efficiency of Ni nanoparticles supported on hierarchical porous nitrogen-doped carbon for hydrogenolysis of kraft lignin in flow and batch systems. *ACS Sustain Chem Eng*, 5: 2415–2420.

Lavoie JM, Baré W, and Bilodeau M (2011). Depolymerization of steam-treated lignin for the production of green chemicals. *Bioresour Technol*, 102: 4917–4920.

Lee W, Wang Z, Wu RJ, and Bhan A (2014). Selective vapor-phase hydrodeoxygenation of anisole to benzene on molybdenum carbide catalysts. *J Catal*, 319: 44–53.

Li C, Zhao X, Wang A, Huber GW, and Zhang T (2015). Catalytic transformation of lignin for the production of chemicals and fuels. *Chem Rev*, 115: 11559–11624.

Li S, Li W, Zhang Q, Shu R, Wang H, Xin H, and Ma L (2018). Lignin-first depolymerization of native corn stover with an unsupported MoS2 catalyst. *RSC Adv*, 8: 1361–1370.

Li Y, Cai Z, Liao M, Long J, Zhao W, Chen Y, and Li X (2017). Catalytic depolymerization of organosolv sugarcane bagasse lignin in cooperative ionic liquid pairs. *Catal Today*, 298: 168–174.

Li ZM, Long JX, Zeng Q, Wu YH, Cao Ml, Liu SJ, and Li XH (2020). Production of methyl p-hydroxycinnamate by selective tailoring of herbaceous lignin using metal-based deep eutectic solvents (DES) as catalyst. *Ind Eng Chem Res*, 59: 17328–17337.

Liguori L and Barth T (2011). Palladium-Nafion SAC-13 catalysed depolymerisation of lignin to phenols in formic acid and water. *J Anal Appl Pyrolysis*, 92(2): 477–484.

Limarta SO, Ha JM, Park YK, Lee H, Suh DJ, and Jae J (2018). Efficient depolymerization of lignin in supercritical ethanol by a combination of metal and base catalysts. *J. Ind. Eng. Chem*, 57: 45–54.

Long J, Zhang Q, Wang T, Zhang X, Xu Y, Ma L (2014). An efficient and economical process for lignin depolymerization in biomass-derived solvent tetrahydrofuran. *Bioresour. Technol*, 154: 10–17.

Lundquist K (1976). Low-molecular weight lignin hydrolysis products. *Applied Polymer Symposium*. JohnWiley & Sons, Inc., Hoboken, NJ, USA, pp. 1393–1407.

Ma H, Li H, Zhao W, Li L, Liu S, Long J, and Li X (2019). Selective depolymerization of lignin catalyzed by nickel supported on zirconium phosphate. *Green Chem*, 21: 658–668.

Ma R, Guo M, Lin Kt, Hebert VR, Zhang J, Wolcott MP, Quintero M, Ramasamy KK, Chen X, and Zhang X (2016). Peracetic acid depolymerization of biorefinery lignin for production of selective monomeric phenolic compounds. *Chem A Eur J*, 22: 10884–10891.

Ma X, Cui K, Hao W, Ma R, Tian Y, and Li Y (2015). Alumina supported molybdenum catalyst for lignin valorization: effect of reduction temperature. *Bioresour Technol*, Sep; 192: 17–22.

MacFarlane DR, Kar M, and Pringle JM (2017). *Fundamentals of Ionic Liquids: From Chemistry to Applications*. John Wiley & Son, Australia.

Mahmood N, Yuan Z, Schmidt J, and Xu C C (2013). Production of polyols via direct hydrolysis of kraft lignin: effect of process parameters. *Bioresour Technol*, 139: 13–20.

Mahmood N, Yuan Z, Schmidt J, and Xu CC (2015). Hydrolytic depolymerization of hydrolysis lignin: effects of catalysts and solvents. *Bioresour Technol*, 190: 416–419.

Marszałł MP and Kaliszan R (2007). Application of ionic liquids in liquid chromatography. *Crit Rev Anal Chem*, 37: 127–140.

Martin A, Patil PT, and Armbruster U (2013). *Catalytic hydroprocessing of lignin in supercritical ethanol*. Proceedings of the III. Iberoamerican Conference Supercritical Fluids Cartagena de Indias, Cartagena de Indias, Colombia, 1–5 April 2013.

Massoth FE, Politzer P, Concha MC, Murray JS, Jakowski J, and Simons J (2006). Catalytic hydrodeoxygenation of methyl-substituted phenols: correlations of kinetic parameters with molecular properties. *J Phys Chem B*, 110: 14283–14291.

McFarland EW and Metiu H (2013). Catalysis by doped oxides. *Chem Rev*, 113: 4391–4427.

Mensah M, Tia R, Adei E, and de Leeuw NH (2022). A DFT mechanistic study on base-catalyzed cleavage of the β-O-4 ether linkage in lignin: implications for selective lignin depolymerization. *Front Chem*, Feb; 17(10): 793759.

Meshgini M and Sarkanen Kyosti V (1989). Synthesis and kinetics of acid-catalyzed hydrolysis of some α-aryl ether lignin model compounds, Holzforschung. *Int J Biology, Chemistry, Physics and Technology of Wood*, 43: 239.

Miller J, Evans L, Littlewolf A, and Trudell D (1999). Batch microreactor studies of lignin and lignin model compound depolymerization by bases in alcohol solvents. *Fuel*, 78: 1363–1366.

Miyafuji H, Miyata K, Saka S, Ueda F, and Mori M (2009). Reaction behavior of wood in an ionic liquid, 1-ethyl- 3-methylimidazolium chloride. *J Wood Sci*, 55: 215–219.

Mohtar SS, Busu TTM, Noor AM, Shaari N, Yusoff NA, Bustam MA, Mutalib MA, and Mat HB (2015). Extraction and characterization of lignin from oil palm biomass via ionic liquid dissolution and nontoxic aluminium potassium sulfate dodecahydrate precipitation processes. *Bioresour Technol*, 192: 212–218.

Molinari V, Giordano C, Antonietti M, and Esposito D (2014). Titanium nitride-nickel nanocomposite as heterogeneous catalyst for the hydrogenolysis of aryl ethers. *J Am Chem Soc*, 136: 1758–61.

Mortensen PM, de Carvalho HWP, Grunwaldt J-D, Jensen PA, and Jensen AD (2015). Activity and stability of Mo2C/ZrO2 as catalyst for hydrodeoxygenation of mixtures of phenol and 1-octanol. *J Catal*, 328: 208–215.

Muley PD, Mobley JK, Tong X, Novak B, Stevens J, Moldovan D, Shi J, and Boldor D (2019). Rapid microwave-assisted biomass delignification and lignin depolymerization in deep eutectic solvents. *Energy Convers Manag*, 196: 1080–1088.

Nandiwale KY, Danby AM, Ramanathan A, Chaudhari RV, and Subramaniam B (2018). Dual function lewis acid catalyzed depolymerization of industrial corn stover lignin into stable monomeric phenols. *ACS Sustain Chem Eng*, 7: 1362–1371.

Nolte MW, Zhang J, and Shanks BH (2016). Ex situ hydrodeoxygenation in biomass pyrolysis using molybdenum oxide and low pressure hydrogen. *Green Chem*, 18: 34–138.

Numan-Al-Mobin AM, Kolla P, Dixon D, and Smirnova A (2016). Effect of water–carbon dioxide ratio on the selectivity of phenolic compounds produced from alkali lignin in sub-and supercritical fluid mixtures. *Fuel*, 185: 26–33.

Ogawa S and Miyafuji H (2015). Reaction behavior of milled wood lignin in an ionic liquid, 1-ethyl-3-methylimidazolium chloride. *J Wood Sci*, 61: 285–291.

Okuda K, Umetsu M, Takami S, and Adschiri T (2004). Disassembly of lignin and chemical recovery—rapid depolymerization of lignin without char formation in water-- phenol mixtures. *Fuel Process Technol*, 85: 803–13.

Oregui Bengoechea M, Hertzberg A, Mileti'c N, Arias PL, and Barth T (2015). Simultaneous catalytic de-polymerization and hydrodeoxygenation of lignin in water/ formic acid media with Rh/Al2O3, Ru/Al2O3 and Pd/Al2O3 as bifunctional catalysts. *J Anal Appl Pyrolysis*, 113: 713–722.

Oregui-Bengoechea M, Gandarias I, Arias PL, and Barth T (2018). Solvent and catalyst effect in the formic acid aided lignin-to-liquids. *Bioresour Technol*, 270: 529–536.

Paterson RJ (2012). *Lignin: Properties and Applications in Biotechnology and Bioenergy*. Nova Science Publishers, New York.

Paananen H, Eronen E, Makinen M, Janis J, Suvanto M, Pakkanen TT (2020). Base-catalyzed oxidative depolymerization of softwood kraft lignin. Ind. *Crops Prod*, 152, 112473.

Pham TPT, Cho C-W, and Yun Y-S (2010). Environmental fate and toxicity of ionic liquids: a review. *Water Res*, 44: 352–372.

Plechkova NV and Seddon KR (2008). Applications of ionic liquids in the chemical industry. *Chem Soc Rev*, 37: 123–150.

Prado R, Erdocia X, and Labidi J (2013). Lignin extraction and purification with ionic liquids. *J Chem Technol Biotechnol*, 88: 1248–1257.

Prado R, Erdocia X, and Labidi J (2016). Study of the influence of reutilization ionic liquid on lignin extraction. *J Clean Prod*, 111: 125–132.

Prasomsri T, Nimmanwudipong T, and Román-Leshkov Y (2013). Effective hydrodeoxygenation of biomass-derived oxygenates into unsaturated hydrocarbons by MoO3 using low H2 pressures. *Energy Environ Sci*, 6: 1732–1738.

Prasomsri T, Shetty M, Murugappan K, and Román-Leshkov Y (2014). Insights into the catalytic activity and surface modification of MoO3 during the hydrodeoxygenation of lignin-derived model compounds into aromatic hydrocarbons under low hydrogen pressures. *Energy Environ Sci*, 7: 2660–2669.

Pu Y, Jiang N, and Ragauskas AJ (2007). Ionic liquid as a green solvent for lignin. *J Wood Chem Technol*, 27: 23–33.

Qiu S, Li M, Huang Y, and Fang Y (2018). Catalytic hydrotreatment of Kraft lignin over NiW/ SiC: effective depolymerization and catalyst regeneration. *Ind Eng Chem Res*, 57: 2023–2030.

Rahimi A, Ulbrich A, Coon JJ, and Stahl SS (2014). Formic-acid-induced depolymerization of oxidized lignin to aromatics. *Nature*, 515: 249.

Ranga C, Lødeng R, Alexiadis VI, Rajkhowa T, Bjørkan H, Chytil S, Svenum I, Walmsley JC, Detavernier C, Poelman H, Voort PV, and Thybaut JW (2018). Effect of composition and preparation of supported MoO3 catalysts for anisole hydrodeoxygenation. *Chem Eng J*, 335: 120–132.

Rashid T, Kait CF, Regupathi I, and Murugesan T (2016). Dissolution of Kraft lignin using protic ionic liquids and characterization. *Ind Crop Prod*, 84: 284–293.

Reverchon E (2002). Micro and nano-particles produced by supercritical fluid assisted techniques: present status and perspectives. *Chem Eng Trans*, 2: 1–10.

Roberts VM, Stein V, Reiner T, Lemonidou A, Li X, and Lercher JA (2011). Towards quantitative catalytic lignin depolymerization. *Chem Eur J*, 17: 5939–5948.

Roy R, Rahman MS, Amit TA, and Jadhav B (2022). Recent advances in lignin depolymerization techniques: a comparative overview of traditional and greener approaches. *Biomass*, 2: 130–154. https://doi.org/10.3390/biomass2030009.

Rui M, Wenyue H, Xiaolei M, Ye T, and Yongdan L (2014). Catalytic ethanolysis of Kraft lignin into high-value small-molecular chemicals over a nanostructured α-molybdenum carbide catalyst. *Angew Chem Int Ed*, 53: 7310–7315.

Ruiz PE, Frederick BG, DeSisto WJ, Austin RN, Radovic LR, Garcia R, Escalona N, and Wheeler MC (2012). Guaiacol hydrodeoxygenation on MoS2 catalysts: influence of activated carbon supports. *Catal Commun*, 27: 44–48.

Saisu M, Sato T, Watanabe M, Adschiri T, and Arai K (2003). Conversion of lignin with supercritical water– phenol mixtures. *Energy Fuels*, 17: 922–8.

Sasaki M and Goto M (2008). Recovery of phenolic compounds through the decomposition of lignin in near and supercritical water. *Chem Eng Processing Process Intensif*, 47: 1609–1619.

Sathitsuksanoh N, Holtman KM., Yelle DJ, Morgan T, Stavila V, Pelton J, Blanch H, Simmons BA, and George A (2014). Lignin fate and characterization during ionic liquid biomass pretreatment for renewable chemicals and fuels production. *Green Chem*, 16: 1236–1247.

Schüth F, Rinaldi R, Meine N, Käldström M, Hilgert J, and Kaufman-Rechulski MD (2014). Mechanocatalytic depolymerization of cellulose and raw biomass and downstream processing of the products. *Catal Today*, 234: 24–30.

Shen X-J, Chen T, Wang H-M, Mei Q, Yue F, Sun S, Wen J-L, Yuan T-Q, and Sun R-C (2019). Structural and morphological transformations of lignin macromolecules during bio-based deep eutectic solvent (DES) pretreatment. *ACS Sustain Chem Eng*, 8: 2130–2137.

Shetty M, Murugappan K, Prasomsri T, Green WH, and Roman-Leshkov Y (2015). Reactivity and stability investigation of supported molybdenum oxide catalysts for the hydrodeoxygenation (HDO) of m-cresol. *J Catal*, 331: 86–97.

Shu R, Xu Y, Ma L, Zhang Q, Wang C, and Chen Y (2018). Controllable production of guaiacols and phenols from lignin depolymerization using Pd/C catalyst cooperated with metal chloride. *Chem Eng J*, 338: 457-464. https://doi.org/10.1016/j.cej.2018.01.002.

Shuai L and Luterbacher J (2016). Organic solvent effects in biomass conversion reactions. *ChemSusChem*, 9: 133–55.

Singh SK, Nandeshwar K, and Ekhe JD (2016). Thermochemical lignin depolymerization and conversion to aromatics in subcritical methanol: effects of catalytic conditions. *New J Chem*, 40(4): 3677–3685.

Solt P, Rosiger B, Konnerth J, Van Herwijnen HW (2018). Lignin phenol formaldehyde resoles using base-catalysed depolymerized Kraft lignin. *Polymers*, 10: 1162.

Song Q, Wang F, Cai J, Wang Y, Zhang J, Yu W, and Xu J (2013). Lignin depolymerization (LDP) in alcohol over nickel-based catalysts via a fragmentation–hydrogenolysis process. *Energy Environ Sci*, 6: 994–1007.

Song Q, Wang F, and Xu J (2012). Hydrogenolysis of lignosulfonate into phenols over heterogeneous nickel catalysts. *Chem Comm*, 48(56): 7019–7021.

Stark A (2010). Ionic liquids in the biorefinery: a critical assessment of their potential. *Energ Environ Sci*, 4(1): 19–32.

Stärk K, Taccardi N, Bösmann A, and Wasserscheid P (2010). Oxidative depolymerization of lignin in ionic liquids. *ChemSusChem*, 3: 719–723.

Sturgeon MR, Kim S, Lawrence K, Paton RS, Chmely SC, Nimlos M, Foust TD, and Beckham GT (2014). A mechanistic investigation of acid-catalyzed cleavage of aryl-ether linkages: implications for lignin depolymerization in acidic environments. *ACS Sustain Chem Eng*, 2: 472–485.

Sullivan MM and Bhan A (2016). Acetone hydrodeoxygenation over bifunctional metallic– acidic molybdenum carbide catalysts. *ACS Catal*, 6: 1145–1152.

Sullivan MM, Chen CJ, and Bhan A (2016). Catalytic deoxygenation on transition metal carbide catalysts. *Catal Sci Technol*, 6: 602–616.

Sun Y-C, Xu J-K, Xu F, Sun R-C, and Jones GL (2014). Dissolution, regeneration and characterisation of formic acid and Alcell lignin in ionic liquid-based systems. *Rsc Adv*, 4: 2743–2755.

Takami S, Okuda K, Man X, Umetsu M, Ohara S, and Adschiri T (2012). Kinetic study on the selective production of 2- (Hydroxybenzyl)-4-methylphenol from organosolv lignin in a mixture of supercritical water and p-cresol. *Ind Eng Chem Res*, 51(13): 4804–4808.

Tan SS, MacFarlane DR, Upfal J, Edye LA, Doherty WO, Patti AF, Pringle JM, and Scott JL (2009). Extraction of lignin from lignocellulose at atmospheric pressure using alkylbenzenesulfonate ionic liquid. *Green Chem*, 11: 339–345.

Tang D, Huang X, Tang W, and Jin Y (2021). Lignin-to-chemicals: application of catalytic hydrogenolysis of lignin to produce phenols and terephthalic acid via metal-based catalysts. *Int J Biol Macromol*, 190: 72–85.

Thring R (1994). Alkaline degradation of ALCELL®lignin. *Biomass Bioenergy*, 7: 125–130.

Toledano A, Serrano L, and Labidi J (2012). Organosolv lignin depolymerization with different base catalysts. *J Chem Technol Biotechnol*, 87: 1593–1599.

Toledano A, Serrano L, and Labidi J (2014a). Improving base catalyzed lignin depolymerization by avoiding lignin repolymerization. *Fuel*, 116: 617–624.

Toledano A, Serrano L, Pineda A, Romero AA, Luque R, and Labidi J (2014b). Microwave-assisted depolymerisation of organosolv lignin via mild hydrogen-free hydrogenolysis: catalyst screening. *Appl Catal B, Environmental*, 145: 43–55.

Tolesa LD, Gupta BS, and Lee M-J (2017). The chemistry of ammonium-based ionic liquids in depolymerization process of lignin. *J Mol Liq*, 248: 227–234.

Underkofler KA, Teixeira RE, Pietsch SA, Knapp KG, and Raines RT (2015). Separation of lignin from corn stover hydrolysate with quantitative recovery of ionic liquid. *ACS Sustain Chem Eng*, 3: 606–613.

Vangeel T, Renders T, Van Aelst K, Cooreman E, Van Den Bosch S, Van Den Bossche G, Koelewijn SF, Courtin CM, and Sels BF (2019). Reductive catalytic fractionation of black locust bark. *Green Chem*, 21: 5841–5851.

Vekariya RL (2017). A review of ionic liquids: applications towards catalytic organic transformations. *J Mol Liq*, 227: 44–60.

Wahyudiono W, Machmudah S, and Goto M (2008). Recovery of phenolic compounds through the decomposition of lignin in near and supercritical water. *Chem Eng Process*, 47(9–10): 1609–1619.

Wahyudiono W, Machmudah S, and Goto M (2013). Utilization of sub and supercritical water reactions in resource recovery of biomass wastes. *Eng J*, 17(1): 1–12.

Walsh JM, Ikonomou GD, and Donohue MD (1987). Supercritical phase behavior: the entrainer effect. *Fluid Phase Equilib*, 33: 295–314. https://doi.org/10.1016/0378-3812(87)85042-2.

Wang H, Block LE, and Rogers RD (2014). Catalytic conversion of biomass in ionic liquids. *Catalysis in Ionic Liquids: From Catalyst Synthesis to Application*, Hardacre C and Parvulescu V (ed.). The Royal Society of Chemistry's, London, UK.

Wang H, Pu Y, Ragauskas A, and Yang B (2019). From lignin to valuable products–strategies, challenges, and prospects. *Bioresour Technol*, 271: 449–461.

Wang H, Tucker M, and Ji Y (2013). Recent development in chemical depolymerization of lignin: a review. *J Appl Chem*, 9: 838645.

Wang Q, Guan S, and Shen D (2017). Experimental and kinetic study on lignin depolymerization in water/formic acid system. *Int J Mol Sci*, 18: 2082.

Wang X, Wang N, Nguyen TT, and Qian EW (2018). Catalytic depolymerization of lignin in ionic liquid using a continuous flow fixed-bed reaction system. *Ind Eng Chem Res*, 57: 16995–17002.

Wanmolee W, Daorattanachai P, and Laosiripojana N (2016). Depolymerization of organosolv lignin to valuable chemicals over homogeneous and heterogeneous acid catalysts. *Energy Procedia*, 100: 173–177.

Weldemhret TG, Bañares AB, Ramos KRM, Lee W-K, Nisola GM, Valdehuesa KNG, and Chung W-J (2020). Current advances in ionic liquid-based pre-treatment and depolymerization of macroalgal biomass. *Renew Energ*, 152: 283–299.

Welton T (2011). Ionic liquids in green chemistry. *Green Chem*, 13: 225.

Wen J-L, Yuan T-Q, Sun S-L, Xu F, and Sun R-C (2014). Understanding the chemical transformations of lignin during ionic liquid pretreatment. *Green Chem*, 16: 181–190.

Whiffen VML and Smith KJ (2010). Hydrodeoxygenation of 4-methylphenol over unsupported MoP, MoS2, and MoOx catalysts. *Energy Fuel*, 24: 4728–4737.

Wu H, Song J, Xie C, Wu C, Chen C, and Han B (2018). Efficient and mild transfer hydrogenolytic cleavage of aromatic ether bonds in lignin-derived compounds over Ru/C. *ACS Sustain Chem Eng*, 6: 2872–2877.

Xiao L, Wang S, Li H, Li Z, Shi Z, Xiao L, Sun R, Fang Y, and Song G (2017). Catalytic hydrogenolysis of lignins into phenolic compounds over carbon nanotube supported molybdenum oxide. *ACS Catal*, 7: 7535–7542.

Xu C, Arancon RAD, Labidi J, and Luque R (2014). Lignin depolymerisation strategies: towards valuable chemicals and fuels. *Chem Soc Rev*, 43: 7485–7500.

Xu W, Miller SJ, Agrawal PK, and Jones CW (2012). Depolymerization and hydrodeoxygenation of switchgrass lignin with formic acid. *ChemSusChem*, 5(4): 667–675.

Yan B, Lin X, Chen Z, Cai Q, and Zhang S (2021). Selective production of phenolic monomers via high efficient lignin depolymerization with a carbon based nickel-iron-molybdenum carbide catalyst under mild conditions. *Bioresour Technol*, 321: 124503.

Yan N, Zhao C, Dyson PJ, Wang C, Liu LT, and Kou Y (2008). Selective degradation of wood lignin over noble-metal catalysts in a two-step process. *ChemSusChem*, 1: 626–629.

Yang J, Zhao L, Liu S, Wang Y, and Dai L (2016). High-quality bio-oil from one-pot catalytic hydrocracking of kraft lignin over supported noble metal catalysts in isopropanol system. *Bioresour Technol*, 212: 302–310.

Yang T, Wu K, Li B, Du C, Wang J, and Li R (2021). Conversion of lignin into phenolic-rich oil by two-step liquefaction in subsupercritical ethanol system assisted by carbon dioxide. *J Energy Inst*, 94: 329–336.

Yao SG, Mobley JK, Ralph J, Crocker M, Parkin S, Selegue JP, and Meier MS (2018). Mechanochemical treatment facilitates two-step oxidative depolymerization of kraft lignin. *ACS Sustain Chem Eng*, 6: 5990–5998.

Ye Y, Zhang Y, Fan J, and Chang J (2012). Selective production of 4-ethylphenolics from lignin via mild hydrolysis. *Bioresour Technol*, 118: 648–651.

Yokoyama T (2015a). Revisiting the mechanism of β-O-4 bond cleavage during acidolysis of lignin. Part 6: a review. *J Wood Chem Technol*, 35: 27–42.

Yokoyama T (2015b). Role of counter anion in the chemical reaction of wood components under acidic conditions. *Mokuzai Gakkaishi*, 61: 217–225.

Yong TL-K and Matsumura Y (2013). Kinetic analysis of lignin hydrothermal conversion in sub-and supercritical water. *Ind Eng Chem Res*, 52: 5626–5639.

Yoshikawa T, Shinohara S, Yagi T, Ryumon N, Nakasaka Y, and Tago T (2012). Production of phenols from lignin via depolymerization and catalytic cracking. *Fuel Process Technol*, 108: 69–75.

Yu Q, Song Z, Chen X, Fan J, Clark JH, Wang Z, Sun Y, and Yuan Z (2020). A methanol–choline chloride based deep eutectic solvent enhances the catalytic oxidation of lignin into acetovanillone and acetic acid. *Green Chem*, 22: 6415–6423.

Yuan Z, Cheng S, Leitch M, and Xu CC (2010). Hydrolytic degradation of alkaline lignin in hot-compressed water and ethanol. *Bioresour Technol*, 101: 9308–9313.

Zakzeski J, Bruijnincx PCA, Jongerius AL, and Weckhuysen BM (2010). The catalytic valorization of lignin for the production of renewable chemicals. *Chem Rev*, 110(6): 3552–3599.

Zhai Y, Li C, Xu G, Ma Y, Liu X, and Zhang Y (2017). Depolymerization of lignin via a nonprecious Ni–Fe alloy catalyst supported on activated carbon. *Green Chem*, 19: 1895–1903.

Zhang J, Asakura H, Rijn JV, Yang J, Duchesne PN, Zhang B, Chen X, Zhang P, Saeys M, and Yan N (2014b). Highly efficient, NiAu-catalyzed hydrogenolysis of lignin into phenolic chemicals. *Green Chem*, 16: 2432–2437.

Zhang J, Teo J, Chen X, Asakura H, Tanaka T, Teramura K, and Yan N (2014a). A series of NiM (M=Ru, Rh, and Pd) bimetallic catalysts for effective lignin hydrogenolysis in water. *ACS Catal*, 4: 1574–1483.

Zhang J-w, Cai Y, Lu G-p, and Cai C (2016). Facile and selective hydrogenolysis of β-O-4 linkages in lignin catalyzed by Pd-Ni bimetallic nanoparticles supported on ZrO2. *Green Chem*, 18: 6229–35.

Zhang S, Sun J, Zhang X, Xin J, Miao Q, and Wang J (2014a). Ionic liquid-based green processes for energy production. *Chem Soc Rev*, 43: 7838–7869.

Zhang X, Tang W, Zhang Q, Li Y, Chen L, Xu Y, Wang C, and Ma L (2018a). Production of hydrocarbon fuels from heavy fraction of bio-oil through hydrodeoxygenative upgrading with Ru-based catalyst. *Fuel*, 215: 825–834.

Zhang X, Tang W, Zhang Q, Wang T, and Ma L (2018b). Hydrodeoxygenation of lignin-derived phenoic compounds to hydrocarbon fuel over supported Ni-based catalysts. *Appl Energy*, 227: 73–79.

Zhang X, Zhang Q, Long J, Xu Y, Wang T, Ma L, and Li Y (2014b). Phenolics production through catalytic depolymerization of alkali lignin with metal chlorides. *BioRes*, 9(2): 3347–3360.

Zhang ZC (2013). Catalytic transformation of carbohydrates and lignin in ionic liquids. *Wiley Interdiscip Rev Energy Environ*, 2: 655–672.

Zhu XH, Peng C, Chen H, Chen Q, Zhao ZK, Zheng Q, and Xie H (2018). Opportunities of ionic liquids for lignin utilization from biorefinery. *Chem Select*, 3(27): 7945–7962. https://doi.org/10.1002/slct.201801393.

3.4 Oxidative Depolymerization of Lignin

A promising strategy for conversion of lignin into monomers is oxidative depolymerization (Störk et al., 2010; Sun et al., 2018; Zaid, 2019; Zhou et al., 2022). Compared to acid or base catalyst methods, this method uses less forceful conditions which cleave the ether linkages of lignin. Acid or base catalyst methods are expensive, necessitate relatively severe reaction conditions, are difficult to manage, and are harmful to the environment.

Despite numerous relevant studies on oxidation protocols, most of which focus on lignin model compounds, oxidants and/or oxidizing protocols are not widely used for lignin depolymerization. Lignin has a lot of hydroxyl groups and because of this, oxidative cracking might be a fascinating choice to consider. When it comes to the depolymerization and degradation of lignin, this is one of the most common methods (Das et al., 2017; Rahimi et al., 2014; Störk et al., 2010).

Expanding its use holds promise due to the abundance of hydroxyl groups in its structure. Oxidative methods, such as oxygen, hydrogen peroxide, or peroxyacids, are already widely used in the paper industry for pulp bleaching, so lignin depolymerization may expand significantly in the future. Furthermore, the majority of current oxidation techniques are probably environment friendly and carried out in mild conditions. But, in order to avoid excessive oxidation of the substrate to gaseous products, oxidation methods require sufficient selectivity.

Oxidative depolymerization produces carbonyl derivatives that can be used either "as is" or modified chemically to be valuable. One of the main lignin oxidation products, vanillin, has a market that is already very developed and can be used, for instance, as a fragrance ingredient, a platform compound for higher value added product and polymer applications, or a food flavoring agent (Bjørsvik and Minisci, 1999; Calvo-Flores and Dobado, 2010; Fache et al., 2016; Roberts et al., 2011). In basic media with copper catalysts, Borregaard developed a method to transform sulfite lignin into vanillin (8%wt) (Bjørsvik and Minisci, 1999; McClelland et al., 2019; Rodrigues Pinto et al., 2012; Rødsrud et al., 2012; Vidal, 2000). A number of oxidants, including oxygen, hydrogen peroxide, ozone, percarbonate, and others, have been reported (Cronin et al., 2017; Danby et al., 2018; Maluenda et al., 2015) with or without catalysts. Both homogeneous and heterogeneous catalysts are used as catalysts (Schutyser et al., 2018a, 2018b; Xu et al., 2014; Zakzeski et al., 2010). In general, it was discovered that yield increased with the increase in base concentration when concentrating on catalyst-free lignin depolymerization in simple media. According to the botanical source of the lignin, vanillin yields ranging from 1 to 6.8 weight % have been reported, showing that the nature of the lignin is crucial to the process. Native lignin, also referred to as poplar sawdust, was used by Schutyser et al. (2018a) to produce aromatic yields of about 30 weight % (vanillin: 7% by weight).

New approaches are being developed to study the capability of oxidative depolymerization of lignin in ambient conditions at a lower cost (Ahmad et al., 2021). It is typically used for producing phenolic derivatives by utilizing oxidants like oxygen, hydrogen peroxide, nitrobenzene and metallic oxide, which aid in the preservation of the lignin aromatic rings (Figueiredo et al., 2018; Laurichesse and Avérous, 2014). A wide range of subsequent reactions, including aromatic ring hydroxylation, phenol oxidation, benzylic acid oxidation, demethylation and ring-opening reactions are caused by the reaction, which is typically linked to either the transfer of electrons or the extraction of hydrogen atoms from lignin (Li and Takkellapati, 2018).

The depolymerization reaction employs a wide range of oxidants, including, hydrogen peroxide, chlorine, hypochlorite, permanganate, nitrobenzene and copper oxide (Kishimoto et al., 2003; Smith et al., 1989; Wu and Heitz, 1995). The oxidative degradation method preserves the aromatic character while selectively breaking down the ether linkages (Ragauskas et al., 2014).

Strong oxidants are able to break the links between the aromatic moieties of lignin more effectively than catalytic methods under milder conditions (Roy et al., 2021b). As a result, the success of oxidative depolymerization depends on selecting the right oxidant. Table 3.17 (Roy et al., 2022) lists the conditions of the reaction as well as the main results of oxidative depolymerization of lignin. For the past few decades, lignin has been depolymerized with hydrogen peroxide, an important oxidant. During lignin depolymerization, it acts as an environmentally friendly and economically advantageous catalyst (Crestini et al., 2006).

Alkali lignin can be depolymerized into monomers by using hydrogen peroxide at a concentration of at least 1% (Hasegawa et al., 2011). The β-O-4 and β-1 linkages can be selectively broken down by hydrogen peroxide at low temperatures, according to other studies (Jennings et al., 2017). Despite having a higher rate of degradation than other methods, hydrogen peroxide yields fewer monophenolic compounds. Hydrogen peroxide can break more side chains and ether bonds in the presence of a copper oxide-type metallic catalyst (as Cu^{2+}). Additionally, hydrogen peroxide has a significantly increased capacity for oxidation when ferric sulfate (as Fe^{3+}) is present, which may result in an

Table 3.17 Reaction conditions and the products of oxidative depolymerization of lignin.

Coconut husk lignin
Nitrobenzene, 160 °C for 2.5 h
Vanilin
Ngadi et al., 2014.
Wheat straw lignin
Hydrogen peroxide, CuO, and $Fe_2(SO_4)_3$, 180 °C for 90 min
4-hydroxybenzaldehyde and vanillin
Ouyang et al., 2014
Corn Stover lignin
Molecular Oxygen, 160 °C for 0.5 h at 1.5 MPa
Vanillin, vanillic acid, p-coumaric acid, and acetovanillone
Lyu et al., 2018
Kraft lignin
$CuSO_4$/water/octanol 170 °C for 60 min and 5 MPa vanillin
Bjelić et al., 2018
Lignosulfonate lignin
Cu-Mn/Al_2O_3, 160 °C for 40 min
Vanillin, p-hydroxybenzaldehyde, vanillic acid, and p-hydroxybenzoic acid
Abdelaziz et al., 2019.
Alkali lignin
Co/TiO_2, Co/ZrO_2, Co/CeO_2, 140 °C for 1 h, 15 bars
Guaiacol, syringol, homovanillic acid, vanillin, and 2-methoxy-4-vinylphenol,
Kumar et al., 2020
Corn stover lignin
Wet air (O_2) 200 °C for 2 h, 500 psi
Vanillin
Irmak et al., 2020.

Roy et al. (2022) / MDPI / CC BY 4.0.

increase in the production of monophenols. Additionally, acid can significantly improve the efficiency of hydrogen peroxide-based depolymerization (Jennings et al., 2017; Zhang et al., 2017).

Oxidative depolymerization can be carried out without the use of toxic metallic materials. It is believed that molecular oxygen and wet air are oxidants that are less harmful and benign than metallic or conventional oxidants. At low temperatures and pressures, wet air oxidation of lignin has been shown to significantly enhance the production of phenolic monomers (Irmak et al., 2020). In moderate reaction environments, lignin can be effectively converted into aromatic compounds through molecular oxygen depolymerization (Lyu et al., 2018).

Oxidative cracking might be a fascinating option to think about because lignin contains a lot of hydroxyl groups. The primary byproducts of the oxidative degradation of lignin are carboxylic acids and aromatic aldehydes (Rahimi et al., 2013). For this purpose, hydrogen peroxide and metal oxides have been considered. Vanillin is a main byproduct of the oxidative depolymerization of lignin. The yields range from 5 to 15 weight % depending upon the source of the lignin. The majority of protocols that have been reported make use of homogeneous conditions like mineral or alkali acids and hydrogen peroxide, homogeneous heteropolyacid catalysts, and sometimes even metals like Cu^{2+}, Fe^{3+}, and so on under high pressures and temperatures. Because they produce radicals that cause partial re-polymerization and the formation of more complex lignin structures, in a few cases, oxidative methods for lignin may not be the best. Intriguingly, a study by Rahimi et al. (2013) outlines the future direction of oxidative depolymerization of lignin. For more information on oxidative protocols, a practical organocatalytic method based on the chemoselective aerobic oxidation of secondary benzylic alcohols was reported for lignin model compounds (with an extension to aspen lignin with a S:G ratio of 2: 1).

The catalytic system of mineral acids like nitric acid and hydrochloric acid (10 mol% each) and 4-acetamido TEMPO (5 mol%) was able to selectively oxidize a variety of lignin model compounds, including significant acids and ketones, under mild reaction conditions (like vanillin, vanillic, and veratric acids). For a specific benzylic ketone that could completely convert the starting material into veratric acid and guaiacol, with a yield of 88% and 42% respectively, an alkali-hydrogen peroxide system was used to combine oxidations with C–C bond cleavage. NMR data demonstrated that the majority of S and G units were successfully oxidized to their benzylic ketone analogues during the selective oxidation of aspen lignin, which was also included in the protocol. Unoxidized beta-S ether units were only found in a minuscule amount (less than 10%). This protocol may be suitable for industrial processes on a larger scale due to the metal-free system's simplicity, milder reaction conditions, and the ability to scale up (as demonstrated for reactions up to 10 g).

In summary, high temperatures and pressures are required for base-catalyzed depolymerization reactions. Depolymerization's difficulty is exacerbated by the fact that it typically results in the production of a significant quantity of gases at much higher temperatures, in addition to the numerous side products that avoid the formation of monomeric aromatic hydrocarbons. The acid-catalyzed process, on the other hand, appears to be a gentle and kinetically preferred method. Additionally, the method may facilitate the fractionation of lignin, hemicellulose, and cellulose by destroying the cavities that surround cellulose in lignocellulosic materials. Even though only tannins were tested for the acid-catalyzed process's efficacy, an integrated biorefinery concept may be able to take advantage of these acid-assisted reaction characteristics (Mosier et al., 2005; Ragauskas et al., 2014). The complex structure of lignin can theoretically be broken down into C–O and C–C bonds by adding an acid. Literature reviews have yet to show that it can depolymerize lignin, a more complicated feedstock. Under ozonolysis conditions, oxidative depolymerization techniques can be helpful in producing aromatic chemicals with structural stability and acyclic organic acids. Product yields, lignin recombination and repolymerization, and product separation and/or isolation feasibility particularly for smaller amounts of generated products still require significant attention, despite the most promising reports (Rahimi et al., 2013). The highest yields of lignin-obtained oxidative products have been reported to be between 10 weight % and 11 weight %, while vanillin-containing products typically yield between 3 weight % and 5% by weight (Das et al., 2012).

Aromatic aldehydes (like vanillin and syringaldehyde) and their corresponding acids (such as vanillic acid and syringic acid) are the primary byproducts of oxidative depolymerization of lignin (Vangeel et al., 2018). According to Li and Takkellapati (2018), oxidative depolymerization is an excellent method for breaking C–C bonds (such as 5–5′ and α–5′) in kraft lignin with a higher C–C content. Processing conditions, particularly pH and oxidant selection, have a significant impact on the lignin oxidation reaction mechanism, product yield, and distribution.

Numerous studies have examined the oxidative depolymerization of lignin by hydrogen peroxide in acidic as well as alkaline conditions. According to Xiang and Lee (2000), for achieving the similar degree of depolymerization (roughly 98%) of precipitated hardwood lignin, under strongly alkaline conditions, lower temperatures (80 to 90 °C) are required, while higher temperatures (130 to 160 °C) are required under acidic conditions. Formic, oxalic, and acetic acids dominate product streams, while vanillin and syringaldehyde were only detected in trace amounts. This suggests that if aromatic aldehyde and the acids that go along with it are what you want, hydrogen peroxide might not be the best choice (Xiang and Lee, 2000).

Nitrobenzene is thought to be a more efficient oxidant than hydrogen peroxide for producing aldehydes like vanillin, syringaldehyde, and p-hydroxybenzaldehyde as well as their respective acids like vanillic acid, syringic acid, and p-hydroxybenzoic acid (Min et al., 2015). Additionally, due to the carcinogenicity of nitrobenzene, transition metals and metallic oxides are sometimes used as alternative oxidants to achieve the highest yield of a product obtained from hardwood and softwood lignin by alkaline nitrobenzene oxidation (Figueiredo et al., 2018).

The mechanism of oxidation of β-O-4 linkages is shown in Figure 3.26 (Chio et al., 2019). Figure 3.27 shows oxidative depolymerization mechanism of lignin (Roy et al., 2021a).

Hydrogen peroxide and potassium permanganate are also frequently used as oxidants for chemical oxidation in industry and laboratories because of their higher availability, lower cost, and ease of production (Das et al., 2017; Nyamunda et al., 2013). It has been extensively researched for many years how to depolymerize or degrade lignin using hydrogen peroxide. For the purpose of selective lignin oxidation, hydrogen peroxide typically reacts with a suitable catalyst, such as acid or metallic catalysts (Crestini et al., 2006; Jennings et al., 2017; Zhang et al., 2017). Hydrogen peroxide was used by Hasegawa et al. (2011) to oxidize various type of lignin. HPLC analysis showed that at 200 °C for 5 minutes, 0.1% diluted hydrogen peroxide can depolymerize alkali lignin and produce 45 weight % of succinic, formic, and acetic acids. Similarly, the organosolv lignin can also be converted to 20 weight % of these three organic acids by diluting hydrogen peroxide (Hasegawa et al., 2011). Hydrogen peroxide can specifically oxidise the β-O-4 and β-1 linkages, according to Jennings et al. (2017). They also demonstrate that a reduction in reaction temperature and time may improve reaction. But, hydrogen peroxide also contributes to excessive oxidation. This results in the ring-opening of aromatic or phenolic compounds and their transformation into alkylic compounds (Fernández-Rodríguez et al., 2017; Wang et al., 2018). Additionally, excessive oxidation makes a product difficult to control and less specific.

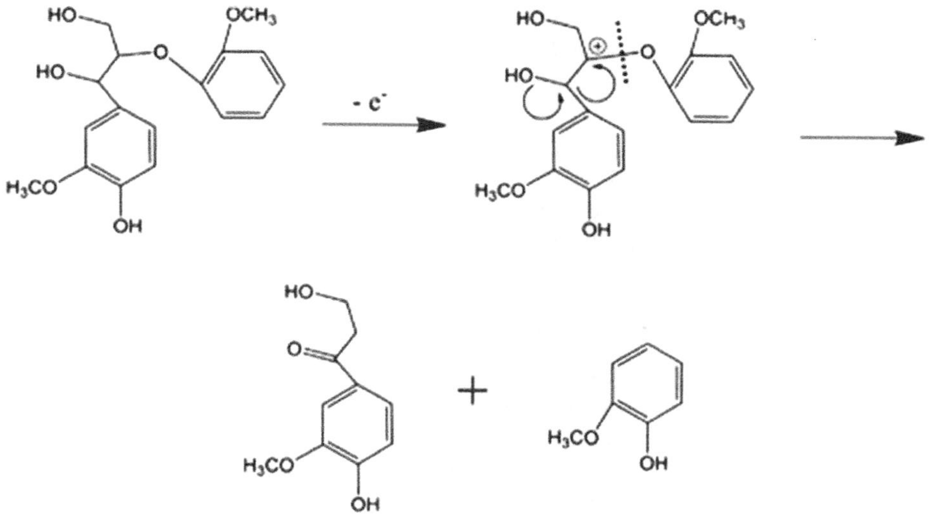

Figure 3.26 The mechanism of cleaving the β-O-4 linkages with the oxidant. The Figure is showing one of the potential products. Actual products can be varied based on the oxidative degree. Reproduced with permission from Chio et al. (2019) / ELSEVIER.

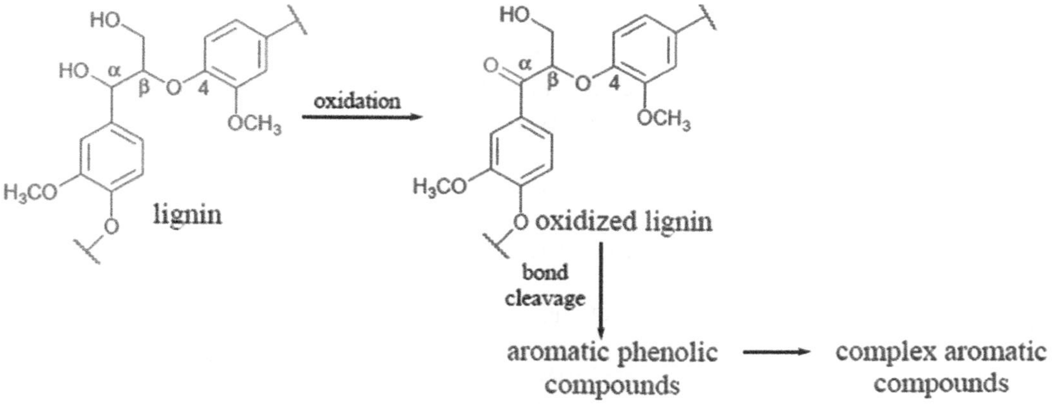

Figure 3.27 Oxidative depolymerization mechanism of lignin. Roy et al. (2021a) / ELSEVIER/ CC BY 4.0.

Bibliography

Abdelaziz OY, Meier S, Prothmann J, Turner C, Riisager A, and Hulteberg CP (2019). Oxidative depolymerisation of lignosulphonate lignin into low-molecular- weight products with Cu–Mn/_-Al2O3. *Top Catal*, 62: 639–648.

Ahmad Z, Paleologou M, and Xu CC (2021). Oxidative depolymerization of lignin using nitric acid under ambient conditions. *Ind Crops Prod*, 170: 113757. https://doi.org/10.1016/j.indcrop.2021.113757.

Bjeli´c S, Garbuio L, Arturi KR, van Bokhoven JA, Jeschke G, and Vogel F, (2018). Oxidative biphasic depolymerization (BPD) of kraft lignin at low pH. *Chemistry Select*, 3: 11680–11686.

Bjørsvik H-R and Minisci F (1999). Fine chemicals from lignosulfonates. 1. Synthesis of vanillin by oxidation of lignosulfonates. *Org Proc Res Develop*, 3(5): 330–340.

Calvo-Flores FG and Dobado JA (2010). Lignin as renewable raw material. *Chem Sus Chem*, 3(11): 1227–1235.

Chio C, Sain MM, and Qin W (2019). Lignin utilization: a review of lignin depolymerization from various aspects. *Renewable Sustainable Energy Rev*, 107: 232–249.

Crestini C, Caponi MC, Argyropoulos DS, and Saladino R (2006). Immobilized methyltrioxo rhenium (MTO)/H2O2 systems for the oxidation of lignin and lignin model compounds. *Bioorganic Med Chem*, 14: 5292–5302.

Cronin DJ, Zhang X, Bartley J, and Doherty WOS (2017). Lignin depolymerization to dicarboxylic acids with sodium percarbonate. *ACS Sus Chem Eng*, 5(7): 6253–6260.

Danby AM, Lundin MD, and Subramaniam B (2018). Valorization of grass lignins: swift and selective recovery of pendant aromatic groups with ozone. *ACS Sus Chem Engineer*, 6(1): 71–76.

Das L, Kolar P, Sharma-Shivappa R, Classen JJ, and Osborne JA (2017). Catalytic valorization of lignin using niobium oxide. *Waste Biomass- Valor*, 8: 2673–2680.

Das L, Kolar P, and Shivappa RS (2012). Heterogeneous catalytic oxidation of lignin into value-added chemicals. *Biofuels*, 3(2): 155–166.

Fache M, Boutevin B, and Caillol S (2016). Vanillin production from lignin and its use as a renewable chemical. *ACS Sus Chem Engineer*, 4(1): 35–46.

Fernández-Rodríguez J, Erdocia X, de Hoyos PL, Alriols MG, and Labidi J (2017). Small phenolic compounds production from kraft black liquor by lignin depolymerization with different catalytic agents. *Chem Eng*, 57.

Figueiredo P, Lintinen K, Hirvonen JT, Kostiainen MA, and Santos HA (2018). Properties and chemical modifications of lignin: towards lignin-based nanomaterials for biomedical applications. *Prog Mater Sci*, 93: 233–269. https://doi.org/10.1016/j.pmatsci.2017.12.001.

Hasegawa I, Inoue Y, Muranaka Y, Yasukawa T, and Mae K (2011). Selective production of organic acids and depolymerization of lignin by hydrothermal oxidation with diluted hydrogen peroxide. *Energy Fuels*, 25: 791–796.

Irmak S, Kang J, and Wilkins M (2020). Depolymerization of lignin by wet air oxidation. *Bioresour Technol Rep*, 9: 100377.

Jennings JA, Parkin S, Munson E, Delaney SP, Calahan JL, Isaacs M, Hong K, and Crocker M (2017). Regioselective Baeyer–Villiger oxidation of lignin model compounds with tin beta zeolite catalyst and hydrogen peroxide. *RSC Adv*, 7: 25987–25997.

Kishimoto T, Kadla JF, Chang H-m, and Jameel H (2003). The reactions of lignin model compounds with hydrogen peroxide at low pH. *Holzforschung*, 57: 52–88.

Kumar A, Biswas B, and Bhaskar T (2020). Effect of cobalt on titania, ceria and zirconia oxide supported catalysts on the oxidative depolymerization of prot and alkali lignin. *Bioresour Technol*, 299: 122589.

Laurichesse S and Avérous L (2014). Chemical modification of lignins: towards biobased polymers. *Prog Polym Sci*, 39(7): 1266–1290. https://doi.org/10.1016/j.progpolymsci.2013.11.004.

Li T and Takkellapati S (2018). The current and emerging sources of technical lignins and their applications. *Biofuels Bioprod Bioref*, 12(5): 756–787.

Lyu G, Yoo CG, and Pan X (2018). Alkaline oxidative cracking for effective depolymerization of biorefining lignin to mono-aromatic compounds and organic acids with molecular oxygen. *Biomass Bioenergy*, 108: 7–14.

Maluenda I, Chen M-T, Guest D, Mark Roe S, Turner ML, and Navarro O (2015). Room temperature, solvent less telomerization of isoprene with alcohols using (N-heterocyclic carbene)–palladium catalysts. *Cat Sci Tech*, 5(3): 1447–1451.

McClelland DJ, Galebach PH, Motagamwala AH, Wittrig AM, Karlen SD, Buchanan JS, Dumesic JA, and Huber GW (2019). Supercritical methanol depolymerization and hydrodeoxygenation of lignin and biomass over reduced copper porous metal oxides. *Green Chem*, 21(11): 2988–3005.

Min D, Xiang Z, Liu J, Jameel H, Chiang V, Jin Y, and Chang HM (2015). Improved protocol for alkaline nitrobenzene oxidation of woody and non-woody biomass. *J Wood Chem Technol*, 35(1): 52–61. https://doi.org/10.1080/02773813.2014.902965.

Mosier N, Wyman C, Dale B, Elander R, Lee YY, Holtzapple M, and Ladisch M (2005). Features of promising technologies for pretreatment of lignocellulosic biomass. *Bioresour Technol*, 96: 673–686.

Ngadi N, Halim NAA, and Ibrahim MNM (2014). Isolation and characterization of vanillin from coconut husk lignin via alkaline nitrobenzene oxidation. *J Teknol*, 67: 4.

Nyamunda BC, Chigondo F, Moyo M, Guyo U, Shumba M, and Nharingo T (2013). Hydrogen peroxide as an oxidant for organic reactions. *J At Mol*, 3: 23.

Ouyang X-P, Tan Y-D, and Qiu X-Q (2014). Oxidative degradation of lignin for producing monophenolic compounds. *J Fuel Chem Technol*, 42: 677–682.

Ragauskas AJ, Beckham GT, Biddy MJ, Chandra R, Chen F, Davis MF, Davison BH, Dixon RA, Gilna P, and Keller M (2014). Lignin valorization: improving lignin processing in the biorefinery. *Science*, 344: 1246843.

Rahimi A, Azarpira A, Kim H, Ralph J, and Stahl SS (2013). Chemoselective metal-free aerobic alcohol oxidation in lignin. *J Am Chem Soc*, 135: 6415–6418.

Rahimi A, Ulbrich A, Coon JJ, and Stahl SS (2014). Formic-acid-induced depolymerization of oxidized lignin to aromatics. *Nature*, 515: 249–252.

Roberts VM, Stein V, Reiner T, Lemonidou A, Li X, and Lercher JA (2011). Towards quantitative catalytic lignin depolymerization. *Chem Eur J*, 17(21): 5939–5948.

Rodrigues Pinto PC, Borges da Silva EA, and Rodrigues AE (2012). Lignin as source of fine chemicals: vanillin and syringaldehyde. *Biomass Conversion*, Baskar C, Baskar S, and Dhillon RS (eds.). Springer, Berlin Heidelberg, pp. 381–420.

Rødsrud G, Lersch M, and Sjöde A (2012). History and future of world's most advanced biorefinery in operation. *Biomass Bioenergy*, 46(0): 46–59.

Roy PS, Garnier G, Allais F, and Saito K (2021a). Effective lignin utilization strategy: major depolymerization technologies, purification process and production of valuable material. *Chem Lett*, 50(6): 1123–1130. https://doi.org/10.1246/cl.200873.

Roy R, Jadhav B, Rahman MS, and Raynie DE (2021b). Characterization of residue from catalytic hydrothermal depolymerization of lignin. *Curr Res Green Sustain Chem*, 4: 100052.

Roy R, Rahman MS, Amit TA, and Jadhav B (2022). Recent advances in lignin depolymerization techniques: a comparative overview of traditional and greener approaches. *Biomass*, 2: 130–154. https://doi.org/10.3390/biomass2030009.

Schutyser W, Kruger JS, Robinson AM, Katahira R, Brandner DG, Cleveland NS, Mittal A, Peterson DJ, Meilan R, Román-Leshkov Y, and Beckham GT (2018b). Revisiting alkaline aerobic lignin oxidation. *Green Chem*, 20(16): 3828–3844.

Schutyser W, Renders T, Van den Bosch S, Koelewijn SF, Beckham GT, and Sels BF (2018a). Chemicals from lignin: an interplay of lignocellulose fractionation, depolymerisation, and upgrading. *Chem Soc Rev*, 47(3): 852–908.

Smith C, Utley JH, Petrescu M, and Viertler H (1989). Biomass electrochemistry: anodic oxidation of an organo-solv lignin in the presence of nitroaromatics. *J Appl Electrochem*, 19: 535–539.

Stärk K, Taccardi N, Bösmann A, and Wasserscheid P (2010). Oxidative depolymerization of lignin in ionic liquids. *ChemSusChem*, 3: 719–723.

Sun Z, Fridrich B, de Santi A, Elangovan S, and Barta K (2018). Bright side of lignin depolymerization: toward new platform chemicals. *Chem Rev*, 118(2): 614–678.

Vangeel T, Schutyser W, Renders T, and Sels BF (2018). Perspective on lignin oxidation: advances, challenges, and future directions. *Top Curr Chem (Cham)*, 376(4): 30–16.

Vidal J-P (2000). Vanillin. *Kirk-Othmer Encyclopedia of Chemical Technology*, Othmer K. (ed.). John Wiley & Sons, Inc, New Jersey.

Wang Q, Tian D, Hu J, Shen F, Yang G, Zhang Y et al. (2018). Fates of hemicellulose, lignin and cellulose in concentrated phosphoric acid with hydrogen peroxide (PHP) pretreatment. *RSC Adv*, 8: 12714–12723.

Wu G and Heitz M (1995). Catalytic mechanism of Cu^{2+} and Fe^{3+} in alkaline O_2 oxidation of lignin. *J Wood Chem Technol*, 15: 189–202.

Xiang Q and Lee YY (2000). Oxidative cracking of precipitated hardwood lignin by hydrogen peroxide. *Abab*, 84-86: 153–162. https:doi.org/10.1385/abab:84-86:1-9:153.

Xu C, Arancon RA, Labidi J, and Luque R (2014). Lignin depolymerisation strategies: towards valuable chemicals and fuels. *Chem Soc Rev*, Nov 21; 43(22): 7485–7500.

Zaid A (2019). *Highly efficient depolymerization of kraft lignin (KL) and hydrolysis lignin (HL) via hydrolysis and oxidation*. Electronic Thesis and Dissertation Repository. 6171. https://ir.lib.uwo.ca/etd/6171 (accessed December, 2021).

Zakzeski J, Bruijnincx PCA, Jongerius AL, and Weckhuysen BM (2010). The catalytic valorization of lignin for the production of renewable chemicals. *Chem Rev*, 110(6): 3552–3599.

Zhang C, Li H, Lu J, Zhang X, MacArthur KE, Heggen M, and Wang F (2017). Promoting lignin depolymerization and restraining the condensation via an oxidation- hydrogenation strategy. *ACS Catal*, 7: 3419–3429.

Zhou N, Thilakarathna WPDW, He QS, and Rupasinghe HPV (2022). A review: depolymerization of lignin to generate high-value bio-products: opportunities, challenges, and prospects. *Front Energy Res*, 9: 758744.

3.5 Microwave-aided Depolymerization

The one-of-a-kind heating potential of microwave-aided technology makes it appealing (Asmadi et al., 2011). When comparison is made to traditional heating methods, this is a more rapid alternative for degrading samples. The speeding up of chemical reactions, which results in shorter reaction times under milder conditions, is one advantage of microwave heating technologies over traditional heating techniques (Aguilar-Reynosa et al., 2017; de la Hoz et al., 2005; Dhar and Vinu, 2017).

Microwave-aided depolymerization of lignin is recently receiving more attention than conventional heating processes because of its higher heating efficiency and quick heating (Cederholm et al., 2020; Liu et al., 2017; Tayier et al., 2017; Tsodikov et al., 2018; Xiaoli et al., 2012; Zhou et al., 2019; Zou et al., 2018). The use of the microwave in the process of converting biomass or lignin has been the subject of decades of research (Beneroso et al., 2017; de la Hoz et al., 2005; Goñi and Montgomery, 2000). Monomeric and oligomeric products are formed efficiently by solvolysis of lignin with the aid of microwave in a variety of organic solvents, oxidants, acids, and bases (Dong and Xiong, 2014; Duan et al., 2018a, 2018b, 2018c; Kim and Park, 2013; Ouyang et al., 2014). The biomass molecule and lignin can be penetrated by high-energy electromagnetic radiation from a microwave. Ionic conduction and rotation of polar molecules are two potential effects of these radiations, which can also generate a significant amount of heat (Yunpu et al., 2016). As a result, in comparison to conventional heating, microwave can reduce reaction times and prevent surface overheating by avoiding the physical contact between the material and the heating source (de la Hoz et al., 2005; Lam et al., 2017). Additionally, the microwave has been suggested as a low-cost method for heating and pyrolyzing biomass (Duan et al., 2017a, 2017b; Lam et al., 2017; Liew et al., 2019).

The lignin is broken down into value-added products using microwave electromagnetic radiation (Beneroso et al., 2017). The materials and the heat source do not need to come into physical contact with one another in this method. It has the potential to generate a significant amount of heat by rotating the polar molecules and causing ionic conduction (Lam et al., 2017; Yunpu et al., 2016). Biomass and lignin can be quickly and economically transformed into valuable products using microwave-assisted technology (Liew et al., 2019).

Table 3.18 summarizes the process conditions and the results of microwave-assisted depolymerization of lignin (Roy et al., 2022).

Numerous studies have revealed that microwave-aided depolymerization outperforms conventional technology in several ways. It is quick, very effective, uniform, selective, and good for the environment (Li et al., 2012). Due to its capability of instantaneous start and stop, for large-scale industries, this method's heating system is advantageous. Polar materials preferentially absorb energy, allow raw materials to penetrate the interior, significantly shorten processing times, and ultimately increase reaction rates (Zlotorzynski, 1995).

Non-thermal effects can be added to reactions using this technology, speeding up the process (Lidström et al., 2001).

Gedye et al. (1988) have shown that the reaction rates of microwave-aided reactions were significantly higher than those of traditional methods. By using a metal catalyst at lower temperatures to break the condensed bonds of lignin, microwave-aided depolymerization can increase the selectivity of the reaction.

Zhu et al. (2017) used ferric sulfate as the catalyst in this method for depolymerizing lignin into phenolic monomers by selectively breaking the $C\alpha$-$C\beta$ linkages. According to their research, the microwave method was able to selectively cleave the $C\alpha$-$C\beta$ bonds in phenolic as well as non-phenolic dimers, with the phenolic dimer having a substantially higher cleavage rate than the

Table 3.18 Reaction conditions and the products of microwave-aided depolymerization of lignin.

Kraft lignin
900–1240 K, 1.5–2.7 kW
Guaiacols
Farag et al., 2014
Olive tree pruning lignin
Formic acid solvent, 400 W, 30 min, 140 °C
Diethyl phthalate
Toledano et al., 2014
Softwood kraft Lignin
1.5 kW, 800 s
Guaiacol, 4-methylguaiacol
Fu et al., 2014
Black liquor lignin
Formic acid solvent, 600 W, 130 °C, 30 min
2-propanone, 1-(4-hydroxy-3-methoxyphenyl), ethanone, 1-(4-hydroxy-3,5-dimethoxyphenyl), and dibutylphthalate
Shen et al., 2015
Wheat straw lignin
10 weight % H_2SO_4 + 10% by weight phenol, 100–180 °C, 10–60 min, 300 W
Mono phenolic compounds
Ouyang et al., 2015
Wheat alkali lignin
100–160 °C, 0.5–120 min, 400 W
Benzylic alcohols
Zhu et al., 2016.
Black liquor lignin
100–180 °C, 5–60 min, 600 W
Ethanone, 1-(4-hydroxy-3-methoxyphenyl) and ethanone, 1-(4-hydroxy-3,5-dimethoxy phenyl)
Liu et al., 2017
Organosolv lignin
80 °C, 10 min, 80 W
vanillic acid and vanillin, 4-hydroxybenzaldehyde
Dai et al., 2018
Kraft lignin
110–150 °C, 10 min, 1.60 kW
Tetra phenolic compounds
Muley et al., 2019

Table 3.18 (Continued)

Birch sawdust lignin
150–210 °C, 15 min, 1800 W
4-propyl and 4-propenyl syringol and guaiacol
Liu et al., 2021
Kraft lignin
225 °C, 1 h 500–1000 W
Vanillin, isoeugenol-2, and homovanillic acid
Agarwal et al., 2021

Roy et al. (2022) / MDPI / CC BY 4.0.

non-phenolic dimer. Additionally, the metal catalyst prevented the formation of char and significantly reduced the activation energy of the depolymerization reaction (Xu et al., 2012).

Microwave-assisted depolymerization may also play a significant role in the solvolysis or liquefaction of lignin because of its high efficiency and fast heating rate. Using this method, in ionic liquids, and at a reduced temperature, lignin can be dissolved quickly (Merino et al., 2018).

Furthermore, 1-butyl-3-methylimidazolium hydrogen sulfate, which is an ionic liquid, can be reused five times without any loss of its catalytic activity (Pan et al., 2014). Because of this, it works well as a catalyst for depolymerizing lignin.

Liew et al. (2019) demonstrated that micro-wave pyrolysis could be a method for conversion of biomass into high-value, high-quality active carbon at a low cost. In addition, microwave has fewer mechanical units and can be precisely controlled throughout the process (Dai et al., 2017). According to previous research, using a microwave as a catalyst for a reaction can increase its selectivity. The effectiveness of microwave-assisted depolymerization is further compared to that of conventional heating by employing ferric sulfide as the catalyst. According to the findings the condense linkages Cα-Cβ can be specifically broken by microwave at temperatures below 160 °C, increasing the ratio of soluble fraction from 67% to 86% (Zhu et al., 2017).

The microwave can also be used as a heater for pyrolysis or pre-treatment prior to pyrolysis, despite the fact that it is generally utilized as a heating method for depolymerization of lignin in conjunction with other catalysts. The methoxyl group can be removed and weak links, like the ether bond, can be broken by using the microwave for pre-treatment. As a result, the formation of char during pyrolysis can be reduced. Additionally, Duan et al. (2018b) show that at 200 °C, 60 minutes, microwave-assisted pyrolysis of alkali lignin can decrease yields of guaiacols such as creosol and eugenol from 36.56% to 22.36% and increase yields of phenolic compounds such as catechol, 2-methyl phenol from 3.81% to 14.15%.

In addition to being used for depolymerizing lignin, the microwave can also help to liquefy or solubilize lignin by increasing the rate of heating. Ionic liquid can easily dissolve lignin at a lower temperature and in less time than with conventional heating methods because of the high heating rate (Merino et al., 2018).

The electromagnetic field may also have an impact on the chemical transformation and produce unique thermal effects that are impossible to achieve with conventional heating (Mazo et al., 2012). In addition, the microwave-converted biopolyols can be directly used in the polyurethane foam production process (Gosz et al., 2018; Xue et al., 2015). These findings demonstrate that the use of microwaves for liquefaction of lignin and production of biopolyol has a significant industrial application potential.

The electromagnetic spectrum includes higher frequency electromagnetic radiation with frequencies between 300 MHz and 300 GHz. Long-chain organic molecules' polar parts take these in and start making high-frequency oscillations inside the molecules (Zlotorzynski, 1995). Consequently, heat energy is directly converted from microwave energy at the molecular level (Toledano et al., 2014). This makes it more likely that the structure of lignin's C–C and β-O-4 ether bonds will be easily cleaved, lowering the activation energy (Ouyang et al., 2015). Over the past few decades, it has been shown that microwave heating technology can increase chemical reactions by reducing the activation energy and utilizing a novel internal non-thermal heating phenomenon to increase heat energy efficiency (Dong et al., 2014; Gedye et al., 1988; Ouyang et al., 2015). For instance, microwave irradiation markedly increased the degradation of lignin model substances like guaiacol and benzyl phenyl ether in ionic liquids based on imidazolium (Pan et al., 2014). Microwave heating may selectively contribute to the cleavage of 96.3% of the C–C bonds in organosolv lignin, primarily vanillin, methyl vanillate, syringaldehyde, and methyl syringate, while only generating a small number of aromatic monomers (Zhu et al., 2017). The benefits of microwave heating technology over other traditional thermochemical procedures based on the depolymerization of lignin include high energy efficiency, speed, low cost, selectivity, portability, and concern for the environment. However, in contrast to conventional thermochemical methods, microwaves frequently exhibit localized heating effects because of the lignin molecule's non-uniform structure (Yunpu et al., 2016). Organic compounds are subjected to microwave irradiation, which generates plasma that predominantly degrades C–H bonds rather than C–C bonds, as opposed to convective heating (Tsodikov et al., 2016). In contrast to convective heating, which transfers heat from the external to the internal surface of the biomass and results in excessive heat loss, microwave heating transfers heat generated from inside to outside of the biomass (Xie et al., 2018). Based on the conditions of the reaction, the microwave-aided depolymerization of lignin can be divided into solvolysis and pyrolysis. Solvolysis takes place at a lower temperature (200 °C) whereas pyrolysis takes place at a higher temperature (> 400 °C) (Dhar and Vinu, 2017).

The microwave-aided catalytic solvolysis of lignin is receiving a growing amount of attention in the recent years (Tayier et al., 2017; Tsodikov et al., 2018; Zou et al., 2018). It has been shown that microwave-aided solvolysis of lignin in different types of organic solvents, oxidants, acids, and bases effectively promotes the generation of monomeric and oligomeric products (Dong et al., 2014; Duan et al., 2018a, 2018b, 2018c; Kim and Park, 2013; Ouyang et al., 2014).

Under mild reaction conditions, microwave-assisted lignin degradation with oxidants like hydrogen peroxide, copper oxide, and ferric sulfate produced monophenolic compounds with yields of 11.86 and degradation rates of 90.88%, respectively (Ouyang et al., 2014). However, because of the opening of aromatic rings, hydrogen peroxide as a single oxidant achieves a high rate of lignin degradation while producing a low yield of monophenolic compounds. The side chain cleavage that is facilitated by the presence of cuprous ion and ferric ion in the oxidant system also improves hydrogen peroxide's oxidation capacity, which results in a higher yield of monophenolic compounds. Products are produced under microwave irradiation with low methoxyl group and high phenolic hydroxyl group content through soda lignin's oxidative degradation in a more constrained and lower molecular weight distribution as opposed to traditional heating (Ouyang et al., 2010).The microwave degradation of alkali lignin produced substantial quantity of phenolic compounds (20 weight %) which contained syringaldehyde, acetosyringone, guaiacol, anisole, and lignin dimers in the presence of organic solvents like dimethyl sulfoxide and dimethyl formamide. Under moderate reaction conditions, only 11% by weight of phenolic compounds were produced when ethylene glycol was present (Dhar and Vinu, 2017). Methyl alcohol and ethyl alcohol had higher conversion rates (84.86 and 84.22%) than butyl alcohol, ethanediol, and isopropyl alcohol during the microwave-aided solvolysis of organosolv lignin that is catalyzed by sulfuric acid (Duan et al., 2018c). This was ascribed to the methyl alcohol and ethyl alcohol's low molecular weight, which encouraged high permeability and fluidity. In addition, the bio-oil produced had a substantially lower molecular weight than other alcohols.

The depolymerization of lignin into aromatic compounds in isopropyl alcohol was studied by Liu et al. (2017). The reaction took 30 minutes at 120 °C, with a 45 weight %yield of liquid products. The production of oligomers, in contrast to monomers, initially decreased before significantly increasing as the temperature reached 120 °C. This demonstrates that lignin undergoes depolymerization to form oligomers, which are then broken down into monomers by hydrogenation.

The microwave-aided depolymerization of lignin is affected differently by various metal chlorides. In comparison to a number of different metal chloride catalysts, such as magnesium chloride, aluminium chloride, ferric chloride, and zinc chloride, manganese chloride catalyzes lignin in the hydrochloric acid and formic acid system, producing a total monomeric yield of 48.7% at mild reaction conditions (160 °C for 30 minutes) that includes 23.5% guaiacyl-type, 11.9% syringyl-type and 14.8% H-type monomer compounds. The order of the catalytic activity of various metal chlorides for producing phenolic monomers is as follows: manganese chloride < aluminium chloride < ferric chloride < zinc chloride < manganese chloride.

In contrast to alkali and alkaline metal chlorides, which are better at producing specific products only, microwave-aided lignin depolymerization yields a wider range of products than transition metal chloride catalysts. Ferric chloride, for instance, outperforms zinc chloride and manganese chloride in the production of monomeric compounds of the G, H, and S types. Multivalent element chlorides, like chromium chloride and manganese chloride, showed outstanding catalytic activity for the microwave-aided breakdown of lignin model compounds out of 14 different types of metal chlorides, including alkali, alkaline, and transition metal chlorides (vanillyl alcohol, 4-hydroxybenzyl alcohol, and 2-phenoxy-1-phenylethanol) (Pan et al., 2015).

When these multivalent elements react with hydrogen peroxide, they oxidize to a higher oxidation state and produce hydroxyl radicals. While aromatic radicals are produced when ions in a higher oxidation state capture electrons from alcohols, hydroxyl radicals react with the alcoholic group. Aldehydes are produced when aromatic radicals and hydroxyl radicals combine. Sulfuric acid has the highest catalytic activity and the quickest rate of thermal decomposition among several acids (phosphoric acid, formic acid, hydrochloric acid, and sulfuric acid) in the microwave-assisted depolymerization of organosolv lignin. Additionally, it produces liquids with low molecular weights (Tayier et al., 2017).

It is common knowledge that the rate of acid-catalyzed β-O-4 bond cleavage in lignin is significantly sped up when a phenolic hydroxyl group is present and that the distribution of the product also changes when a methoxyl group is present on the phenyl ring (Sturgeon et al., 2014). As the temperature of microwave rises from 100–200 °C, the number average molecular weight and weight average molecular weight of the lignin degradation products decrease, efficiently promoting depolymerization of lignin. However, lignin depolymerization is minimally affected by exceeding a predetermined threshold for the reaction time (Duan et al., 2017a, 2017b). Lignin depolymerizes through oxidative cleavage under the influence of low-power microwave radiation, with simultaneous repolymerization and depolymerization reactions. Higher temperatures and a longer reaction time during microwave-aided lignin depolymerization encourage the recondensation of degraded lignin fragments. Therefore, avoiding recondensation reactions is essential for achieving a higher yield of monophenolic aromatic compounds (Ouyang et al., 2015).

During the microwave-assisted lignin degradation, certain materials with high dielectric loss tangents, like carbon, metal nanoparticles, and iron, cobalt, and nickel carbides, are frequently added to make efficient use of microwave irradiation because lignin does not have enough dielectric losses to reach the cracking temperature (Tsodikov et al., 2016; Wen et al., 2011).

Nickel-containing nanoparticles and nanosized iron dispersed on the surface of the lignin significantly enhance microwave absorption as well as higher rate of dehydrogenation during dry reforming of the lignin (Tsodikov et al., 2018, 2017).

Syngas and hydrogen are produced by intensive carbon dioxide reforming. Under microwave activation, functionalized mesoporous SBA-15 has been extensively utilized as an effective catalyst to oxidize lignin and its model compounds (Badamali et al., 2009, 2013). The microwave-aided oxidation of lignin is catalyzed by the surface hydroxyl groups and internal pore structure, as well as by silanol groups acting as active sites (Badamali et al., 2013).

The type of metal catalyst, hydrogen-donating solvent, and microwave irradiation all have an impact on the nature and composition of the phenolic compounds produced. Nickel (10 weight %) nanoparticles over mesoporous Al-SBA-15, for instance, successfully undergo depolymerization of lignin and produce the least biochar and residual lignin content compared to other metal nanoparticles. (Toledano et al., 2014).

Furthermore, after a brief microwave irradiation period of normally 30 minutes, the mesoporous catalyst supported by nickel nanoparticles and Al-SBA-15 prevents repolymerization and yields bio-oil at a maximum of 30 weight %. However, the highest bio-oil yield of up to 82.88 weight % is obtained by catalytic solvolysis of lignin over modified HUSY in formic acid at 130 °C and the same microwave irradiation time (Shen et al., 2015). It is common knowledge that the HUSY catalyst's acidic sites (Si/Al ratio) and proper pore structure (average 2.45 nm) encourage the formation of aromatic monomers. A novel microwave-assisted two-step method for depolymerizing lignin has recently been developed (Zhu et al., 2016). In it, the first step is to methylate the benzylic alcohols that are present in the lignin. The hydrogenolysis of the β-O-4 bond on a palladium/carbon catalyst substantially enhanced the selectivity of products and reduced the amount of oxygen in the aromatic monomers.

In comparison to conventional pyrolysis, microwave lignin pyrolysis is more swift, economical, energy-efficient, and highly selective. According to a numerical analysis of the microwave-aided pyrolysis of lignin, the reaction zone when the sample is treated with microwaves extends from the centre to the outside (Gadkari et al., 2017). The microwave pyrolysis of biomass, like wood, microalgae, and sewage sludge, is receiving a lot of attention (Huang et al., 2017; Luo et al., 2017; Zhou et al., 2018) but there have been very few studies on microwave-assisted lignin pyrolysis (Bu et al., 2014; Duan et al., 2017a, 2017b; Fan et al., 2017).

As a microwave absorber, activated charcoal is frequently utilized to raise the pyrolytic temperature. Nearly 87% of the bio-oil contained hydrocarbons, esters, phenols, and guaiacols produced during the catalytic microwave pyrolysis of lignin with activated carbon (Bu et al., 2014). The highest possible concentration of phenol and phenolic compounds was achieved under ideal circumstances at 550 °C and 2.18 h-1 WHSV (weight hourly space velocity). By raising the pyrolysis temperature during the microwave pyrolysis of kraft lignin, the overall bio-oil yield increased (Farag et al., 2014). Lignin is a biomass devoid of hydrogen with a very low ratio of hydrogen to carbon (H/Ceff), which limits the production of advanced bio-oil and results in an excessive amount of coke being produced during pyrolysis (Zhang et al., 2015).

Microwave-aided co-pyrolysis of lignin with hydrogen-rich materials, like low-density, polypropylene, polyethylene and soapstock (waste oils/fats from animal fat and vegetable oil), has attracted a lot of interest for improving the quality of bio-oil (2018a, 2018b; Duan et al., 2017a, 2018c, 2017b; Fan et al., 2017). While simultaneously reducing char formation from 32.44 to 24.35 weight %, co-pyrolysis of acid-treated lignin and soapstock under microwave irradiation for 60 minutes at a temperature of 150 °C produces a bio-oil with a lot of gasoline hydrocarbons and aromatics (Duan et al., 2018a).

Through the thermal depolymerization of soapstock, hydrogen remains available for oxygenates obtained from lignin during co-pyrolysis. The dehydrogenation and condensation of aromatic hydrocarbons are slowed down as a result, thereby preventing the formation of char and speeding up the production of aromatic compounds (Fan et al., 2017). The microwave-aided pre-treatment of lignin in the presence of acid produces two different types of free radicals and intermediates of the carbenium ion during demethoxylation and the breaking of β-O-4 bonds. The microwave pyrolysis of acid-pretreated lignin yields more monophenols and fewer guaiacols when compared to untreated lignin (Duan et al., 2018b). Maximum bio-oil yield was achieved by microwave-aided catalytic copyrolysis of lignin using low-density polyethylene in a 3:1 ratio over HZSM-5 and magnesium oxide catalyst at 500 °C (Fan et al., 2017). By complete conversion of methoxyl phenols into phenol and alkylated phenols, LDPE improves the bio-oil's quality and serves as a good hydrogen donor in this situation. Alkylated phenols are also reduced when the HZSM-5 to magnesium oxide catalyst ratio is increased, while aromatic compounds are produced in greater quantities. This is because the HZSM-5 pores favor aromatic ring structures with six members (Fan et al., 2018). During co-pyrolysis, HZSM-5 pores serve as catalytic sites for the aromatization of pyrolytic vapors that result in the formation of aromatic compounds. During microwave-aided pyrolysis of lignin's catalytic upgradation of vapors, the ratio of HZSM-5 to lignin was increased from 0 to 0.3 (Fan et al., 2018). The selectivity of methoxyl phenols diminished as a result and aromatics increased from 1.1 to 41.4%. Catalytic upgradation of pyrolytic vapors from the microwave pyrolysis of lignin over Co/ZSM-5 increased the quality as well as quantity of bio-oil, and the bio-oil mostly contained furans, phenols, ketones, and guaiacols, accounting for more than 85% of the bio-oil (Xie et al., 2018). In a microwave reactor, the co-pyrolysis of lignin with 1,4-butanediol reduces the average molecular weight of the liquid products while increasing the yield of bio-oil (Tarves Paul et al., 2017).

Additionally, the production of methoxy-phenols (guaiacols, syringols) is significantly preferred to that of non-methoxylated alkylphenols (phenol, cresol, etc.) when diol is added as a co-reactant during microwave-aided lignin pyrolysis.

Dai et al. (2018) looked into the possibility of hydroxyl radical production by oxidative depolymerization of lignin in the presence of copper sulfate and hydrogen peroxide. Within seven minutes, this method was found to effectively depolymerize lignin (Table 3.19). Copper salts, which are effectively soluble in water, enable depolymerization in water (as a solvent). Vanillin, vanillic acid, and 4-hydroxybenzaldehyde are the primary monomers that can be produced by employing this strategy. Many valuable compounds, including phenol, guaiacol, syringol, and catechol, can be extracted as phenol-mixtures from the oil using switchable hydrophilicity solvents, which is an intriguing method of microwave-aided pyrolysis of lignin. In the presence or absence of carbon dioxide, for instance, these solvents can alter their properties (Fu et al., 2014). The basicity of these solvents makes it possible to extract phenols, which are weak acids, from lignin. These phenols can then be recovered from these solvents by altering their properties. But, its commercial application has been constrained by the non-uniform heating caused by microwave irradiation and the low bulk diffusion rate.

Figure 3.28 shows microwave-aided catalytic depolymerization of lignin from birch sawdust to produce phenolic monomers using a hydrogen-free strategy (Liu et al., 2021).

Table 3.19 Depolymerization of lignin under microwave irradiation.

Lignin		
Vanillin	4-hydroxy Benzaldehyde	Syringaldehyde
Vanillic acid	4-hydroxy Benzoic acid	2-hydroxy-3-(4-Hydroxyphenyl) propanal

Based on Roy et al. (2021)

Figure 3.29 shows depolymerization of lignin to phenols via microwave-aided solvolysis process (Dhar and Vinu, 2017).

Figure 3.30 shows microwave processing of lignin in green solvents (Cederholm et al., 2020).

Figure 3.31 shows ultrasonic and microwave aided organosolv pre-treatment of pine wood for the production of pyrolytic sugars and phenols (Yang et al., 2020).

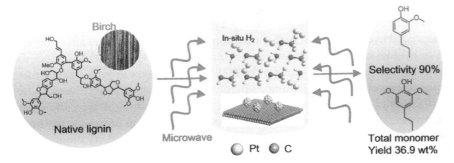

Figure 3.28 Microwave-assisted catalytic depolymerization of lignin from birch sawdust to produce phenolic monomers utilizing a hydrogen-free strategy. Reproduced with permission from Liu et al. (2021) / ELSEVIER.

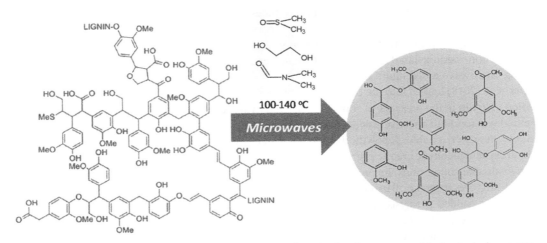

Figure 3.29 Understanding lignin depolymerization to phenols via microwave-assisted solvolysis process. Reproduced with permission from Dhar and Vinu (2017) / ELSEVIER.

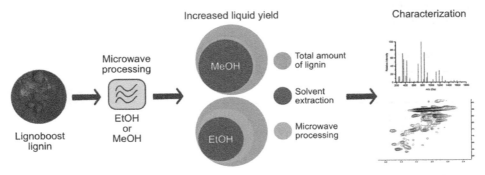

Figure 3.30 Microwave processing of lignin in green solvents: a high-yield process to narrow-dispersity oligomers Cederholm et al. (2020) / ELSEVIER / CC BY 4.0.

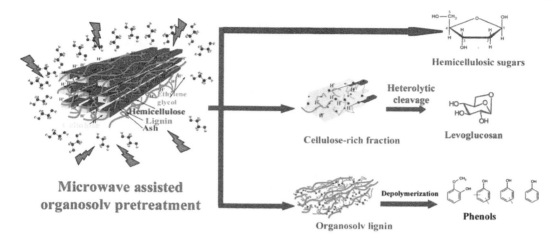

Figure 3.31 Ultrasonic and microwave assisted organosolv pre-treatment of pine wood for producing pyrolytic sugars and phenols. Adapted from Yang et al. (2020).

Bibliography

Agarwal A, Jo YT, and Park JH (2021). Hybrid microwave-ultrasound assisted catalyst-free depolymerization of Kraft lignin to bio-oil. *Ind Crops Prod*, 162: 113300.

Aguilar-Reynosa A, Romaní A, Rodríguez-Jasso RM, Aguilar CN, Garrote G, and Ruiz HA (2017). Microwave heating processing as alternative of pretreatment in second-generation biorefinery: an overview. *Energy Convers Manag*, 136: 50–65.

Asmadi M, Kawamoto H, and Saka S (2011). Gas-and solid/liquid-phase reactions during pyrolysis of softwood and hardwood lignins. *J Anal Appl Pyrolysis*, 92: 417–425.

Badamali SK, Luque R, Clark JH, and Breeden SW (2009). Microwave assisted oxidation of a lignin model phenolic monomer using Co(salen)/SBA-15. *Catal Commun*, 10: 1010–1013.

Badamali SK, Luque R, Clark JH, and Breeden SW (2013). Unprecedented oxidative properties of mesoporous silica materials: towards microwave-assisted oxidation of lignin model compounds. *Catal Commun*, 31: 1–4.

Beneroso D, Monti T, Kostas E, and Robinson J (2017). Microwave pyrolysis of biomass for bio-oil production: scalable processing concepts. *Chem Eng J*, 316: 481–498.

Bu Q, Lei H, Wang L, Wei Y, Zhu L, Zhang X, Liu Y, Yadavalli G, and Tang J (2014). Biobased phenols and fuel production from catalytic microwave pyrolysis of lignin by activated carbons. *Bioresour Technol*, 162: 142–147.

Cederholm L, Xu Y, Tagami A, Sevastyanova O, Odelius K, and Hakkarainen M (2020). Microwave processing of lignin in green solvents: a high-yield process to narrow-dispersity oligomers. *Ind Crops Prod*, 145: 112152.

Dai J, Styles GN, Patti AF, and Saito K (2018). CuSO4/H2O2-catalyzed lignin depolymerization under the irradiation of microwaves. *ACS Omega*, 3: 10433–10441.

Dai L, Fan L, Duan D, Ruan R, Wang Y, Liu Y, Zhou Y, Yu Z, Liu Y, and Jiang L (2017). Production of hydrocarbon-rich bio-oil from soapstock via fast microwave-assisted catalytic pyrolysis. *J Anal Appl Pyrolysis*, 125: 356–362.

de la Hoz A, Diaz-Ortiz A, and Moreno A (2005). Microwaves in organic synthesis. Thermal and non-thermal microwave effects. *Chem Soc Rev*, 34: 164–178.

Dhar P and Vinu R (2017). Understanding lignin depolymerization to phenols via microwave-assisted solvolysis process. *J Environ Chem Eng*, 5(2017): 4759–4768.

Dong C, Feng C, Liu Q, Shen D, and Xiao R (2014). Mechanism on microwave-assisted acidic solvolysis of black-liquor lignin. *Bioresour Technol Jun*, 162: 136–141.

Dong Q and Xiong Y (2014). Kinetics study on conventional and microwave pyrolysis of moso bamboo. *Bioresour Technol*, 171: 127–131.

Duan D, Ruan R, Lei H, Liu Y, Wang Y, Zhang Y, Zhao Y, Dai L, Wu Q, and Zhang S (2018a). Microwave-assisted co-pyrolysis of pretreated lignin and soapstock for upgrading liquid oil: effect of pretreatment parameters on pyrolysis behavior. *Bioresour Technol Jun*, 258: 98–104.

Duan D, Ruan R, Wang Y, Liu Y, Dai L, Zhao Y, Zhou Y, and Wu Q (2018b). Microwave-assisted acid pretreatment of alkali lignin: effect on characteristics and pyrolysis behavior. *Bioresour Technol*, 2018 Mar; 251: 57–62.

Duan D, Wang Y, Dai L, Ruan R, Zhao Y, Fan L, Tayier M, and Liu Y (2017a). Ex-situ catalytic co-pyrolysis of lignin and polypropylene to upgrade bio-oil quality by microwave heating. *Bioresour Technol*, 2017 Oct; 241: 207–213.

Duan D, Wang Y, Ruan R, Tayier M, Dai L, Zhao Y, and Liu Y (2018c). Comparative study on various alcohols solvolysis of organosolv lignin using microwave energy: physicochemical and morphological properties. *Chem Eng Process*, 126: 38–44.

Duan D, Zhao Y, Fan L, Dai L, Lv J, Ruan R, Wang Y, and Liu Y (2017b). Low-power microwave radiation-assisted depolymerization of ethanol organosolv lignin in ethanol/formic acid mixtures. *BioRes*, 12(3): 5308–5320.

Fan L, Chen P, Zhang Y, Liu S, Liu Y, Wang Y, Dai L, and Ruan R (2017). Fast microwave-assisted catalytic co-pyrolysis of lignin and low-density polyethylene with HZSM-5 and MgO for improved bio-oil yield and quality. *Bioresour Technol*, 2017 Feb; 225: 199–205.

Fan L, Chen P, Zhou N, Liu S, Zhang Y, Liu Y, Wang Y, Omar MM, Peng P, Addy M, Cheng Y, and Ruan RR (2018). In-situ and ex-situ catalytic upgrading of vapors from microwave-assisted pyrolysis of lignin. *Bioresource Technol*, 247: 851–858.

Farag S, Fu D, Jessop PG, and Chaouki J (2014). Detailed compositional analysis and structural investigation of a bio-oil from microwave pyrolysis of kraft lignin. *J Anal Appl Pyrolysis*, 109: 249–257.

Fu D, Farag S, Chaouki J, and Jessop PG (2014). Extraction of phenols from lignin microwave-pyrolysis oil using a switchable hydrophilicity solvent. *Bioresour Technol*, 154: 101–108.

Gadkari S, Fidalgo B, and Gu S (2017). Numerical investigation of microwave-assisted pyrolysis of lignin. *Fuel Process Technol*, 156: 473–484.

Gedye RN, Smith FE, and Westaway KC (1988). The rapid synthesis of organic compounds in microwave ovens. *Can J Chem*, 66: 17–26.

Goñi MA and Montgomery S (2000). Alkaline CuO oxidation with a microwave digestion system: lignin analyses of geochemical samples. *Anal Chem*, 72: 3116–3121.

Gosz K, Kosmela P, Hejna A, Gajowiec G, and Piszczyk Ł (2018). Biopolyols obtained via microwave-assisted liquefaction of lignin: structure, rheological, physical and thermal properties. *Wood Sci Technol*, 52: 599–617.

Huang F, Tahmasebi A, Maliutina K, and Yu J (2017). Formation of nitrogen-containing compounds during microwave pyrolysis of microalgae: product distribution and reaction pathways. *Bioresour Technol*, Dec; 245(Pt A): 1067–1074.

Kim HG and Park Y (2013). Manageable conversion of lignin to phenolic chemicals using a microwave reactor in the presence of potassium hydroxide. *Ind Eng Chem Res*, 52: 10059–10062.

Lam SS, Mahari WAW, Jusoh A, Chong CT, Lee CL, and Chase HA (2017). Pyrolysis using microwave absorbents as reaction bed: an improved approach to transform used frying oil into biofuel product with desirable properties. *J Clean Prod*, 147: 263–272.

Li M-F, Sun S-N, Xu F, and Sun R-C (2012). Microwave-assisted organic acid extraction of lignin from bamboo: structure and antioxidant activity investigation. *Food Chem*, 134: 1392–1398.

Lidström P, Tierney J, Watheyb B, and Westmana J (2001). Microwave assisted organic synthesis a review. *Tetrahedron*, 57: 9225–9283.

Liew RK, Chai C, Yek PNY, Phang XY, Chong MY, Nam WL, Su MH, Lam WH, Ma NL, and Lam SS (2019). Innovative production of highly porous carbon for industrial effluent remediation via microwave vacuum pyrolysis plus sodium-potassium hydroxide mixture activation. *J Clean Prod*, 208: 1436–1445.

Liu Q, Li P, Liu N, and Shen D (2017). Lignin depolymerization to aromatic monomers and oligomers in isopropanol assisted by microwave heating. *Polym Degrad Stab*, 135: 54–60.

Liu X, Bouxin FP, Fan J, Budarin VL, Hu C, and Clark JH (2021). Microwave-assisted catalytic depolymerization of lignin from birch sawdust to produce phenolic monomers utilizing a hydrogen-free strategy. *J Hazard Mater*, 402: 123490.

Luo H, Bao L, Kong L, and Sun Y (2017). Low temperature microwave-assisted pyrolysis of wood sawdust for phenolic rich compounds: kinetics and dielectric properties analysis. *Bioresource Technol*, Aug; 238: 109–115.

Mazo P, Estenoz D, Sponton M, and Rios L (2012). Kinetics of the Transesterification of castor oil with maleic anhydride using conventional and microwave heating. *J Am Oil Chem Soc*, 89: 1355–1361.

Merino O, Fundora-Galano G, Luque R, and Martínez-Palou R (2018). Understanding microwave-assisted lignin solubilization in protic ionic liquids with multiaromatic imidazolium cations. *ACS Sustain Chem Eng*, 6: 4122–4129.

Muley PD, Mobley JK, Tong X, Novak B, Stevens J, Moldovan D, Shi J, and Boldor D (2019). Rapid microwave-assisted biomass delignification and lignin depolymerization in deep eutectic solvents. *Energy Convers Manag*, 196: 1080–1088.

Ouyang X, Zhu G, Huang X, and Qiu X (2015). Microwave assisted liquefaction of wheat straw alkali lignin for the production of monophenolic compounds. *J Energ Chem*, 24: 72–76.

Ouyang XP, Lin ZX, Deng YH, Yang DJ, and Qiu XQ (2010). Oxidative degradation of soda lignin assisted by microwave irradiation. *Chin J Chem Eng*, 18(2010): 695–702.

Ouyang X-P, Tan Y-D, and Qiu X-Q (2014). Oxidative degradation of lignin for producing monophenolic compounds. *J Fuel Chem Technol*, 42: 677–682.

Pan J, Fu J, and Lu X (2015). Microwave-assisted oxidative degradation of lignin model compounds with metal salts. *Energy & Fuels*, 29: 4503–4509.

Pan JY, Fu J, Deng SG, and Lu XY (2014). Microwave-assisted degradation of lignin model compounds in imidazolium-based ionic liquids. *Energy Fuel*, 28: 1380–1386.

Roy PS, Garnier G, Allais F, and Saito K (2021). Effective lignin utilization strategy: major depolymerization technologies, purification process and production of valuable material. *Chem Letters*, 50(6): 1123–1130. https://doi.org/10.1246/cl.200873.

Roy R, Rahman MS, Amit TA, and Jadhav B (2022). Recent advances in lignin depolymerization techniques: a comparative overview of traditional and greener approaches. *Biomass*, 2: 130–154. https://doi.org/10.3390/biomass2030009.

Shen D, Liu N, Dong C, Xiao R, and Gu S (2015). Catalytic solvolysis of lignin with the modified HUSYs in formic acid assisted by microwave heating. *Chem Eng J*, 270: 641–647.

Sturgeon M, Kim S, Lawrence K, Paton RS, Chmely SC, Nimlos MR, Foust TD, and Beckham GT (2014). A mechanistic investigation of acid-catalyzed cleavage of aryl-ether linkages: implications for lignin depolymerization in acidic environments. *ACS Sustain Chem Eng*, 2: 472–485.

Tarves Paul C, Mullen Charles A, Strahan Gary D, and Boateng Akwasi A (2017). Depolymerization of lignin via co-pyrolysis with 1,4-butanediol in a microwave reactor. *ACS Sustain Chem Eng*, 5(2017): 988–994.

Tayier M, Duan D, Zhao Y, Ruan R, Wang Y, and Liu Y (2017). Catalytic effects of various acids on microwave-assisted depolymerization of organosolv lignin. *Bioresources*, 13: 412–424.

Toledano A, Serrano L, Pineda A, Romero AA, Luque R, and Labidi J (2014). Microwave-assisted depolymerisation of organosolv lignin via mild hydrogen-free hydrogenolysis: catalyst screening. *Appl Catal B Environ*, 145: 43–55.

Tsodikov MV, Éllert OG, Nikolaev SA, Arapova OV, Bukhtenko OV, Maksimov YV, Kirdyankin DI, and Vasil'kov A (2018). Fe-containing nanoparticles used as effective catalysts of lignin reforming to syngas and hydrogen assisted by microwave irradiation. *J Nanopart Res*, 20: 1–15.

Tsodikov MV, Éllert OG, Nikolaev SA, Arapova OV, Konstantinov GI, Bukhtenko OV, and Vasil'kov A (2017). The role of nanosized nickel particles in microwave-assisted dry reforming of lignin. *Chem Eng J*, 309: 628–637.

Tsodikov MV, Konstantinov GI, Chistyakov AV, Arapova OV, and Perederii MA (2016). Utilization of petroleum residues under microwave irradiation. *Chem Eng J*, 292: 315–320.

Wen F, Zhang F, and Liu Z (2011). Investigation on microwave absorption properties for multiwalled carbon nanotubes/Fe/Co/Ni nanopowders as light weight absorbers. *J Phys Chem C*, 115: 14025–14030.

Xiaoli G, Cheng K, Ming H, Shi Y, and Zhongzheng L (2012). La-Modified SBA-15/H2O2 systems for the microwave assisted oxidation of organosolv beech wood lignin. *Maderas-Ciencia Y Tecnologia*, 14(2012): 31–42.

Xie W, Liang J, Morgan HM, Zhang X, Wang K, Mao H, and Bu Q (2018). Ex-situ catalytic microwave pyrolysis of lignin over Co/ZSM-5 to upgrade bio-oil. *J Anal Appl Pyrolysis*, 132(Complete): 163–170.

Xu W, Miller SJ, Agrawal PK, and Jones CW (2012). Depolymerization and hydrodeoxygenation of switchgrass lignin with formic acid. *ChemSusChem*, 5: 667–675.

Xue BL, Wen JL, and Sun RC (2015). Producing lignin-based polyols through microwave-assisted liquefaction for rigid polyurethane foam production. *Materials*, 8: 586–599.

Yang X, Cui C, Zheng A, Zhao Z, Chenyang W, Shengpeng X, Zhen H, and Guoqiang W (2020). Ultrasonic and microwave assisted organosolv pretreatment of pine wood for producing pyrolytic sugars and phenols. *Ind Crops Prod*, 1 December 2020, 157: 112921.

Yunpu W, Leilei D, Liangliang F, Shaoqi S, Yuhuan L, and Roger R (2016). Review of microwave-assisted lignin conversion for renewable fuels and chemicals. *J Anal Appl Pyrolysis*, 119: 104–113.

Zhang H, Xiao R, Nie J, Jin B, Shao S, and Xiao G (2015). Catalytic pyrolysis of black-liquor lignin by co-feeding with different plastics in a fluidized bed reactor. *Bioresour Technol Sep*, 192: 68–74.

Zhou J, Liu S, Zhou N, Fan L, Zhang Y, Peng P, Anderson E, Ding K, Wang Y, Liu Y, Chen P, and Ruan R (2018). Development and application of a continuous fast microwave pyrolysis system for sewage sludge utilization. *Bioresour Technol*, 2018 May; 256: 295–301.

Zhou M, Sharma BK, Li J, Zhao J, Xu J, and Jiang J (2019). Catalytic valorization of lignin to liquid fuels over solid acid catalyst assisted by microwave heating. *Fuel*, 239: 239–244.

Zhu G, Jin D, Zhao L, Ouyang X, Chen C, and Qiu X (2017). Microwave-assisted selective cleavage of CαCβ bond for lignin depolymerization. *Fuel Process Technol*, 161: 155–161.

Zhu G, Qiu X, Zhao Y, Qian Y, Pang Y, and Ouyang X (2016). Depolymerization of lignin by microwave-assisted methylation of benzylic alcohols. *Bioresour Technol*, 218: 718–722.

Zlotorzynski A (1995). The application of microwave radiation to analytical and environmental chemistry. *Crit Rev Anal Chem*, 2: 43–76.

Zou R, Zhao Y, Wang Y, Duan D, Fan L, Dai L, Liu Y, and Rongsheng R (2018). Microwave-assisted depolymerization of lignin with metal chloride in a hydrochloric acid and formic acid system. *Bioresources*, 13: 3704–3719.

3.6 Electrochemical Lignin Depolymerization

Recently, lignin raw material has been electrochemically depolymerized as a technique for creating valuable chemicals from its monomeric complement of components, a waste valorization process that has the potential to be environmentally friendly. Due to its robustness and economic viability, this technology for valorization is one of the most industrially promising ones (Agrawal et al., 2018). Electrochemical depolymerization, in contrast to the thermochemical process, uses less energy and frequently only needs milder reaction conditions (Marino et al., 2016). However, low yield and a lack of product selectivity remains a significant obstacle. Additionally, the electrochemical process's sustainability may be enhanced by using renewable energy as a power source (Stiefel et al., 2016a). The most common electrolyte for electrochemical lignin depolymerization has been water.

Depolymerization is possible through electrochemical processes in which electrons act as the reagent at lower temperatures and pressures. As a result, the conversions can be regarded as low-priced, reagent-free, and environment friendly procedures that can be conducted in conditions of milder reaction.

Figure 3.32 shows electrochemical/radical mechanisms during lignin degradation (Dier et al., 2017).

The range of produced products as well as the rates of electrochemical conversion are significantly influenced by particular process parameters, including temperature, the nature of the electrolyte and the electrodes. Electrochemical depolymerization involves a direct exchange of

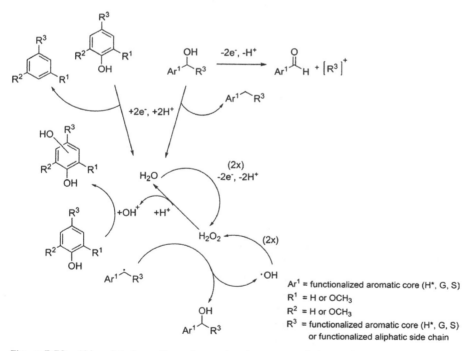

Figure 3.32 Abbreviated reaction scheme showing proposed electrochemical/radical mechanisms during lignin degradation. The numbers in parentheses give the equivalents of raw material needed for the reaction. Aromatic core units are defined as follows: (H*) 4-hydroxybenzyl, (G) 3-methoxy-4-hydroxybenzyl, (S) 3,5-dimethoxy-4- hydroxybenzyl. Dier et al. (2017). Creative Commons Attribution 4.0 International License https://doi.org/10.1038/s41598-017-05316-x.

electrons between the working electrode and the lignin component while the catalyst remains immobile on the surface of the electrode. By controlling the oxidation potential and time, electrochemical pathways can limit the oxidation products to a certain range of molecular weight (Movil-Cabrera et al., 2016).

Rates and ranges of produced products from electrochemical depolymerization are strongly influenced by the temperature, electrode material, current density, and electrolyte (Gao et al., 2017; Hao et al., 2015; Schmitt et al., 2015; Stiefel et al., 2016a, 2016b). Metals cannot be used as electrode materials because lignin cannot be dissolved and an electrically conducting solution cannot be obtained without a high pH. Additionally, products undergo excessive oxidation to undesirable organic acids and carbon dioxide at high pH (Lu et al., 2003). Lignin was depolymerized completely in less than four hours when polymeric nanoporous membranes and an electrochemical reactor were combined under normal conditions (Stiefel et al., 2016a). It is possible to remove the desired depolymerization products on the spot without causing excessive oxidation (Stiefel et al., 2015). Pt, Au, Ni, Cu, Cu/Ni-Mo-Co, DSA-O_2, Ti/TiO_2NT/PhO_2, RuO_2IrO_2/Ti, Pb/PbO_2, IrO_2, and so on are some of the electrodes examined for the possibility of electrochemical lignin oxidation (Cai et al., 2018; Hao et al., 2015; Pan et al., 2012; Parpot et al., 2000; Tolba et al., 2010; Zhu et al., 2014).

The kinetics of depolymerization are significantly influenced by the electrode's porosity (Stiefel et al., 2016a). For the electrochemical depolymerization of lignin, non-aqueous solvent systems are frequently utilized because parasitic oxygen evolution reduces the amount of current used in aqueous solutions (Reichert et al., 2012).

Aprotic and protic solvent systems of ionic liquids like triethylammonium methanesulfonate and 1-ethyl-3-methylimidazolium trifluoromethanesulfonate have been used as electrolytes that can be used again for the extensive electrochemical depolymerization of lignin (Dier et al., 2017).

An environmentally friendly alternative to other methods of lignin depolymerization is provided by the nearly complete recovery of the ionic electrolyte material obtained in the above process. Deep eutectic solvents, in addition to ionic liquids, have also been utilized for the electrochemical degradation of lignin because they are non-toxic, biodegradable, non-flammable, and inexpensive (Marino et al., 2016). The lignin molecule was found to be cleaved by electrochemically generated reactive oxygen species in aqueous electrolytes (Wang et al., 2017).

Aprotic ionic liquid 1-Butyl-3-methylimidazolium tetrafluoroborate [BMIM]BF4 was used to study the degradation of lignin model compound (p-benzyloxyl phenol) to determine the type of reactive oxygen species (% oxygen, % hydroperoxyl radical, or hydrogen peroxide) involved in the bond cleavage reaction. Hydroperoxyl radical transformed electrochemically generated superoxide radical (% oxygen) into hydrogen peroxide. The resulting hydrogen peroxide attacked the 4-C atom on the opposite side of the hydroxyl group, breaking the PBP molecule's alkyl-O-aryl bond (Wang et al., 2018).

By comprehending the kinetics of lignin depolymerization, one could control both the quantity and quality of the electrochemical degradation products.

Bawareth et al. (2018) proposed a model that can predict the distribution of lignin's molecular weight as a function of electrochemical processing time. This opens up the electrochemical lignin depolymerization process to a variety of commercial applications and provides an understanding of the kinetics of the process (Ragauskas et al., 2014).

Schmitt et al. (2015) utilized nickel and cobalt-based foams as electrodes and an anion exchange resin to adsorb vanillin at significant levels to develop a system for selective vanillin synthesis. More than half of the raw materials in other studies were broken down by polymer through unintentional conversions (Shao et al., 2014; Zhu et al., 2014).

In a short period of time, Stiefel et al. (2016a), were successful in reducing the initial lignin precursor's molecular weight by more than 93%. However, the non-self-destructive water-based system's constant cell voltage of 1.23 V limited the range of electrochemical reactions. The electrochemical voltage window of the electrolyte was extended by room temperature ionic liquids (RTIL) allowing for additional electrochemical reactions before the electrolyte began to degrade (Chatel and Rogers, 2014; Li et al., 2016, 2015; Shamsipur et al., 2010; Xu et al., 2014). RTIL have been utilized for the catalytic depolymerization of lignin. The RTIL's potential as electrolytes for the decomposition of lignin has also been enhanced by its thermal stability, capacity to stimulate free radical reactions, energy efficiency for processing biomass, and the capacity to dissolve a significant quantity of raw lignin (Brandt et al., 2013; Dier et al., 2016; Eshtaya et al., 2016; Hart et al., 2015; Kulkarni et al., 2007; Prado et al., 2013; Reichert et al., 2012; Yang et al., 2015).

For lower cell voltages (1–1.5 V), triethylammonium methanesulfonate as RTIL produced electrochemical decomposition rates of 3 to 6% (w/w), but these rates were found to increase to 20% (w/w) when the ionic liquid reached its upper limit (1.7 V) (Dier et al., 2016; Reichert et al., 2012). Along with typical breakdown reactions like β-O-4 bond cleavages, previously unidentified reductive reactions were also seen. After identifying the degradation products that resulted, the reductive mechanisms of dehydroxylation, demethoxylation, and hydrogenation were proposed.

Levulinic acid was investigated by da Cruz et al. (2022) as a possible medium for the reductive electrochemical depolymerization of lignin macromolecules. Levulinic acid was recognised by the US Department of Energy as one of the top 12 value-added chemicals from biomass because it is created during the hydrothermal process of lignocellulosic biomass. The aketone group in the structure of levulinic acid (Figure 3.33) is responsible for the substance's good ability to dissolve lignin. The hydrophilic and hydrophobic groups can both be resolved by this structure. The disaggregation of lignin macromolecules in solution is also made easier by the weakening and breaking of some intermolecular interactions. Additionally, the higher content of carbonyl groups makes it possible for electrochemical applications to benefit from improved electrical and protonic conductivities. In addition, the recovered lignin has a higher concentration of carbonyl groups, which may also play a significant role in increasing its electrical conductivity (Melro et al., 2020). Both of these outcomes make the utilization of levulinic acid as a long-lasting solvent for the electrocatalyzed fractionation of lignin advantageous. Both lignin's effective solvent and the reaction medium for the complex polyphenol's reductive electrochemical depolymerization can be levulinic acid.

da Cruz et al. (2022) found that monomers and dimers derived from lignin were produced after 20 hours of lignin depolymerization in levulinic acid. Aryl ether and phenol groups dominated the compounds produced as a result of this procedure. A crucial step in producing renewable chemicals for the fuel, lubricant, and coating industries is the extraction of these compounds from lignin. Utilizing a solvent based on biomass and copper, an inexpensive transition metal, as an electrocatalyst, these researchers showed a simple method for producing aromatic and phenolic compounds from an abundant industrial side-stream. Incorporating nature's biodegradation design into the production of functional biogenic materials, this method may hold promise.

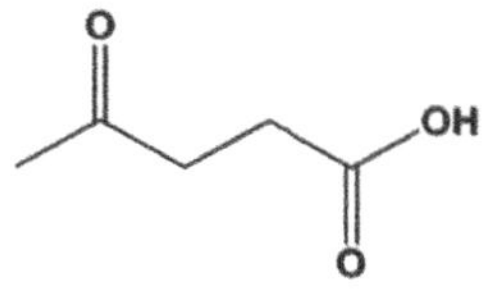

Direct injection high-resolution MS revealed the main levulinic acid-derived products of kraft lignin depolymerization (Figure 3.34).

Figure 3.33 A representation of the structure of levulinic acid.

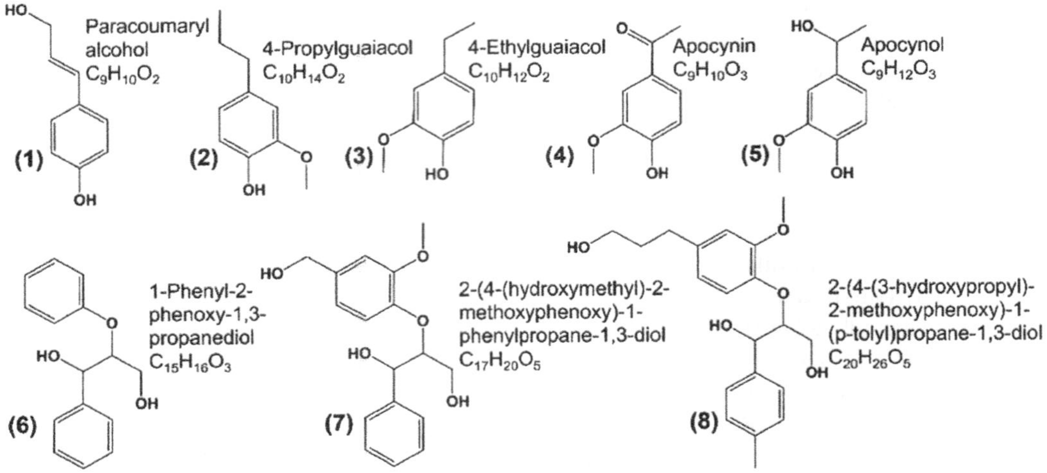

Figure 3.34 Main levulinic acid-derived products of kraft lignin depolymerization identified by direct injection high-resolution MS. da. Reproduced with permission da Cruz et al. (2022) / John Wiley & Sons.

Bibliography

Agrawal A, Rana M, and Park JH (2018). Advancement in technologies for the depolymerization of lignin. *Fuel Process Technol*, 181: 115–132.

Bawareth B, Di Marino D, Nijhuis TA, and Wessling M (2018). Unravelling electrochemical lignin depolymerization. *ACS Sustain Chem Eng*, 6: 7565–7573.

Brandt A, Gräsvik J, Hallett JP, and Welton T (2013). Deconstruction of lignocellulosic biomass with ionic liquids. *Green Chem*, 15: 550–583.

Cai P, Fan H, Cao S, Qi J, Zhang S, and Li G (2018). Electrochemical conversion of corn stover lignin to biomass-based chemicals between Cu/NiMoCo cathode and Pb/ PbO2 anode in alkali solution. *Electrochim Acta*, 264: 128–139.

Chatel G and Rogers RD (2014). Oxidation of lignin using ionic liquids – An innovative strategy to produce renewable chemicals. *ACS Sustain Chem Eng*, 2: 322–339.

da Cruz MGA, Gueret R, Chen J, Piątek J, Beele B, Sipponen MH, Frauscher M, Budnyk S, Rodrigues BVM, and Slabon A (2022). Electrochemical depolymerization of lignin in a biomass-based solvent. *ChemSusChem*, 2022 Aug 5; 15(15): e202200718.

Dier TKF, Egele K, Fossog V, Hempelmann R, and Volmer DA (2016). Enhanced mass defect filtering to simplify and classify complex mixtures of lignin degradation products. *Anal Chem*, 88: 1328–1335.

Dier TKF, Rauber D, Durneata D, Hempelmann R, and Volmer DA (2017). Sustainable electrochemical depolymerization of lignin in reusable ionic liquids. *Sci Rep*, 2017 Jul 11; 7(1): 5041.

Eshtaya M, Ejigu A, Stephens G, Walsh DA, Chena GZ, and Croft AK (2016). Developing energy efficient lignin biomass processing – Towards understanding mediator behaviour in ionic liquids. *Faraday Discuss*, 190: 127–145.

Gao WJ, Lam CM, Sun BG, Little RD, and Zeng CC (2017). Selective electrochemical CO bond cleavage of β-O-4 lignin model compounds mediated by iodide ion. *Tetrahedron*, 73(2017): 2447–2454.

Hao X, Quansheng Y, Dan S, Honghui Y, Jidong L, Jiang-tao F, and Wei YJ (2015). Fabrication and characterization of PbO2 electrode modified with [Fe(CN)6](3-) and its application on electrochemical degradation of alkali lignin. *J Hazard Mater*, 286: 509–516.

Hart WES, Harper JB, and Aldous L (2015). The effect of changing the components of an ionic liquid upon the solubility of lignin. *Green Chem*, 17: 214–218.

Kulkarni PS, Branco LC, Crespo JG, Nunes MC, Raymundo A, and Afonso CA (2007). Comparison of physicochemical properties of new ionic liquids based on imidazolium, quaternary ammonium, and guanidinium cations. *Chemistry*, 13(30): 8478–8488.

Li C, Zhao X, Wang A, Huber GW, and Zhang T (2015). Catalytic transformation of lignin for the production of chemicals and fuels. *Chem Rev*, 115: 11559–11624.

Li Q, Jiang J, Li G, Zhao W, Zhao X, and Mu T (2016). The electrochemical stability of ionic liquids and deep eutectic solvents. *Sci China Chem*, 59: 1–7.

Lu Z, Tu B, and Chen F (2003). Electro-degradation of sodium lignosulfonate. *J Wood Chem Technol*, 23: 261–277.

Marino DD, Stöckmann D, Kriescher S, Stiefel S, and Wessling M (2016). Electrochemical depolymerisation of lignin in a deep eutectic solvent. *Green Chem*, 18: 6021–6028.

Melro E, Filipe A, Valente AJM, Antunes FE, Romano A, Norgren M, and Medronho B (2020). Levulinic acid: a novel sustainable solvent for lignin dissolution. *Int J Biol Macromol*, 164: 3454–3461.

Movil-Cabrera O, Rodriguez-Silva A, Arroyo-Torres C, and Staser JA (2016). Electrochemical conversion of lignin to useful chemicals. *Biomass Bioenergy*, 88: 89–96.

Pan K, Tian M, Jiang ZH, Kjartanson B, and Chen AC (2012). Electrochemical oxidation of lignin at lead dioxide nanoparticles photoelectrodeposited on TiO2 nanotube arrays. *Electrochim Acta*, 60: 147–153.

Parpot P, Bettencourt AP, Carvalho AM, and Belgsir EM (2000). Biomass conversion: attempted electrooxidation of lignin for vanillin production. *J Appl Electrochem*, 30(2000): 727–731.

Prado R, Erdocia X, and Labidi J (2013). Lignin extraction and purification with ionic liquids. *J Chem Technol Biotechnol*, 88: 1248–1257.

Ragauskas AJ, Beckham GT, Biddy MJ, Chandra R, Chen F, Davis MF, Davison BH, Dixon RA, Gilna P, and Keller M (2014). Lignin valorization: improving lignin processing in the biorefinery. *Science*, 344: 1246843.

Reichert E, Wintringer R, Volmer DA, and Hempelmann R (2012). Electro-catalytic oxidative cleavage of lignin in a protic ionic liquid. *Phys Chem Chem Phys*, 14: 5214–5221.

Schmitt D, Regenbrecht C, Hartmer M, Stecker F, and Waldvogel SR (2015). Highly selective generation of vanillin by anodic degradation of lignin: a combined approach of electrochemistry and product isolation by adsorption. *Beilstein J Org Chem* Apr 13, 11: 473–480.

Shamsipur M, Beigi AAM, Teymouri M, Pourmortazavi SM, and Irandoust M (2010). Physical and electrochemical properties of ionic liquids 1-ethyl-3-methylimidazolium tetrafluoroborate, 1-butyl-3-methylimidazolium trifluoromethanesulfonate and 1-butyl-1-methylpyrrolidinium bis(trifluoromethylsulfonyl)imide. *J Mol Liq*, 157: 43–50.

Shao D, Liang J, Cui X, Xu H, and Yan W (2014). Electrochemical oxidation of lignin by two typical electrodes: Ti/SbSnO2 and Ti/PbO2. *Chem Eng J*, 244: 288–295.

Stiefel J, Lolsberg J, Kipshagen L, Moller-Gulland R, and Wessling M (2015). Controlled depolymerization of lignin in an electrochemical membrane reactor. *Electrochem Commun*, 61(2015): 49–52.

Stiefel S, Marks C, Schmidt T, Hanisch S, Spalding G, and Wessling M (2016b). Overcoming lignin heterogeneity: reliably characterizing the cleavage of technical lignin. *Green Chem*, 18: 531–540.

Stiefel S, Schmitz A, Peters J, Di Marino D, and Wessling M (2016a). An integrated electrochemical process to convert lignin to value-added products under mild conditions. *Green Chem*, 18: 4999–5007.

Tolba R, Tian M, Wen J, Jiang Z, and Chen A (2010). Electrochemical oxidation of lignin at IrO2-based oxide electrodes. *J Electroanal Chem*, 649: 9–15.

Wang L, Chen Y, Liu S, Jiang H, Wang L, Sunab Y, and Wan P (2017). Study on the cleavage of alkyl-O-aryl bonds by in situ generated hydroxyl radicals on an ORR cathode. *RSC Adv*, 7: 51419–51425.

Wang L, Liu S, Jiang H, Chen Y, Wang L, Duan G, Sun Y, Chen Y, and Wan P (2018). Electrochemical generation of ROS in ionic liquid for the degradation of lignin model compound. *J Electrochem Soc*, 165: H705–H710.

Xu C, Arneil R, Arancon D, Labidi J, and Luque R (2014). Lignin depolymerisation strategies: towards valuable chemicals and fuels. *Chem Soc Rev*, 43: 7485–7500.

Yang Y, Fan H, Song J, Meng Q, Zhou H, Wu L, Yang G, and Han B (2015). Free radical reaction promoted by ionic liquid: a route for metal-free oxidation depolymerization of lignin model compound and lignin. *Chem Commun*, 51: 4028–4031.

Zhu HB, Wang L, Chen Y, Li G, Li H, Tang Y, and Wan P (2014). Electrochemical depolymerization of lignin into renewable aromatic compounds in a non-diaphragm electrolytic cell. *RSC Adv*, 4: 29917–29924.

3.7 Reductive De-polymerization of Lignin

Reductive approaches typically make use of hydrogen gas or hydrogen-donor solvents to produce fuels and bio-oils (Ahmad, 2019; Akhtar and Amin, 2011; Bu et al., 2012; Mohan et al., 2006; Sun et al., 2018; Xiu and Shahbazi, 2012; Zacher et al., 2014).

The first works on structural elucidation used reductive treatment of lignin. Aliphatic compounds, primarily 4-propylcyclohexanol, were then isolated and characterized. CuCr catalysts were used to treat lignin in relatively severe conditions for reaction (250–260 °C, 220–240 bar) (Cooke et al., 1941; Godard et al., 1941; Harris et al., 1938).

Reductive methods necessitate catalysts that can selectively break C–O bonds, resulting in depolymerization (Chauvier and Cantat, 2017; Zaheer and Kempe, 2015).

One of the first reductive systems, at temperatures of 140–180 °C, with a high isolated yield for specific aromatic compounds has been described (Barta et al., 2014). In this study, organosolv lignin obtained from candlenut shells was depolymerized into aromatic monomers using copper-doped porous metal oxide and hydrogen gas at 50 bar. The main product was 4-propanolcatechol at 140 °C. It was isolated using column chromatography to isolate this. The yield was 43.3% and the yield of total monomers was 63.7%.

Another efficient method for producing chemical feedstocks and fuels from biomass or coal is the selective hydrogenolysis of aromatic carbon–oxygen bonds in aryl ethers, which results in the reductive depolymerization of lignin. According to Sergeev and Hartwig (2011), an NHC ligand and Ni (COD)2—bis(cyclooctadiene) nickel were able to cleave diarylethers very effectively in the presence of a base at a temperature of 80–100 °C under hydrogen atmosphere, of a 1 bar yielding excellent arene and phenol products.

A variety of hydride donors, such as aluminum hydride donors, silane reducing agents, and hydrogen gas with the addition of trimethyl aluminium, benzylic ethers and aryl-alkyl ethers, can be utilized, as stated by Sergeev and Hartwig (2011). For electron-removing substituents in aryl substrates, these catalytic systems were found to be more active. Due to the fact that formic acid is able to get decomposed into hydrogen gas, it was hypothesized that depolymerization of lignin with the use of formic acid as a solvent or catalyst also utilized the reductive depolymerization mechanism. For the reductive degradation of lignin, it has been demonstrated that formic acid, which is produced in situ from solvents that donate hydrogen, is more efficient than hydrogen gas sources from the outside (Huang et al., 2014). To selectively split C–O bonds during reductive depolymerization, effective catalysts are required.

Depolymerization-derived aromatic monomers undergo stepwise reductive deoxygenation and typically reduce the complexity of product mixtures and increase selectivity for particular aromatic compounds. Due to these unique characteristics, the approach makes use of structure elucidation (Barta et al., 2010).

Pd/C and formic acid as a reducing agent was used by Galkin et al. (2014) under mild reaction conditions (80 °C in air) for developing an effective catalyst system for the breakage of C–O bonds in lignin β-O-4 linkages of model compounds. Remarkably, this group of researchers discovered that the redox neutral cleavage of the β-O-4 linkage could be accelerated by adding catalytic amounts of a hydrogen source (such as formic acid, ammonium formate, 2-propanol, and sodium borohydride). Moreover, Barta et al. (2014) under relatively benign reaction conditions, reported an efficient lignin depolymerization reductive system with a higher isolated yield for particular aromatic compounds at 140–180 °C.

Monsigny et al. (2018) used hydrosilanes (R_3SiH) as a reductant and an iridium-based Brookhart's catalyst to efficiently reduce lignin model compounds and also hardwood and softwood lignins. Compared to $B(C_6F_5)_3$/hydrosilane system, this catalyst allowed for a convergent reductive depolymerization of wood lignins into isolable mono-aromatics, which was more stable and selective. A highly effective route into lignin solubilization is provided by the oxidative depolymerization of lignin. The isolation of pure aromatics remains a challenge with this strategy (Zakzeski et al., 2010). By substituting C–H bonds for C–O bonds, mild reductive routes, on the other hand, are extremely uncommon and would converge on monoaromatics derivatives of the constituent monolignols. As a result, this strategy requires additional research (Galkin et al., 2014).

A commercially available Pd/C catalyst can selectively cleave complex β-O-4 model compounds, resulting in acetophenone or ethylsubstituted arenes and phenols. The secondary alcohol was dehydrogenated in the β-O-4 moieties, and then the hydrogen produced in the first step was used to hydrogenolyze the alkyl C–O bond. When the method was applied to organosolv lignin, the generated hydrogen had to be supplemented because of the presence of olefins in natural lignin samples. *Miscanthus giganteus* acetonesolv lignins produced 12–15% combined yields of seven main products under optimal conditions, and pine lignin yielded 9% of alkyl-substituted phenols (4-ethylphenol M6P, 4-ethylsyringol M6S, 4-ethylguaiacol M6G) (Gao et al., 2016).

Bamboo lignin was depolymerized without using any external hydrogen source (Jiang et al., 2015). Raney nickel and zeolites increased the yield of phenolic monomers, primarily 4-hydroxy-3,5-dimethoxy-benzeneacetic acid M8S, 4-propylguaiacol M7G, and 4-allyl-2,6-dimethoxyphenol M9S (12.9% to 27.9%), as well as the bio-oil yield of more than 60 weight % under optimal conditions (270 °C, 1 atm nitrogen). Because this catalyst combination increased depolymerization efficiency while decreasing the production of undesirable high-molecular-weight polymers, these researchers came to the conclusion that there was a synergistic effect.

Konnerth et al. (2015) investigated the effect of pH levels between 1 and 14 on the reductive depolymerization of lignin by utilizing a Ni_7Au_3 catalyst and water. According to the findings, there was a positive correlation between the rate of hydrogenolysis and rising pH values. The addition of sodium hydroxide to organosolv lignin made from birch sawdust, a Ni_7Au_3 catalyst, and 10 bar hydrogen at 160 °C increased the total monomer yield from 7.6% to 10.9%. The most important products were 4-propanolsyringol (M10S), 4-propylguaiacol (M7G) and 4-propanolguaiacol (M10G). Depolymerization became more selective when a base was added. After being characterized using, ultraviolet-visible, transmission electron microscopy, and X-ray photoelectron spectroscopy, it was discovered that the catalyst itself remained chemically and structurally unchanged. The authors came to the conclusion that the basic reaction medium increases selectivity because the base makes it difficult for the bulky aromatic ring to coordinate with the catalyst, which prevents arene hydrogenation. Using sodium hydroxide as an additive significantly speeds up the depolymerization of organosolv lignin into aromatic monomers, which is important.

Singh and Ekhe (2015) looked into how solid acids affect depolymerization. These researchers created a one-pot method for producing alkyl phenols with a Cu/Mo-loaded ZSM-5 catalyst by using water as a cosolvent and methyl alcohol as a hydrogen donor. After seven hours at 220 °C, kraft lignin had almost completely transformed (>95%) and only produced a small quantity of char (0.5%). A GC-MS/FID was used to analyze the products. Cu/Mo-ZSM-5 was the catalyst for the reaction, which resulted in the production of 3-methoxy- 2,5,6-trimethyl phenol (M11) as the major product and higher selectivity (70.3%).

Si et al. (2017) discovered that treating woody biomass at 170–200 °C with water and tetrahydrofurfuryl alcohol (THFA) produces a high yield of lignin (77.4%) and 92.8% of good-quality cellulose simultaneously. Without the use of acid, high-quality lignin with a high retention of β-O-4 linkages was produced, making it suitable for catalytic treatment. which resulted in a high yield of aromatic monomers. A total monomer yield of 14.7% (primarily M10S and M10G) was achieved through hydrogenolysis at 220 °C using nickel/C.

Xiao et al. (2017) demonstrated a low-cost nanostructured MoOx/CNT catalyst that, in terms of activity, compatibility and reusability, with biomass feedstock, is on par with catalysts based on precious metals. Enzymatic mild acidolysis lignins produced a high yield of aromatic products (up to 47%). Unexpectedly, high yields of unsaturated monomeric phenols (M18S, M18G) were obtained.

Feghali et al. (2015) found that high yields of clearly defined aromatic products could be achieved by the first demonstration of lignin's reductive depolymerization at room temperature under metal-free conditions. Hydrosilanes were used as reductants in place of hydrogen gas, and $B(C_6F_5)_3$ served as a Lewis acid catalyst. This adaptable strategy worked well with various lignin species extracted using the formacell method, including 15 gymnosperm and angiosperm woods. The isolated yield ranged from 0.5 to 2.4 weight % for lignocellulose and 7 to 24 weight % for lignin, depending on the type of wood. These aromatic products (M12G, M12S, M13G, and M13S) were obtained with exceptional selectivity. For determining whether or not the depolymerization step was successful, estimating the maximum yield of monoaromatics from lignins is crucial.

A catalytic method developed by Shao et al. (2017) produced liquid aromatic hydrocarbons from lignin in a yield of 35.5 weight % and enabled the complete removal of oxygen. Surprisingly, when birch lignin was used, a nearly quantitative carbon yield was seen. The selectivity was as high as 71 weight % for arenes (methylbenzene M14, ethylbenzene M15, and propylbenzene M16). In water at 250 °C, organosolv lignin was hydrodeoxygenated directly over a porous Ru/Nb_2O_5 catalyst to produce arenes. The existence of an active Nb_2O_5 species was confirmed by combining calculations using density functional theory (DFT) and inelastic neutron scattering (INS). Strong adsorption, selective phenol activation, and a synergistic effect between the Ru and NbOx species were credited with the catalytic activity.

Luo et al. (2017) discovered that aliphatic alcohol moieties (CαH−OH) in lignin can serve as the hydrogen donor in addition to hydrogen gas or other reducing reagents. On $ZnIn_2S_4$, Lignin- β-O-4 linkages were first dehydrated to form a "hydrogen pool" and then hydrogen from the "hydrogen pool" was used to hydrogenolyze the C–O bond in close proximity. From organosolv lignin, this method yielded p-hydroxy acetophenone derivatives in a yield of 10% and phenols in a yield of 71–91% during the conversion of lignin β-O-4 models.

Figure 3.35 shows reductive depolymerization of kraft lignin to produce aromatics (Mankar et al., 2021).

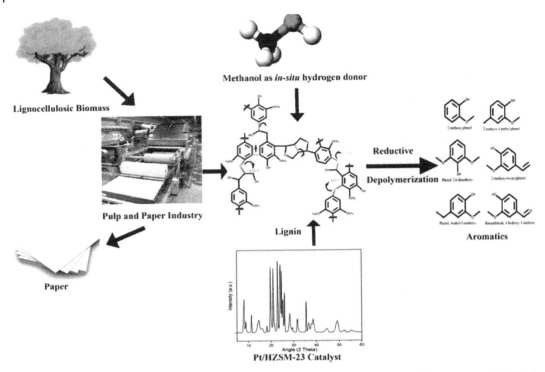

Figure 3.35 Reductive depolymerization of kraft lignin to produce aromatics. Akshay R. Mankar et al. 2021 / Reproduced with permission from Elsevier.

Bibliography

Ahmad Z (2019). Highly efficient depolymerization of kraft lignin (KL) and hydrolysis lignin (HL) via hydrolysis and oxidation. *Electronic Thesis and Dissertation Repository*. 6171. https://ir.lib.uwo.ca/etd/6171.

Akhtar J and Amin NAS (2011). A review on process conditions for optimum Bio-oil yield in hydrothermal liquefaction of biomass. *Renewable Sustainable Energy Rev*, 15: 1615–1624.

Barta K, Matson TD, Fettig ML, Scott SL, Iretskii AV, and Ford PC (2010). Catalytic disassembly of an organosolv lignin via hydrogen transfer from supercritical methanol. *Green Chem*, 12(9): 1640–1647.

Barta K, Warner GR, Beach ES, and Anastas PT (2014). Depolymerization of organosolv lignin to aromatic compounds over Cu-doped porous metal oxides. *Green Chem*, 16(1): 191–196.

Bu Q, Lei H, Zacher AH, Wang L, Ren S, Liang J, Wei Y, Liu Y, Tang J, Zhang Q, and Ruan R (2012). A review of catalytic hydrodeoxygenation of lignin-derived phenols from biomass pyrolysis. *Bioresour Technol*, Nov; 124: 470–477.

Chauvier C and Cantat T (2017). A view point on chemical reductions of Carbon–Oxygen bonds in renewable feedstocks including CO2 and biomass. *ACS Catal*, 7: 2107–2115.

Cooke LM, Mccarthy JL, and Hibbert H (1941). Studies on lignin and related compounds. LXI. Hydrogenation of ethanolysis fractions from maple wood (Part 2). *J Am Chem Soc*, 63: 3056–3061.

Feghali E, Carrot G, Thuéry P, Genre C, and Cantat T (2015). Convergent reductive depolymerization of wood lignin to isolated phenol derivatives by metal-free catalytic hydrosilylation. *Energy Environ Sci*, 8: 2734–2743.

Galkin MV, Sawadjoon S, Rohde V, Dawange M, and Samec JSM (2014). Mild heterogeneous palladium-catalyzed cleavage of β-O-4'- ether linkages of lignin model compounds and native lignin in air. *ChemCatChem*, 6: 179–184.

Gao F, Webb JD, Sorek H, Wemmer DE, and Hartwig JF (2016). Fragmentation of lignin samples with commercial Pd/C under ambient pressure of hydrogen. *ACS Catal*, 6: 7385–7392.

Godard HP, Mccarthy JL, and Hibbert H (1941). Studies on lignin and related compounds. LXII. High pressure hydrogenation of wood using copper chromite catalyst (Part 1). *J Am Chem Soc*, 63: 3061–3066.

Harris EE, D'Ianni J, and Adkins H (1938). Reaction of hardwood lignin with hydrogen. *J Am Chem Soc*, 60: 1467–1470.

Huang S, Mahmood N, Tymchyshyn M, Yuan Z, and Xu CC (2014). Reductive depolymerization of kraft lignin for chemicals and fuels using formic acid as an in-situ hydrogen source. *Bioresource Technol*, 171: 95–102.

Jiang Y, Li Z, Tang X, Sun Y, Zeng X, Liu S, and Lin L (2015). Depolymerization of cellulolytic enzyme lignin for the production of monomeric phenols over raney ni and acidic zeolite catalysts. *Energy Fuels*, 29: 1662–1668.

Konnerth H, Zhang J, Ma D, Prechtl MHG, and Yan N (2015). Base promoted hydrogenolysis of lignin model compounds and organosolv lignin over metal catalysts in water. *Chem Eng Sci*, 123: 155–163.

Luo N, Wang M, Li H, Zhang J, Hou T, Chen H, Zhang X, Lu J, and Wang F (2017). Visible-light-driven self-hydrogen transfer hydrogenolysis of lignin models and extracts into phenolic products. *ACS Catal*, 7: 4571–4580.

Mankar AR, Ahmad E, and Pant KK (2021). Insights into reductive depolymerization of Kraft lignin to produce aromatics in the presence of Pt/HZSM-23 catalyst. *Mater Sci Enery Technol*, 4: 341–348.

Mohan D, Pittman CU, and Steele PH (2006). Pyrolysis of Wood/biomass for Bio-oil: a critical review. *Energy Fuels*, 20: 848–889.

Monsigny L, Feghali E, Berthet JC, and Cantat T (2018). Efficient reductive depolymerization of hardwood and softwood lignins with Brookhart's iridium (iii) catalyst and hydrosilanes. *Green Chem*, 20(9): 1981–1986.

Sergeev AG and Hartwig JF (2011). Selective, nickel-catalyzed hydrogenolysis of aryl ethers. *Science*, 332(6028): 439–443.

Shao Y, Xia Q, Dong L, Liu X, Han X, Parker SF, Cheng Y, Daemen LL, Ramirez-Cuesta AJ, Yang S, and Wang Y (2017). Selective production of arenes via direct lignin upgrading over a niobium-based catalyst. *Nat Commun*, 8: 16104.

Si X, Lu F, Chen J, Lu R, Huang Q, Jiang H, Taarning E, and Xu J (2017). A strategy for generating high-quality cellulose and lignin simultaneously from woody biomass. *Green Chem*, 19: 4849–4857.

Singh SK and Ekhe JD (2015). Cu–Mo doped zeolite ZSM-5 catalyzed conversion of lignin to alkyl phenols with high selectivity. *Catal Sci Technol*, 5: 2117–2124.

Sun Z, Fridrich B, de Santi A, Elangovan S, and Barta K (2018). Bright side of lignin depolymerization: toward new platform chemicals. *Chem Rev*, Jan 24; 118(2): 614–678.

Xiao L-P, Wang S, Li H, Li Z, Shi Z-J, Xiao L, Sun R-C, Fang Y, and Song G (2017). Catalytic hydrogenolysis of lignins into phenolic compounds over carbon nanotube supported molybdenum oxide. *ACS Catal*, 7: 7535–7542.

Xiu S and Shahbazi A (2012). Bio-oil production and upgrading research: a review. *Renewable Sustainable Energy Rev*, 16: 4406–4414.

Zacher AH, Olarte MV, Santosa DM, Elliott DC, and Jones SB (2014). A review and perspective of recent bio-oil hydrotreating research. *Green Chem*, 16: 491–515.

Zaheer M and Kempe R (2015). Catalytic hydrogenolysis of aryl ethers: a key step in lignin valorization to valuable chemicals. *ACS Catal*, 5: 1675–1684.

Zakzeski J, Bruijnincx PC, Jongerius AL, and Weckhuysen BM (2010). The catalytic valorization of lignin for the production of renewable chemicals. *Chem Rev*, 110(6): 3552–3599.

4

Lignin-first Biorefining Process

Abstract

Traditional lignocellulosic biomass utilization techniques struggle to stop the undesirable reactive intermediate condensation in industrial applications during biomass deconstruction. By either selectively catalyzing the conversion of these intermediates to stable derivatives, the lignin-first biorefinery prevents the condensation of reactive intermediates or prevents their formation by modifying the natural building blocks or intermediates. This approach has emerged as one of the most efficient ways to produce novel platform chemicals from lignin because it depolymerizes native lignin effectively. The lignin-first strategy is a catalyst-dependent, heterogeneous process that involves the following three main steps: solvolysis, fractionalization or depolymerization, and reductive stabilization. The obtained monolignol and phenolic units can then be utilized as substrates for producing pharmaceuticals, biobased fuels, polymers, and aromatic chemicals etc. In this chapter, the revolutionary "lignin-first" method for lignocellulosic catalytic valorization is discussed.

Keywords   *Lignin-first biorefinery; Lignocellulosic biomass; Lignin and its derivatives; Depolymerization; Reductive stabilization; Fuels and chemicals*

4.1 Introduction

Lignin is an amorphous polymer with three dimensions made up of methoxylated phenylpropane units and their derivatives. It makes up between 15 and 35 weight % of lignocellulosic biomass and has the potential to be a plentiful renewable resource for the production of fuels and chemicals (Baeyens et al., 2015; Kang et al., 2014; Liu et al., 2015; Xu et al., 2014).

The pulp and paper industry currently produces and processes a significant amount of lignin. Several excellent review papers have reported and compiled the results of several lignin conversion initiatives which include, pyrolysis, gasification, and selective depolymerization (Hanson et al., 2012; Lancefield et al., 2015; Li et al., 2015; Luterbacher et al., 2015; Ma et al., 2014; Ragauskas et al., 2014; Song et al., 2013; Vardon et al., 2015; Wang and Rinaldi, 2013; Zakzeski et al., 2010). However, lignin processing in a biorefinery is still difficult, in part because of the structural change that takes place during pretreatments of biomass (Banerjee et al., 2011; Bussemaker and Zhang, 2011; Dias et al., 2011; Ma et al., 2009; Rahimi et al., 2014; Yan et al., 2009). Depolymerization and use of the resulting lignin are hampered by the subsequent formation of new C–C linkages (Luterbacher et al., 2014).

A brand new idea known as lignin-first refinery, which stands out from the standard cellulose-first refinery, was proposed, and it has been the subject of extensive research (Bridgwater and Cottam, 1991; Ferrini and Rinaldi, 2014; Renders et al., 2017; Weingarten et al., 2012). The idea was to process lignin in its most reactive and usable form.

4.2 The Revolutionary "Lignin-first" Method for Lignocellulosic Catalytic Valorization

The lignin-first approach is a heterogeneous catalyst-dependent process that involves three elementary steps: solvolysis, fractionalization/depolymerization, and reductive stabilization. The obtained monolignol and phenolic units can be subsequently used as feedstocks for the preparation of aromatic chemicals, bio-based fuels, polymers, and drugs (Figure 4.1) (Paone et al., 2020).

Several approaches have been described thus far. Figure 4.2 depicts the chemical reaction mechanism (Luo et al., 2023). Most often, lignin extraction from lignocellulosic raw material with solvent is done in a reductive atmosphere in the presence of a metal catalyst. The most widely used metal catalysts are platinium, palladium, rhodium, ruthenium, and nickel on activated carbon or aluminium supports (Kenny et al., 2022; Sun et al., 2018a). This technique, known as reductive catalytic fractionation (RCF), is quick and produces a lot of lignin.

4.2.1 Reductive Catalytic Fractionation

Hydrodeoxygenation directly for producing chemicals and fuels derived from hydrocarbons with added value is made possible by the presence of lignin derivatives produced by liquid-phase catalytic depolymerization (Yan et al., 2008). The issue of lignin degradation that comes up frequently when using conventional lignin-isolation methods is mitigated by this strategy (Schutyser et al., 2018). During biomass fractionation, one of the additional active stabilization strategies is the proper use of chemistry of protection groups.

Alcohols have been used for nearly a century to fractionate lignocellulosic biomass. However, the benefits of using them in the lignin extraction process have only become apparent in recent years. An alcohol that produces ethers by acting as an external nucleophilic reagent can intercept benzylic carbocation ion intermediates during lignin extraction (Kaiho et al., 2015).

Small molecules like ethylene glycol and aldehydes can also be used to protect reactive intermediates as acetals. As a result, chemically stable lignin can be extracted, which paves the way for additional depolymerization and transformation (Deuss et al., 2015; Shuai et al., 2016). This method

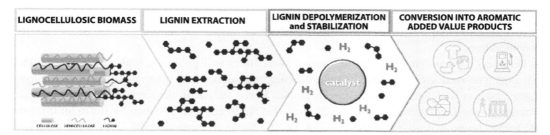

Figure 4.1 The revolution of the "lignin-first" approach: from lignocellulosic biomasses to added value (Paone et al., 2020). Reproduced with permission.

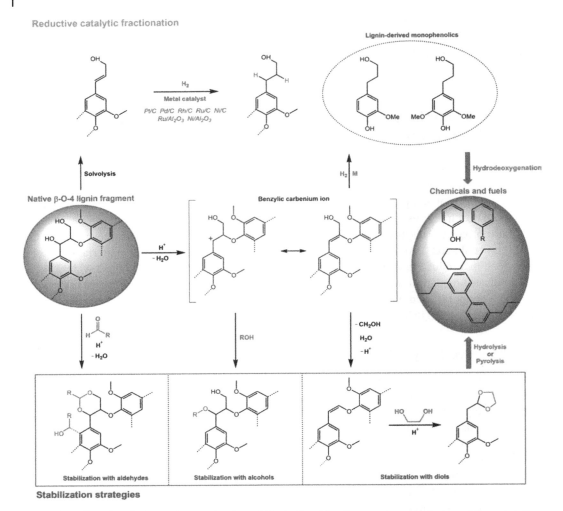

Figure 4.2 The chemical-reaction mechanism of lignin-first biorefinery using solvolysis and the catalytic stabilization of reactive intermediates to stable products or protection-group chemistry and subsequent upgrading. Luo et al. (2023). Creative Commons Attribution (CC BY) license (https://creativecommons.org/licenses/by/4.0).

is more adaptable and easier to control because it physically separates the lignin depolymerization and biomass fractionation processes. However, the addition of protection reagents made up of small molecules may have negative effects on the environment, which necessitates comprehensive consideration.

The recent development of the lignin-first biorefinery will gradually replace these research hotspots with strategies for downstream processing that successfully integrate the use of carbohydrate residues and lignin-degradation products. New catalytic approaches have been studied extensively for the purpose of increasing the value of lignin and its derivatives for the manufacture of chemicals and hydrocarbon fuels (Li et al., 2020a, 2020b; Kalogiannis et al., 2019).

Cascade processes have also been used to make carbohydrate residues more valuable. Examples include the direct conversion of carbohydrates into mixed alcohols and the subsequent upgrade to fuel-range alkanes (Sun et al., 2018b); treatment with ferric choride to simultaneously obtain levulinic acid, 5-hydroxymethylfurfural, and furfural (Chen et al., 2020) and fermentation and saccharification for the production of bioethanol (Van den Bosch et al., 2017).

It has been deemed a significant obstacle to successfully separate the solid catalyst mixture from the delignified pulp. As a result, a viable option might be to rationally design multi-purpose catalysts for the direct catalytic upgradation of solid-residue mixtures (carbohydrates and catalyst).

Several methods for lignin-first refining have been reported; the chemical steps of these methods are depicted in Figure 4.3 (Abu-Omar et al., 2021). The extraction of lignin from biomass using a solvent is the most common approach that donates hydrogen or another reducing agent or in a hydrogen atmosphere in the presence of a transition metal (Bosch et al., 2015; Parsell et al., 2015; Song et al., 2013; Yan et al., 2008).

Regardless of the H-source, this method, which has been referred to as catalytic upstream biorefinery (CUB) or early-stage catalytic conversion of lignin (ECCL) for the process that uses 2-propyl alcohol as a H-donor (Ferrini et al., 2017, 2016; Ferrini and Rinaldi, 2014; Graça et al., 2018; Kennema et al., 2017; Rinaldi et al., 2016, 2018; Wang et al., 2016) is now commonly referred to as reductive catalytic fractionation (RCF) (Figure 4.3a).

As a means of pulping and producing chemicals based on lignin with high yields, RCF under hydrogen pressure was first used or developed in the late 1930s and early 1940s, but it was not yet commercialized (Pepper and Lee, 1969; Pepper and Rahman, 1987; Pepper and Supathna, 1978; Sobolev et al., 1957).

The main functions of the metal catalyst in systems that operate with sugars in the biomass as reducing agents are the reductive stabilization of reactive solvolysis intermediates from lignin and the depolymerization of solubilized lignin oligomers (Galkin and Samec, 2014). As demonstrated by reactor setups which physically separate the catalyst from the solid biomass, no contact between the catalyst and solid biomass is required (Anderson et al., 2017; Kumaniaev et al., 2017; Van den Bosch et al., 2017). Flow-through operations or catalyst baskets have recently been used to separate the solvolysis of biomass from the hydrogenation/hydrogenolysis of lignin intermediates. The yield of monophenolic compounds and hemicelluloses retention may be affected by the solvent combination as well as the possibility of an additional acid or base catalyst being present. Not only does the solvent affect the degree of delignification (solvolysis and lignin extraction), but it also affects the monophenolic selectivity and distribution, as well as the retention of (hemi-)cellulose (Renders et al., 2017). High levels of delignification can be achieved using water or protic solvents with a higher water content, but they will also hydrolyze and solubilize carbohydrates, which can then be hydrogenated by the catalyst or react with the solvent. The intelligent application of protection-group chemicals is one of the additional active stabilization techniques used during acidic fractionation in the lignin-first sphere (Luo et al., 2020; Questell-Santiago et al., 2020).

Through acidolysis of the β-O-4 linkage, two pathways (Lundquist, 1970; Yokoyama, 2014) are involved in the acid-catalyzed depolymerization of lignin. Both the unstable C2 aldehyde (a mixture of G, S, or H depending upon the type of wood) and the ketones known as Hibbert's can be produced through the same pathway (Kulka and Hibbert, 1943; Steeves and Hibbert, 1939). This has been proven with a number of organosolv lignins and diol-assisted fractionation (DAF), a "metal-free" lignin-first process in which direct extraction of C2-acetals from lignocellulose is carried out under carefully selected reaction conditions (De Santi et al., 2020; Deuss et al., 2016, 2017).

Using ethylene glycol, newly generated aldehydes produced by a process can be protected as acetals (Figure 4.3b) (De Santi et al., 2020; Deuss et al., 2015). Another option is aldehyde-assisted fractionation (AAF), which stops the production of benzylic cations by capping the benzylic

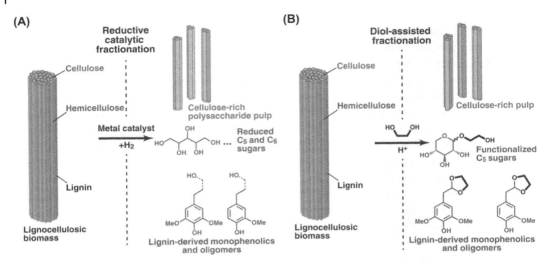

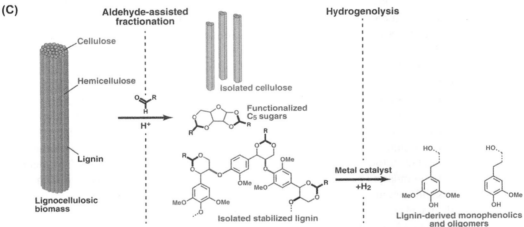

Figure 4.3 Three lignin-first strategies (a) reductive catalytic fractionation (RCF), (b) diol-assisted fractionation (DAF), (c) aldehyde-assisted fractionation (AAF). Abu-Omar et al. (2021). Licensed under Creative Commons Attribution 3.0 Unported Licence.

alcohol. For instance, formaldehyde or other simple aldehydes are combined with the natural 1,3-diol found in the side chains of lignin to produce an acetal in AAF (Figure 4.3c) (Lan et al., 2018; Shuai et al., 2016).

In the future, it will be possible to further depolymerize and transform chemically stabilized lignin using this second method, which requires the use of stoichiometric reagents for capping. New research in this area is constantly being published by the global community working on biomass conversion and the development of the lignin-first concept for fractionating lignocellulosic biomass has substantially increased over the past few years. There are no widely accepted standards for selecting feedstocks, conducting product analyses, or evaluating process performance, which presents a challenge in this new field (Galkin et al., 2017). For instance, narrow product distributions can be achieved through the use of hydrogenolysis reactions (Lan et al., 2018; Pepper and Lee, 1969; Shuai et al., 2016). There are no universally recognized guidelines for selecting feedstocks, analyzing products, or assessing process performance, which presents a challenge in this emerging field (Galkin et al., 2017).

This is a significant limitation because:

1) it is difficult to quantitatively compare laboratory methods and results
2) it becomes difficult, or even impossible, to duplicate the procedures of other research groups
3) there are no standardized lignocellulosic materials available.

Luo et al. (2023) have examined the most recent advancements in RCF and how reactor configurations, catalysts, and solvents affect phenolic monomer yield, delignification degree, and carbohydrate pulp retention. Some reaction systems with high monomer yields are presented in Table 4.1.

Table 4.1 Reductive catalytic fractionation of biomass feedstock.

Feedstock	Catalyst	Solvent	Monomer yield	Sugar Retention	Reference
Miscanthus	Ni/C	Methanol	68 weight %	86 weight %	(Luo et al., 2016)
Corn stover	Ni/C	Methanol	24.5 weight %	76 weight %	(Anderson et al., 2016)
Flax shive	Ru/C	Ethanol	9.5 weight %	Glucan 67.2 weight %	(Kazachenko et al., 2020)
Spruce	Ru/C	Ethanol	30 weight %	Glucan 84.4 weight %	(Taran et al., 2022)
Bamboo	Pd/C	Methanol	32.2 weight %	Glucan 73.4 weight % Xylan 57.4 weight %	(Zhang et al., 2019b)
Eucalyptus	Pd/C	Methanol	49.8 weight %	Glucan 82.5 weight % Xylan 67.8 weight %	(Chen et al., 2020)
Poplar	Pd/C Zn/Pd/C	Methanol/ H2O (7:3) Methanol	43.5 weight % 54 weight %	66.7 weight % 79 weight %	(Parsell et al., 2015; Renders et al., 2016)
Birch	Ru/C	Methanol	51.5% (C-Yield)	81% (C-Yield)	(Van den Bosch et al., 2015a)
	Pd/C	Methanol	49.3% (C-Yield)	89% (C-Yield)	(Van den Bosch et al., 2015b)
	Pd/C	Water	43.8 weight %	55 weight %	(Schutyser et al., 2015)
	Pd/C	Ethanol/ H2O (1:1)	36% (C-Yield)	84.4 weight %	(Galkin et al., 2016)
	Ni/Al2O3 [a]	Methanol	36 weight %	84.9 weight %	(Van den Bosch et al., 2017)
	Pd/ C+H3PO4 [b]	Methanol/ H2O (7:3)	37 weight %	56 weight %	(Kumaniaev et al., 2017)

a) Ni/Al2O3 pellets in catalyst cage.
b) Reaction operated in a flow-through reactor.
Luo et al. (2023). Distributed under the terms and conditions of the Creative Commons Attribution (CC BY) license (https://creativecommons.org/licenses/by/4.0).

The mechanistic studies by Luo et al. (2023) have provided a general comprehension of the RCF processes, which can be summed up as the following three main steps (Renders et al., 2019):

- lignin extraction, which is completely solvent-dependent
- solvolytic depolymerization and catalytic hydrogenolysis
- stabilization, which is regulated by a catalyst that is redox-active.

The type and quantity of products produced by C–O bond hydrogenolysis can be controlled by choosing the right metal (Sun et al., 2018a). It has been demonstrated that lignin depolymerization can be effectively catalyzed by platinium, palladium, rhodium, ruthenium, and nickel all of which are abundant on Earth (Kenny et al., 2022; Van den Bosch et al., 2015a, 2015b; Renders et al., 2018; Luo et al., 2016).

Lignin fraction in the birch RCF was degraded using a ruthenium/carbon catalyst to produce a propyl-substituted phenol compound with a 52% monomer yield. In the subsequent hydrolysis reaction, the carbohydrates were transformed into C2–C6 sugar polyol products, with 95% retention of cellulose and 47% retention of hemicellulose (Van den Bosch et al., 2015a). Palladium/carbon and ruthenium/carbon catalysts that were identical were also compared. As was to be expected, the yields of lignin products produced by the two catalysts were comparable. However, the products had very different chemical structures, and the palladium/carbon catalyst retained more carbohydrate residues and was more selective for lignin monomers with a lot of hydroxyl groups (Van den Bosch et al., 2015b).

Luo et al. (2023) demonstrated that palladium/zinc synergistic catalysis is necessary for lignin conversion in terms of the subsequent hydrodeoxygenation and breakage of β-O-4 linkages (Parsell et al., 2013). Additionally, zinc/palladium/carbon treatment of various biomass materials resulted in the primary conversion of native lignin into two products: dihydroeugenol and 2,6-dimethoxy-4-propylphenol, with yields of 40 to 54% for lignin monomers (Parsell et al., 2015).

Palladium/carbon and ZnII work together in a synergistic manner, according to subsequent mechanistic studies; Cγ–OH removal from the β-O-4 bond could be activated and aided by the addition of zinc+2, according to one hypothesis (Klein et al., 2016).

The creation of inexpensive and readily available catalysts is crucial for industrial applications. Natural birch wood lignin fractions were selectively hydrogenated using a nickel/carbon catalyst to produce dihydroeugenol, 2,6-dimethoxy-4-propylphenol, and a small quantity of propenyl-substituted phenols (Song et al., 2013). Interestingly, the distribution of the monomer product shifted from PG–OH and PS–OH to PG and PS and the Fe-doped bimetallic catalyst removed hydroxyls more effectively than the nickel/carbon catalyst (Zhai et al., 2017).

Li et al. (2012) discovered a new nickel-W2C/AC bimetallic catalyst with a synergistic effect between nickel and W2C that may substantially boost the formation of monomers obtained from lignin. Ethylene glycol and other diol products were produced further from carbohydrates. This catalyst has numerous applications in raw materials like birch, pine, poplar, beech, and others.

Solvent decomposition may break up the lignin–carbohydrate complex between hemicellulose and lignin during the direct catalytic treatment of lignocellulosic raw material, resulting in the removal of lignin from the biomass substrates. Under solvent decomposition, the lignin structure's β-O-4 linkage bond is then severed. The production of soluble lignin fragments, which continue to come into contact with the surface of the catalyst, completes the activation of the β-O-4 linkage bond into a single-molecule compound. During biomass delignification and lignin depolymerization, solvents affect both the retention of carbohydrates and the yield of aromatic monomers (Shuai and Luterbacher, 2016; Zhu et al., 2017).

How various solvents affected the RCF of birch wood has been investigated (Schutyser et al., 2015). The degree of delignification was found to increase with the solvent's polarity. This was due to highly polar solvents' increased lignin-accessibility and improved ability to complete the dissolution of the structure of wood fibers. The delignification efficiencies of methanol and ethylene glycol were the highest among them. The distribution of lignin degradation products in a palladium/carbon catalytic system revealed that as the polarity of the solvent increased, monomers and dimers of degradation products increased whereas oligomer products substantially reduced. This suggests that the breakdown of the lignin oligomer into monomers and dimers can also be sped up by highly polar solvents.

Bartling et al. (2021) carried out a techno-economic examination of the RCF process with a variety of solvents and used ethylene glycol instead of methanol as the solvent. The total pressure in the reactor was significantly reduced as a result of the lower ethylene glycol vapor pressure. In general, capital costs are lower when there is less pressure during RCF. Ethylene glycol, on the other hand, is more expensive and uses more energy to recover solvents than methanol. The sale of bioethanol at a price of USD 2.50 per gallon of gasoline equivalent is supported by the fact that methanol has a higher MSP-monomer fraction at USD 3.63 per kg and ethylene glycol has a lower MSP-monomer fraction at USD 3.07 per kg.

The results of a subsequent study into the effects of various alcohol/water-mixing solvent systems on the RCF revealed that lignin extraction efficiency was significantly improved when water was added in moderate amounts. However, there was less delignification when there was too much water (Renders et al., 2016).

Chen et al. (2016) also proved that addition of water improved the yield of lignin monomers. Importantly, hydrolysis reactions are also carried out on the carbohydrate fraction, which removes approximately 20% of the cellulose and almost all of the hemicelluloses while using pure water as the medium, the lignin fraction is effectively separated and degraded (Schutyser et al., 2015).

Other catalytic systems exhibit solvent-polarity effects in a manner that is comparable. When water was used in place of methanol as the solvent in the ruthenium/carbon system, the yield of phenol monomer dropped from 52% to 25%. However, the carbohydrate fraction also broke down into soluble polyols (Van den Bosch et al., 2015a). This could be explained by the high-temperature auto-ionization of water into H^+ acid ions, which can initiate the hydrolysis of carbohydrates (Sun et al., 2019). When water is used as a solvent, it is also important to consider the reposition of dissolved lignin on the surface of lignocellulosic fibers (Chen et al., 2015). Most importantly, the current direct catalytic reduction method for biomass feedstocks may not function with a pure water system.

Using metal catalysts and biomass, the new method of RCF has been suggested for depolymerizing and stabilizing lignin. However, this typically hinders the recovery of the catalyst. As a result, lignin-first biorefinery flow-through systems were developed (Figure 4.4). For the RCF process in 2017, two research groups introduced flow-through reactors. These reactors separated the catalyst and biomass by filling two distinct beds. By pumping the solvent through the heated biomass bed, partial depolymerization and extraction of the lignin polymer were made possible. The catalyst bed was traversed by a liquid mixture of dissolved lignin fragments for the purpose of further depolymerization and active intermediate stabilization (Anderson et al., 2017; Kumaniaev et al., 2017).

4.2.2 From Phenolic Units to Value-added Products

In the process of being depolymerized, lignocellulosics produce a wide range of substances that are thought to be bio-derived components for the long-term production of valuable chemicals,

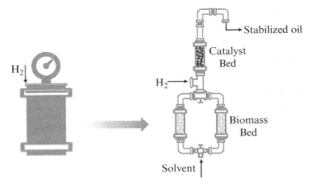

Figure 4.4 The evolution of reactor configurations for reductive catalytic fractionation. Luo et al. (2023) Creative Commons Attribution (CC BY) license (https://creativecommons.org/licenses/by/4.0).

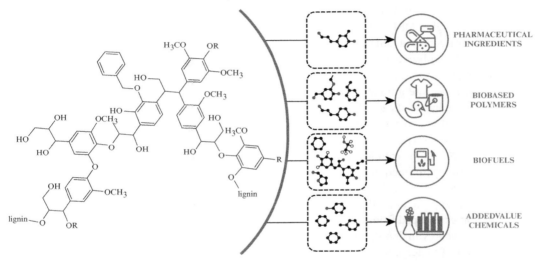

Figure 4.5 A schematic overview of potential added value products of lignin biorefinery (Paone et al., 2020). Reproduced with permission.

polymers, pharmaceuticals, and fuels (Figure 4.5) (Isikgora and Becer, 2015; Sun and Barta, 2018; Zakzeski et al., 2010). Due to the strong connection between lignin and holocellulosic fractions, RCF processes frequently result from the simultaneous partial degradation of hemicellulose which provides numerous polyols and furanics from C5. The solid carbohydrate cellulose pulp, which is naturally high in C5-C6 sugars, can also be made valuable through hydrolysis by enzymes (Ouyang et al., 2018) or a heterogeneously catalyzed hydrogenation/hydrogenolysis process into biofuels after being separated from the reaction medium (Cao et al., 2018; Gumina et al., 2019; Guo et al., 2018; Huang et al., 2018; Zhang et al., 2019a).

For the first time, proof-of-concept evaluation of the RFC carbohydrate pulp for the enzymatic production of ethanol was conducted (Van den Bosch et al., 2017). A "lignin-centered convergent approach" was presented for the reduction of poplar or spruce wood using a phosphidated nickel/silicon dioxide catalyst into aliphatic and aromatic biohydrocarbons (Cao et al., 2018). In light of this, bio-derived fuel additives or lignocellulosics were suggested by Huang et al. (2018) for the direct production of jet fuels.

Businesses like DuPont de Nemours, Covestro, Novamont Avantium, Novamont, Neste, BASF SE, and numerous others have shifted to producing plastics and materials made from renewable monomers in response to the growing importance of using bio-based products as a result of customer attitudes and perceptions. In addition, it is abundantly clear that the modern polymer industry relies heavily on lignin, a natural source of aromatics. Phenylpropanols and propylphenols, two examples of phenolic units produced by deconstruction of lignin, only have a limited direct application in the market (Lan and Luterbacher, 2019). They can be further processed to produce bio-BTX aromatics like xylene, benzene, and toluene, but their final sales costs are still too high when compared to those of compounds that are similar to gasoline.

The global Styrene market stood at approximately 30 million tonnes in 2021 and is expected to grow at a CAGR of 5.64% during the next few years until 2030 (https://finance.yahoo.com/news/global-styrene-market-analysis-report-233000200.html). However, they may have a significant impact in the coming years. After a very interesting contribution, it was shown that a reaction with epichlorohydrin can directly use the lignin oil to make epoxy resins made from biomaterials (Van de Pas et al., 2017).

Biopolymers have been produced with the help of a number of monolignols and phenolics (Esposito & Antonietti, 2015; Hernandez et al., 2016; Llevot et al., 2016; Rajesh Banu et al., 2019; Thakur et al., 2014; Wang et al., 2017). Polyesters can be made by using p-coumaric acid as a precursor. Either vanillin or 4-propylguaiacol can be used to successfully make epoxy resins (Zhao and Abu-omar, 2017). Because it can also be used to make polyurethans, polybenzoxazines, polyaldimines (Schiff base polymers), polyacetals, methacrylates, and other conjugated polymers through its oxidative dimer, divanillin, vanillin is an important intermediate in biobased polymers. Starting monomers include polyesters, polycarbonates, epoxy resins, liquid crystals, optoelectronic conjugated polymers, and bisphenols and their derived precursors (Fache et al., 2015).

Bisphenols are formed when two phenolic units are coupled together. Already, phenols and lignans belong to a significant group of natural products that have pharmacological and nutritional properties. The use of polyphenols and the potential use of ligno-phenols to reduce vascular oxidative stress and/or inflammation, and also the prevention of ischaemic heart disease, have been reported. Sinapyl alcohol's anti-inflammatory and anti-nociceptive properties have also been reported (Spridon, 2018).

Aromatic compounds obtained through the reduction of lignocellulose have also been used as precursors for producing biobased drugs and active pharmaceutical ingredients (APIs). A novel synthetic strategy known as "cleave and couple" has been successfully developed by Katalin Barta's group (Sun and Barta, 2018). C–C and C–N bonds are used in this method to transform aromatic and aliphatic compounds derived from lignocelluloses. The authors specifically suggested that functionalized phenolic monomers could be utilized to produce a number of active pharmaceutical compounds. They also recently presented a synthetic method for making a number of 2-benzazepine derivatives from phenylpropanol, which are frequently used as anti-depressants (Elangovan et al., 2019). This opened up new opportunities for lignin-biorefinery.

4.3 Future Challenges

For a green and sustainable biorefinery, the reductive catalytic fractionation of lignocellulosic biomasses pioneers new territory. There are still numerous scientific and technological hurdles to overcome, despite the fact that we currently possess all of the knowledge necessary to advance the production of aromatics from native lignocellulose under conventional batch conditions. The efficiency and sustainability of the following generation of biorefineries will be further enhanced by the creation of novel one-pot, multistep, continuous flow processes for the direct conversion of lignin fractions into pharmaceuticals. The use of agro-industrial and household wastes as feedstocks, more environmentally friendly hydrogen sources (such as H2O as an in situ hydrogen donor medium or photodriven water splitting), and these processes will all improve biorefineries' overall sustainability and efficiency (Paone et al., 2020).

Bibliography

Abu-Omar MM, Barta K, Beckham GT, Luterbacher J, Ralph J, Rinaldi R, Roman-Leshkov Y, Samec J, Sels B, and Wang F (2021). Guidelines for performing lignin-first biorefining. *Energy Environ Sci*, 14(1). https://doi.org/10.1039/d0ee02870c.

Anderson EM, Katahira R, Reed M, Resch MG, Karp EM, Beckham GT, and Román-Leshkov Y (2016). Reductive catalytic fractionation of corn stover lignin. *ACS Sustain Chem Eng*, 4: 6940–6950.

Anderson EM, Stone ML, Katahira R, Reed M, Beckham GT, and Román-Leshkov Y (2017). Flowthrough reductive catalytic fractionation of biomass. *Joule*, 1: 613–622.

Baeyens J, Kang Q, Appels L, Dewil R, Lv YQ, and Tan TW (2015). Challenges and opportunities in improving the production of bio-ethanol. *Prog Energ Combust Sci*, 47: 60–88.

Banerjee G, Car S, Scott-Craig JS, Hodge DB, and Walton JD (2011). Alkaline peroxide pretreatment of corn stover: effects of biomass, peroxide, and enzyme loading and composition on yields of glucose and xylose. *Biotechnol Biofuels* Jun 9, 4(1): 16.

Bartling AW, Stone ML, Hanes RJ, Bhatt A, Zhang Y, Biddy MJ, Davis R, Kruger JS, Thornburg NE, Luterbacher JS, Rinaldi R, Samec JSM, Sels BF, Román-leshkov Y, and Beckham GT (2021). Techno-economic analysis and life cycle assessment of a biorefinery utilizing reductive catalytic fractionation. *Energy Environ Sci*, 14: 4147–4168.

Bosch SV, Schutyser W, Vanholme R, Driessen T, Koelewijn S, Renders T, Meester BD, Huijgen WJ, Dehaen W, Courtin CM, Lagrain B, Boerjan W, and Sels BF (2015). Reductive lignocellulose fractionation into soluble lignin-derived phenolic monomers and dimers and processable carbohydrate pulps. *Energy Environ Sci*, 8: 1748–1763.

Bridgwater AV and Cottam ML (1991). Opportunities for biomass pyrolysis liquids production and upgrading. *Energy and Fuel*, 6(2): 113–120.

Bussemaker MJ and Zhang D (2011). Effect of ultrasound on lignocellulosic biomass as a pretreatment for biorefinery and biofuel applications. *Ind Eng Chem Res*, 52: 3563–3580.

Cao Z, Dierks M, Clough MT, de Castro IBD, and Rinaldi R (2018). A convergent approach for a deep converting lignin-first biorefinery rendering high-energy-density drop-in fuels. *Joule*, 2: 1118–1133.

Chen H, Fu Y, Wang Z, and Qin M (2015). Degradation and redeposition of the chemical components of aspen wood during hot water extraction. *BioResources*, 10: 3005–3016.

Chen J, Lu F, Si X, Nie X, Chen J, Lu R, and Xu J (2016). High yield production of natural phenolic alcohols from woody biomass using a nickel-based catalyst. *ChemSusChem*, 9: 3353–3360.

Chen X, Zhang K, Xiao LP, Sun RC, and Song G (2020). Total utilization of lignin and carbohydrates in *Eucalyptus grandis*: an integrated biorefinery strategy towards phenolics, levulinic acid, and furfural. *Biotechnol Biofuels*, 13: 2.

De Santi A, Galkin MV, Lahive CW, Deuss PJ, and Barta K (2020). Lignin-first fractionation of softwood lignocellulose using a mild dimethyl carbonate and ethylene glycol organosolv process. *ChemSusChem*, 13: 4468–4477.

Deuss PJ, Lahive CW, Lancefield CS, Westwood NJ, Kamer PC, Barta K, and de Vries JG (2016). Metal triflates for the production of aromatics from lignin. *ChemSusChem* Oct 20, 9(20): 2974–2981.

Deuss PJ, Lancefield CS, Narani A, de Vries JG, Westwood NJ, and Barta K (2017). Phenolic acetals from lignins of varying compositions via iron(iii) triflate catalysed depolymerisation. *Green Chem*, 19: 2774–2782.

Deuss PJ, Scott M, Tran F, Westwood NJ, de Vries JG, and Barta K (2015). Aromatic monomers by in situ conversion of reactive intermediates in the acid-catalyzed depolymerization of lignin. *J Am Chem Soc*, 137: 7456–7467.

Dias MO, Cunha MP, Maciel Filho R, Bonomi A, Jesus CD, and Rossell CE (2011). Simulation of integrated first and second generation bioethanol production from sugarcane: comparison between different biomass pretreatment methods. *J Ind Microbiol Biotechnol*, 38: 955–966.

Elangovan S, Afanasenko A, Haupenthal J, Sun Z, Liu Y, Hirsch AKH, and Barta K (2019). From wood to tetrahydro-2-benzazepines in three waste-free steps: modular synthesis of biologically active lignin-derived scaffolds. *ACS Cent Sci*, 5: 1707–1716.

Esposito D and Antonietti M (2015). Redefining biorefinery: the search for unconventional building blocks for materials. *Chem Soc Rev*, 44: 5821–5835.

Fache M, Boutevin B, and Caillol S (2015). Vanillin, a key-intermediate of biobased polymers. *Eur Polym J*, 68: 488–502.

Ferrini P, Chesi C, Parkin N, and Rinaldi R (2017). Effect of methanol in controlling defunctionalization of the propyl side chain of phenolics from catalytic upstream biorefining. *Faraday Discuss*, 202: 403–413.

Ferrini P, Rezende CA, and Rinaldi R (2016). Catalytic upstream biorefining through hydrogen transfer reactions: understanding the process from the pulp perspective. *ChemSusChem*, 9: 3171–3180.

Ferrini P and Rinaldi R (2014). Catalytic biorefining of plant biomass to non-pyrolytic lignin bio-oil and carbohydrates through hydrogen transfer reactions. *Angew Chem Int Ed Engl* Aug 11, 53(33): 8634–8639.

Galkin MV, Di Francesco D, Edlund U, and Samec JSM (2017). Sustainable sources need reliable standards. *Faraday Discuss*, 202: 281–301.

Galkin MV and Samec JSM (2014). Selective route to 2-propenyl aryls directly from wood by a tandem organosolv and palladium-catalysed transfer hydrogenolysis. *ChemSusChem*, 7: 2154–2158.

Galkin MV, Smit AT, Subbotina E, Artemenko KA, Bergquist J, Huijgen WJ, and Samec JS (2016). Hydrogen-free catalytic fractionation of woody biomass. *ChemSusChem*, 9: 3280–3287.

Graça I, Woodward RT, Kennema MO, and Rinaldi R (2018). Formation and fate of carboxylic acids in the lignin-first biorefining of lignocellulose via H-transfer catalyzed by Raney Ni. *ACS Sustain Chem Eng*, 6: 13408–13419.

Gumina B, Espro C, Galvagno S, Pietropaolo R, and Mauriello F (2019). Bioethanol production from unpretreated cellulose under neutral self sustainable hydrolysis/hydrogenolysis conditions promoted by the heterogeneous Pd/Fe3O4 catalyst. *ACS Omega*, 4: 352–357.

Guo T, Li X, Liu X, Guo Y, and Wang Y (2018). Catalytic transformation of lignocellulosic biomass into arenes, 5-hydroxymethylfurfural, and furfural. *ChemSusChem*, 11: 2758–2765.

Hanson SK, Wu R, and Silks LA (2012). C—C or C—O bond cleavage in a phenolic lignin model compound: selectivity depends on vanadium catalyst. *Angew Chem Int Ed*, 51: 3410–3413.

Hernandez ED, Bassett AW, Sadler JM, La Scala JJ, and Stanzione JF (2016). Synthesis and characterization of bio-based epoxy resins derived from vanillyl alcohol. *ACS Sustain Chem Eng*, 4: 4328–4339.

Huang Y, Duan Y, Qiu S, Wang M, Ju C, Cao H, Fang Y, and Tan T (2018). Lignin-first biorefinery: a reusable catalyst for lignin depolymerization and application of lignin oil to jet fuel aromatics and polyurethane feedstock. *Sustain Energy Fuels*, 2: 637–647.

Isikgora FH and Becer CR (2015). Lignocellulosic biomass: a sustainable platform for the production of bio-based chemicals and polymers. *Polym Chem*, 6: 4497–4559.

Kaiho A, Kogo M, Sakai R, Saito K, and Watanabe T (2015). In situ trapping of enol intermediates with alcohol during acid-catalysed de-polymerisation of lignin in a nonpolar solvent. *Green Chem*, 17: 2780–2783.

Kalogiannis KG, Matsakas L, Lappas AA, Rova U, and Christakopoulos P (2019). Aromatics from beech wood organosolv lignin through thermal and catalytic pyrolysis. *Energies*, 12: 1606.

Kang Q, Appels L, Tan TW, and Dewil R (2014). Bioethanol from lignocellulosic biomass: current findings determine research priorities. *Sci World J*, 2014: 1–13, 298153.

Kazachenko AS, Tarabanko VE, Miroshnikova AV, Sychev VV, Skripnikov AM, Malyar YN, Mikhlin YL, Baryshnikov SV, and Taran OP (2020). Reductive catalytic fractionation of flax shive over Ru/C catalysts. *Catalysts*, 11: 42.

Kennema M, de Castro IBD, Meemken F, and Rinaldi R (2017). Liquid-phase H-transfer from 2-propanol to phenol on Raney Ni: surface processes and inhibition. *ACS Catal*, 7: 2437–2445.

Kenny JK, Brandner DG, Neefe SR, Michener WE, Román-Leshkov Y, Beckham GT, and Medlin JW (2022). Catalyst choice impacts aromatic monomer yields and selectivity in hydrogen-free reductive catalytic fractionation. *React Chem Eng*, 7: 2527–2533.

Klein I, Marcum C, Kenttämaa H, and Abu-Omar MM (2016). Mechanistic investigation of the Zn/Pd/C catalyzed cleavage and hydrodeoxygenation of lignin. *Green Chem*, 18: 2399–2405.

Kulka M and Hibbert H (1943). Studies on lignin and related compounds. LXVII. Isolation and identification of 1-(4- hydroxy-3,5-dimethoxyphenyl)-2-propanone and 1-(4-hydroxy-3-methoxyphenyl)-2-propanone from maple wood ethanolysis products – metabolic changes in lower and higher plants. *J Am Chem Soc*, 65: 1180–1185.

Kumaniaev I, Subbotina E, Sävmarker J, Larhed M, Galkin MV, and Samec JSM (2017). Lignin depolymerization to monophenolic compounds in a flow-through system. *Green Chem*, 19: 5767–5771.

Lan W, Amiri MT, Hunston CM, and Luterbacher JS (2018). Protection group effects during α,γ-diol lignin stabilization promote high-selectivity monomer production. *Angew Chem Int Ed Engl* Jan 26, 57(5): 1356–1360.

Lan W and Luterbacher JS (2019). Preventing lignin condensation to facilitate aromatic monomer production. *Chimia (Aarau)* Aug 21, 73(7): 591–598.

Lancefield CS, Ojo OS, Tran F, and Westwood NJ (2015). Isolation of functionalized phenolic monomers through selective oxidation and C-O bond cleavage of the β-O-4 linkages in lignin. *Angew Chem Int Ed Engl* Jan 2, 54(1): 258–262.

Li C, Zhao X, Wang A, Huber GW, and Zhang T (2015). Catalytic transformation of lignin for the production of chemicals and fuels. *Chem Rev* 2015 Nov 11, 115(21): 11559–11624.

Li C, Zheng M, Wang A, and Zhang T (2012). One-pot catalytic hydrocracking of raw woody biomass into chemicals over supported carbide catalysts: simultaneous conversion of cellulose, hemicellulose and lignin. *Energy Environ Sci*, 5: 6383–6390.

Li L, Dong L, Liu X, Guo Y, and Wang Y (2020a). Selective production of ethylbenzene from lignin oil over FeOx modified Ru/Nb2O5 catalyst. *Appl Catal B Environ*, 260: 118143.

Li S, Luo Z, Wang W, Sun H, Xie J, and Liang X (2020b). Catalytic fast pyrolysis of enzymatic hydrolysis lignin over Lewis-acid catalyst niobium pentoxide and mechanism study. *Bioresour Technol*, 316: 123853.

Liu W-J, Jiang H, and Yu H-Q (2015). Thermochemical conversion of lignin to functional materials: a review and future directions. *Green Chem*, 17: 4888–4907.

Llevot A, Grau E, Carlotti S, Grelier S, and Cramail H (2016). From lignin-derived aromatic compounds to novel biobased polymers macromol. *Rapid Commun.*, 37: 9–28.

Lundquist K (1970). Acid degradation of lignin. II. Separation and identification of low-molecular weight phenols. *Acta Chem Scand*, 24: 889–907.

Luo H, Klein IM, Jiang Y, Zhu H, Liu B, Kenttämaa HI, and Abu-Omar MM (2016). Total utilization of miscanthus biomass, lignin and carbohydrates, using earth abundant nickel catalyst. *ACS Sustain Chem Eng*, 4: 2316–2322.

Luo X, Li Y, Gupta NK, Sels B, Ralph J, and Li S (2020). *Protection Strategies Enable Selective Conversion of Biomass*. N. p., Vol. 59, Germany, p. 11704–11716.

Luo Z, Qian Q, Sun H, Wei Q, Zhou J, and Wang K (2023). Lignin-first biorefinery for converting lignocellulosic biomass into fuels and chemicals. *Energies*, 16: 25.

Luterbacher JS, Azarpira A, Motagamwala AH, Lu F, Ralph J, and Dumesic JA (2015). Lignin monomer production integrated into the γ-valerolactone sugar platform. *Energy Environ Sci*, 8: 2657–2663.

Luterbacher JS, Rand JM, Alonso DM, Han J, Youngquist JT, Maravelias CT, Pfleger BF, and Dumesic JA (2014). Nonenzymatic sugar production from biomass using biomass-derived γ-valerolactone. *Science* Jan 17, 343(6168): 277–280.

Ma H, Liu -W-W, Chen X, Wu Y-J, and Yu Z-L (2009). Enhanced enzymatic saccharification of rice straw by microwave pretreatment. *Bioresource Technol*, 100: 1279–1284.

Ma R, Hao W, Ma X, Tian Y, and Li Y (2014). Catalytic ethanolysis of Kraft lignin into high-value small-molecular chemicals over a nanostructured α-molybdenum carbide catalyst. *Angew Chem Int Ed Engl* Jul 7, 53(28): 7310–7315.

Ouyang X, Huang X, Hendriks BMS, Boot MD, and Hensen EJM (2018). Coupling organosolv fractionation and reductive depolymerization of woody biomass in a two-step catalytic process. *Green Chem*, 10: 2308–2319.

Paone E, Tabanelli T, and Mauriello F (2020). The rise of lignin biorefinery. *Green Sustain Chem*, 24: 1–6.

Parsell T, Yohe S, Degenstein J, Jarrell T, Klein I, Gencer E, Hewetson B, Hurt M, Kim JI, Choudhari H, Saha B, Meilan R, Mosier N, Ribeiro F, Delgass WN, Chapple C, Kenttämaa HI, Agrawal R, and Abu-Omar MM (2015). A synergistic biorefinery based on catalytic conversion of lignin prior to cellulose starting from lignocellulosic biomass. *Green Chem*, 17: 1492–1499.

Parsell TH, Owen BC, Klein I, Jarrell TM, Marcum CL, Haupert LJ, Amundson LM, Kenttämaa HI, Ribeiro F, and Miller JT (2013). Cleavage and hydrodeoxygenation (HDO) of C–O bonds relevant to lignin conversion using Pd/Zn synergistic catalysis. *Chem Sci*, 4: 806–813.

Pepper JM and Lee YW (1969). Lignin and related compounds. I. A comparative study of catalysts for lignin hydrogenolysis. *Can J Chem*, 47: 723–727.

Pepper JM and Rahman MD (1987). Lignin and related compounds. XI. Selective degradation of aspen poplar lignin by catalytic hydrogenolysis. *Cellul Chem Technol*, 21: 233–239.

Pepper JM and Supathna P (1978). Lignin and related compounds. VI. Study of variables affecting hydrogenolysis of spruce wood lignin using a rhodium-on-charcoal catalyst. *Can J Chem*, 56: 899–902.

Questell-Santiago YM, Galkin MV, Barta K, and Luterbacher JS (2020). Stabilization strategies in biomass depolymerization using chemical functionalization. *Nat Rev Chem*, 4: 311–330.

Ragauskas AJ, Beckham GT, Biddy MJ, Chandra R, Chen F, Davis MF, Davison BH, Dixon RA, Gilna P, Keller M, Langan P, Naskar AK, Saddler JN, Tschaplinski TJ, Tuskan GA, and Wyman CE (2014). Lignin valorization: improving lignin processing in the biorefinery. *Science*, 344: 1246843.

Rahimi A, Ulbrich A, Coon JJ, and Stahl SS (2014). Formic-acid-induced depolymerization of oxidized lignin to aromatics. *Nature*, 515: 249–252.

Rajesh Banu R, Kavitha S, Yukesh Kannah R,T, Poornima Devi M, Gunasekaran M, Kim SH, and Kumar G (2019). A review on biopolymer production via lignin valorization. *Bioresour Technol*, 290: 121790.

Renders T, Cooreman E, Van den Bosch S, Schutyser W, Koelewijn SF, Vangeel T, Deneyer A, Van den Bossche G, Courtin CM, and Sels BF (2018). Catalytic lignocellulose biorefining in n-butanol/water: a one-pot approach toward phenolics, polyols, and cellulose. *Green Chem*, 20: 4607–4619.

Renders T, Van den Bosch S, Koelewijn SF, Schutyser W, and Sels BF (2017). Lignin-first biomass fractionation: the advent of active stabilisation strategies. *Energy Environ Sci*, 10: 1551–1557.

Renders T, Van den Bosch S, Vangeel T, Ennaert T, Koelewijn S-F, Van den Bossche G, Courtin CM, Schutyser W, and Sels BF (2016). Synergetic effects of alcohol/water mixing on the catalytic reductive fractionation of poplarwood. *ACS Sustain Chem Eng*, 4: 6894–6904.

Renders T, Van den Bossche G, Vangeel T, Van Aelst K, and Sels B (2019). Reductive catalytic fractionation: state of the art of the lignin-first biorefinery. *Curr Opin Biotechnol*, 56: 193–201.

Rinaldi R, Jastrzebski R, Clough MT, Ralph J, Kennema M, Bruijnincx PC, and Weckhuysen BM (2016). Paving the way for lignin valorisation: recent advances in bioengineering, biorefining and catalysis. *Angew Chem Int Ed Engl* Jul 11, 55(29): 8164–8215.

Rinaldi R, Woodward R, Ferrini P, and Rivera H (2018). Lignin-first biorefining of lignocellulose: the impact of process severity on the uniformity of lignin oil composition. *J Braz Chem Soc*, 30: 479–491.

Schutyser W, Renders T, Van den Bosch S, Koelewijn SF, Beckham GT, and Sels BF (2018). Chemicals from lignin: an interplay of lignocellulose fractionation, depolymerisation, and upgrading. *Chem Soc Rev*, 47: 852–908.

Schutyser W, Van den Bosch S, Renders T, De Boe T, Koelewijn SF, Dewaele A, Ennaert T, Verkinderen O, Goderis B, Courtin CM, and Sels BF (2015). Influence of bio-based solvents on the catalytic reductive fractionation of birch wood. *Green Chem*, 17: 5035–5045.

Shuai L, Amiri MT, Questell-Santiago YM, Héroguel F, Li Y, Kim H, Meilan R, Chapple C, Ralph J, and Luterbacher JS (2016). Formaldehyde stabilization facilitates lignin monomer production during biomass depolymerization. *Science* Oct 21, 354(6310): 329–333.

Shuai L and Luterbacher J (2016). Organic solvent effects in biomass conversion reactions. *ChemSusChem*, 9: 133–155.

Sobolev I, Arlt HG, and Schuerch C (1957). Alkaline hydrogenation pulping. *Ind Eng Chem*, 49: 1399–1400.

Song Q, Wang F, Cai J, Wang Y, Zhang J, Yu W, and Xu J (2013). Lignin depolymerization (LDP) in alcohol over nickel-based catalysts via a fragmentation–hydrogenolysis process. *Energy Environ Sci*, 6: 994–1007.

Spridon I (2018). Biological and pharmaceutical applications of lignin and its derivatives: a minireview cell. *Chem Technol*, 52: 543–550.

Steeves WH and Hibbert H (1939). Studies on lignin and related compounds. XLII. The isolation of a bisulfate soluble "extracted lignin". *J Am Chem Soc*, 61: 2194–2195.

Sun J, Li H, Xiao L-P, Guo X, Fang Y, Sun R-C, and Song G (2019). Fragmentation of Woody lignocellulose into primary monolignols and their derivatives. *ACS Sustain Chem Eng*, 7: 4666–4674.

Sun Z and Barta K (2018). Cleave and couple: toward fully sustainable catalytic conversion of lignocellulose to value added building blocks and fuels. *Chem Commun*, 54: 7725–7745.

Sun Z, Bottari G, Afanasenko A, Stuart MCA, Deuss PJ, Fridrich B, and Barta K (2018b). Complete lignocellulose conversion with integrated catalyst recycling yielding valuable aromatics and fuels. *Nat Catal*, 1: 82–92.

Sun Z, Fridrich B, de Santi A, Elangovan S, and Barta K (2018a). Bright side of lignin depolymerization: toward new platform chemicals. *Chem Rev*, 118: 614–678.

Taran OP, Miroshnikova AV, Baryshnikov SV, Kazachenko AS, Skripnikov AM, Sychev VV, Malyar YN, and Kuznetsov BN (2022). Reductive catalytic fractionation of spruce wood over Ru/C bifunctional catalyst in the medium of ethanol and molecular hydrogen. *Catalysts*, 12: 1384.

Thakur VK, Thakur MK, Raghavan P, and Kessler MR (2014). Progress in green polymer composites from lignin for multifunctional applications: a review. *ACS Sustain Chem Eng*, 2: 1072–1092.

Van de Pas DJ and Torr KM. (2017). Bio-based epoxy resins from deconstructed native softwood lignin. *Biomacromolecules*, 18: 2640–2648.

Van den Bosch S, Renders T, Kennis S, Koelewijn SF, Van den Bossche G, Vangeel T, Deneyer A, Depuydt D, Courtin CM, Thevelein JM, Schutyser W, and Sels BF (2017). Integrating lignin valorization and bio-ethanol production: on the role of Ni-Al2O3 catalyst pellets during lignin-first fractionation. *Green Chem*, 19: 3313–3326.

Van den Bosch S, Schutyser W, Koelewijn SF, Renders T, Courtin CM, and Sels BF (2015a). Tuning the lignin oil OH–content with Ru and Pd catalysts during lignin hydrogenolysis on birch wood. *Chem Commun (Camb)*, 51: 13158–13161.

Van den Bosch S, Schutyser W, Vanholme R, Driessen T, Koelewijn SF, Renders T, De Meester B, Huijgen WJJ, Dehaen W, Courtin CM, Lagrain B, Boerjan W, and Sels BF (2015b). Reductive lignocellulose fractionation into soluble lignin-derived phenolic monomers and dimers and processable carbohydrate pulps. *Energy Environ Sci*, 8: 1748–1763.

Vardon DR, Franden MA, Johnson CW, Karp EM, Guarnieri MT, Linger JG, Salm MJ, Strathmann TJ, and Beckham GT (2015). Adipic acid production from lignin. *Energy Environ Sci*, 8: 617–628.

Wang G, Cao Z, Gu D, Pfänder N, Swertz A-C, Spliethoff B, Bongard H-J, Weidenthaler C, Schmidt W, Rinaldi R, and Schüth F (2016). Nitrogen-doped ordered mesoporous carbon supported bimetallic PtCo nanoparticles for upgrading of biophenolics. *Angew Chem Int Ed Engl*, 55(31): 8850–8855.

Wang S, Ma S, Xu C, Liu Y, Dai J, Wang Z, Liu X, Chen J, Shen X, Wei J, and Zhu J (2017). Vanillin-derived high-performance flame retardant epoxy resins: facile synthesis and properties. *Macromolecules*, 50: 1892–1901.

Wang X and Rinaldi R (2013). A route for lignin and bio-oil conversion: dehydroxylation of phenols into arenes by catalytic tandem reactions. *Angew Chem*, 52: 11499–11503.

Weingarten R, Conner WC, and Huber GW (2012). Production of levulinic acid from cellulose by hydrothermal decomposition combined with aqueous phase dehydration with a solid acid catalyst. *Energy Environ Sci*, 5(6): 7559–7574.

Xu C, Arancon RAD, Labidi J, and Luque R (2014). Lignin depolymerisation strategies: towards valuable chemicals and fuels. *Chem Soc Rev*, 43: 7485–7500.

Yan N, Zhao C, Dyson PJ, Wang C, Liu LT, and Kou Y (2008). Selective degradation of wood lignin over noble-metal catalysts in a two-step process. *ChemSusChem*, 1(7): 626–629.

Yan W, Acharjee TC, Coronella CJ, and Vásquez VR (2009). Thermal pretreatment of lignocellulosic biomass. *Environ Prog Sustain Energy*, 28: 435–440.

Yokoyama T (2014). Revisiting the mechanism of b-O-4 bond cleavage during acidolysis of lignin. Part 6: a review. *J Wood Chem Technol*, 35: 27–42.

Zakzeski J, Bruijnincx PC, Jongerius AL, and Weckhuysen BM (2010). The catalytic valorization of lignin for the production of renewable chemicals. *Chem Rev*, 110: 3552–3599.

Zhai Y, Li C, Xu G, Ma Y, Liu X, and Zhang Y (2017). Depolymerization of lignin via a non-precious Ni–Fe alloy catalyst supported on activated carbon. *Green Chem*, 19: 1895–1903.

Zhang J, Liu J, Kou L, Zhang X, and Tan T (2019a). Bioethanol production from cellulose obtained from the catalytic hydro-deoxygenation (lignin-first refined to aviation fuel) of apple wood. *Fuel*, 250: 245–253.

Zhang K, Li H, Xiao LP, Wang B, Sun RC, and Song G (2019b). Sequential utilization of bamboo biomass through reductive catalytic fractionation of lignin. *Bioresour Technol*, 285: 121335.

Zhao S and Abu-omar MM (2017). Synthesis of renewable thermoset polymers through successive lignin modification using lignin-derived phenols. *ACS Sustain Chem Eng*, 5: 5059–5066.

Zhu S, Guo J, Wang X, Wang J, and Fan W (2017). Alcoholysis: a promising technology for conversion of lignocellulose and platform chemicals. *ChemSusChem*, 10: 2547–2559.

5

Lignin Production

Abstract

Lignin has typically been regarded as a waste product of low value. Syngas, phenolic compounds, carbon fiber, a variety of oxidized products, and multifunctional hydrocarbons are among the high-value products which can be produced from lignin. In response to the growing interest in lignin and its products, numerous novel lignin extraction methods have been developed to produce extremely pure lignin. The majority of these methods for extracting lignin have been investigated at pilot and commercial scale. The methods developed for pilot-scale and commercial-scale lignin production are presented in this chapter.

Keywords *Lignin valorization; Biorefinery; LignoBoost; TMP-Bio; AFEX, PROESA®; BioFlex; Steam explosion; SLRP; SunCarbon; LignoForce; Pilot-scale production; Commercial- scale production*

5.1 Introduction

Lignin has typically been thought of as a low-value waste product. The high-value products that can be made from lignin include syngas, carbon fibre, phenolic compounds, a variety of oxidized products, and multifunctional hydrocarbons (Agrawal et al., 2014; Ľudmila et al., 2015; Luo and Abu-Omar, 2017; Ragauskas et al., 2014). The establishment of new cellulosic biorefineries will result in an excess supply of unsulfoned lignin, despite the fact that the pulp and paper industry is widely acknowledged as a conventional source of lignin. Pulp and paper mills all over the world currently produce 50 to 70 million tons of lignin every year (Luo and Abu-Omar, 2017; Mandlekar et al., 2018). The amount of biofuel required to be produced under the Renewable Fuel Standard (RFS) programme, 60 billion gallons, is expected to rise by 225 million tonnes yearly by 2030 (EPA, 2018). A byproduct of the production of biofuel is lignin. After conversion, 0.225 billion tonnes of biomass remain as a lignin-rich byproduct (Holladay et al., 2007). To produce this amount of biofuel, 0.75 billion tonnes of biomass are required. Less than 2% of the lignin generated by the paper industry is currently used to create specialty chemicals like surfactants, adhesives, dispersants, and other value-added products (Luo and Abu-Omar, 2017). The majority of it is used to generate heat and electricity as low-value fuel. This is true despite the fact that more and more research is being done on the potential for turning lignin into useful materials, chemicals, and fuel. The formation of a co-product derived from lignin with added value can increase the profitability of the paper industry

Table 5.1 Different sources of lignin and their current volume.

Lignin					
Sulfur containing lignin			**Sulfur free lignin**		
Kraft lignin - sulfur content: 1% to 3% - volume: 55 million tons	Lignosulfonates - sulfur content: 3.5% to 8% - volume: 1 million tons	Hydrolyzed lignin - sulfur content: 0–1% - volume: not available	Organosolv lignin - volume: not available	Soda lignin - volume: 6000 kilo tonnes	Lignin from second generation biorefinery process - volume: 100 kilo tonnes

Reproduced from Bajwa et al (2019) / with permission of ELSEVIER

and the second generation of biorefineries while also ensuring the stability and long-term vitality of the biorefineries by valorizing their byproduct lignin. Growth in the lignin market is anticipated to be driven by investments in the creation of carbon fibers based on lignin. In 2022, the lignin market will be worth USD 790 million (https://www.gminsights.com/industry-analysis/lignin-market).

Table 5.1 shows the current volume of various lignin sources (Agrawal et al., 2014; Bruijnincx et al., 2016; Ľudmila et al., 2015; Mandlekar et al., 2018).

5.2 Pilot-scale

Numerous novel lignin extraction methods have been developed in response to the growing interest in lignin and products obtained from lignin in order to produce extremely pure lignin (Salaghi et al., 2021). Most of these techniques for extracting lignin have been tested in small-scale pilot projects and have the potential to be made commercially available. The list of pilot-scale processes studied is presented in Table 5.2 (Salaghi et al., 2021). These technologies typically yield lignin with interesting properties that lend themselves to applications in the downstream chain (Salaghi et al., 2021).

5.2.1 Ammonia Fiber Explosion Lignin

In Michigan, United States, a plant for pilot-scale pretreatment of biomass produced lignin using an ammonia-based fibre explosion (AFEX) process (https://bioeconomy.msu.edu/2015/09/25/a-patent-and-milestone-why-mbis-afex-biomass-pretreatment-is-of-global-importance). Liquid ammonia is

Table 5.2 Lignin production at pilot scale.

Ammonia fiber explosion (AFEX) lignin
Steam explosion process
BioFlex process
German Lignocellulose Feedstock Biorefinery project
PROESA® lignin
FABIOLA™ lignin
Fast pyrolysis lignin
Sequential liquid-lignin recovery and purification (SLRP) technology

Adapted from Salaghi et al. (2021)

used in this method to treat lignocellulosic biomass in a closed vessel for 30 to 60 minutes at a lower temperature of 60 to 90 °C and pressure of 3 MPa or higher (Teymouri et al., 2005). At high temperatures and pressures, when ammonia is present, carbohydrates become active and the crystallinity of the cellulose in biomass changes. In the meantime, some hemicellulose and lignin are removed. After ammonia treatment, filtration can be used to separate the lignin-rich residues and carbohydrate-rich hydrolysate from the biomass slurry. The majority of hemicellulose and cellulose are unaffected by degradation (Moniruzzaman et al., 1997). Conversely, refluxing can turn unhydrolyzed residues into lignin (such as ethanol: water), drying, washing with hexane and dichloromethane, concentration and drying. About 38% of the stover feedstock is turned into lignin-enriched residue.

The fact that the reaction agents, such as ethanol and ammonia, can be recovered and repurposed at the conclusion of the process makes AFEX appealing to industries, which is one major advantage of AFEX lignin production. Because ammonia oxidation can transform lignin that is nitrogen-free into nitrogenous materials, the AFEX lignin is a good material for soil fertilizer lignin (Klinger et al., 2013). The AFEX process can produce high-purity hemicellulose and cellulose in addition to lignin. These polysaccharides are suitable for the production of bioethanol because about 90% of them can be converted into fermentable sugars. However, this process can also lead to the formation of ecotoxic nitrogenous compounds derived from carbohydrates under ammoxidation conditions which may pose a risk when used (Klinger et al., 2013).

5.2.2 Steam Explosion Process

DSM-POET, Emmetsburg, USA, reported using steam explosion to refine biomass into sugars and lignin at a pilot scale (POET-DSM, 2012a). By soaking the biomass in diluted sulfuric acid, it is first softened. In a steam explosion reactor, high-pressure steam is used to treat the biomass impregnated with acid. After cooling, the produced solid is separated through filtration and washed with water. Acetic acid and sugars derived from hemicellulose are abundant in the aqueous hydrolysate solution. Reconcentration is used to get the acid back, and the polysaccharides could be used to make hydrogen or ferment to make bioethanol. In the meantime, the residual solid still contains lignin (POET-DSM 2012b). The resulting lignin has few carbohydrates and impurities from wood extraction. About 4–7% (dry weight) of the lignin is obtained from birchwood (POET-DSM, 2012a).

Compared to the other technical lignins produced, this process's lignin is more like the native lignin. According to Rachele Pucciariello, lignin effectively protected polycaprolactone (PCL) from UV rays (Pucciariello et al., 2008) after successfully combining steam explosion lignin and PCL. However, the treatment at a high temperature results in lignin with a low molecular weight. Using a steam explosion, lignin from birch wood has a polydispersity index of 1.5 and an average molecular weight of only 2250–2620 g mol^{-1} (Sun and Tomkinson, 2000). Steam explosion lignin appears to be a good material for producing phenolic chemicals due to its lower molecular weight. Steam explosion technology has the potential to make full use of biomass materials as lignin can be used to make aromatics and biomass polysaccharides can be used to make bioethanol.

5.2.3 BioFlex Process

An innovative method for pre-treatment of lignocellulosic biomass, the Lixea proprietary BioFlex process, offers unprecedented adaptability of feedstocks. In the Biorenewable Development Centre in York, UK, the first BioFlex biorefining pilot plant was built. (https://www.lixea.co/#bioflex-process). This process can utilize a variety of biomasses, including residues from forestry, agriculture, energy crops, and even wood from construction waste (https://www.filtsep.com/oil-and-gas/news/bioflex-uses-waste-wood-instead-of-crude-oil). Low-cost ionic liquid that can be recycled can be used to recover biomass's cellulose and lignin components, as well as waste wood's byproducts and heavy metals. Ionic liquid is used to process the biomass feedstock at atmospheric pressure and below 200 °C.

This procedure produces no harmful emissions due to the low vapor pressure of ionic liquids. This method was developed on the basis of the ionosolv biorefining process, which makes use of precipitation and filtration to separate any kind of lignocellulosic biomass into the cellulose and lignin biorefining intermediates. The BioFlex process can also be used to recover heavy metal-containing compounds in biomass as a byproduct. Lixea intends to design and construct a plant for industrial demonstration (about 30 000 tons annually) in the near future (https://www.lixea.co/#bioflex-process). These parts can be used as new materials, biochemicals, plastics precursors, or biochemicals once they are each isolated. To be more specific, the resulting ionosolv lignin can be utilized as an alternative to phenols in the production of adhesives and resins (https://www.filtsep.com/chemicals/features/bioflex-turns-wood-into-green-bioproducts-10). The lignin produced by BioFlex lacks sulfur as this process does not use any substance that contains sulfur (https://www.filtsep.com/oil-and-gas/news/bioflex-uses-waste-wood-instead-of-crude-oil; 2019). In comparison to conventional technical lignin, BioFlex lignin is now more competitive as a result and also environmentally friendly. PFA resins, for example, are examples of a suitable sustainable chemical industry. Because it is difficult and costly to synthesize and purify ionic liquids, BioFlex lignin is expensive to make.

5.2.4 German Lignocellulose Feedstock Biorefinery Project

This project was a joint venture of 15 partners with the goal of developing a pilot-scale sustainable and cost-effective biorefinery process for producing cellulose, hemicellulose, and lignin from hardwood of higher quality for use in biomaterials (https://www.yumpu.com/en/document/read/41220925/the-german-lignocellulose-feedstock-biorefinery-project). To begin, alcohol is used to extract the extractives, like triterpene, from biomass. After that, extractive-free biomass is treated using ethanol and water in a pulping process (2–4 hours, 170–180 °C) (50:50). Hot water precipitates carbohydrates and lignin from cooking liquor. Using NaOH and HCl to adjust the pH, the lignin and polysaccharides can be separated as follows: lignin is dissolved in NaOH solution, the cellulose is left in a solid fraction for filtration; the lignin is then removed from the aqueous phase by adding HCl. 80 kg of lignin, or 27% of the feedstock, can be produced from processing approximately 300 kg of wood per week.

The produced lignin is suitable for the production of biofuel and phenolic chemicals because it does not contain any sulfur. The Fraunhofer Institut Chemische Technologie in Pfinztal, Germany, successfully produced a polylactic acid containing 20% lignin. Dynea, with headquarters in Lillestrm, Norway, claimed that wood construction boards could be made from the resulting resin. Additionally, up to 30 weight % of this lignin was utilized to partially replace phenol-based resin. However, the biorefinery of hardwoods like poplar and beech was the sole focus of this project. This method's ability to separate lignin from softwood warrants additional research. A problem with lignin production is the cost, which could limit its use in other areas. It was reported that this project's lignin production cost € 395 per t; whereas it was reported that the conventional Kraft lignin cost approximately USD 250 per t (Abbati de Assis et al., 2018). This project may produce fermentable C6/C5 sugars for bioethanol applications in addition to lignin.

5.2.5 Proesa® Lignin

In a facility in Crescentino, Italy, the Proesa® technology was created for treating cellulosic biomass from agricultural waste and energy crops for separating lignin from hemicelluloses and cellulose (https://www.bio.org/sites/default/files/legacy/bioorg/docs/beta%20renewables%20proesa%20technology%20june%202013_bio_michele_rubino.pdf). In a method known as "smart cooking," steam and water are used to pre-treat biomass to produce biomass slurry without using corrosive chemicals. The slurry is then fermented and hydrolyzed with enzymes and microorganisms. Evaporation is used to collect the bioethanol from the fermentable sugars that are produced, leaving behind lignin in solid residues for applications that add value (Palmisano, 2013). Woody biomass, agricultural wastes, energy crops, and bagasse can all be processed in this pilot plant at a rate of one ton per day. Bioethanol accounts for approximately 20% of the products, while residual lignin and unfermentable sugars account for the remaining 80%.

Currently, the Proesa® lignin is simply burned to recover energy. Proesa® lignin is said to be suitable for value-added applications such as the production of bioplastics and phenolic chemicals like aromatics, terephthalic acid, and phenols. The cooking process only makes use of steam and water, and no additional corrosive chemicals are used so the structural damage to lignin is minimum. As a result, Proesa® lignin is expected to have a structure that is comparable to that of native lignin found in biomass. By treating the obtained polysaccharides with enzymes, this technology could also produce bioethanol of high quality. Proesa® lignin may be heavily contaminated with ash and sugar as a process liquor that only contains water probably cannot separate lignin from biomass. Proesa® lignin production necessitates a lengthy cooking time because of the unique process, which may also increase energy consumption. Production of Proesa® lignin is also more expensive when the enzyme and microorganism are used. The Proesa® lignin's use may be hindered by these drawbacks.

5.2.6 FABIOLA™ Lignin

Ten European partners are working on the UNRAVEL project, which focuses on cutting-edge pretreatment, separation, and conversion techniques for complex lignocellulosic material (http://unravel-bbi.eu/successful-scale-up-of-the-fabiola-process-in-a-lignocellulose-biorefinery-pilot-plant; 2019). The pilot-scale scaling of the FABIOLA™ fractionation process at the Fraunhofer Center for CBP as part of the UNRAVEL project demonstrated significant potential for enhancing lignocellulosic biomass pretreatment. This method yields high-quality, low-cost compounds that can handle a variety of biomass types (Damen et al., 2017). Acetone is used to cook wood and herbaceous biomass in a digester at 140 °C. The pulp is separated from the mixture after filtration to prepare it for further fermentation for making ethanol. A nanofiltration system is used to filter the liquid phase to separate the lignin from the organic solvent. Water then is added to the liquor with more lignin in it. Centrifugation is used to collect lignin, is added to the liquor that already contains more lignin. Filtration and distillation are used to get the acetone solvent back and use it again. The FABIOLA™ process has a delignification rate of 87%. In the future, it is anticipated that this process can recover 95% of lignin with further advancement.

The FABIOLA™ lignin does not contain any sulfur contamination due to the use of sulfur-free pulping liquor in the FABIOLA™ process. This is a significant advantage for applications in biofuels. Due to the relatively low cooking temperature of the FABIOLA™ process (FABIOLA™ cooking at 140 °C, traditional organosolv at 180 °C), FABIOLA™ lignin has a structure that is very similar to that of native lignin and has more β-O-4 linkages than other types of lignin, which is native lignin's primary interlinkage. The FABIOLA™ process also produces xylose, a popular low-calorie sweetening agent. The production cost of FABIOLA™ lignin is high due to the high demand for acetone solvent. Because the cooking process takes at least two hours, FABIOLA™ lignin production uses a lot of energy. Although FABIOLA™ lignin has demonstrated desirable properties, its production's high chemical demand and high energy requirement may prevent it from being used in additional applications.

5.2.7 Fast Pyrolysis Lignin

A significant step has been taken towards the commercialization of a biorefinery based on rapid pyrolysis with the successful launch of a novel pilot-scale thermo-chemical fractionation plant by BTG, Netherlands (https://www.btgworld.com/en/rtd/technologies/bio-materials-chemicals). It is a novel two-step conversion method. Biomass is pyrolyzed in a reactor chamber at temperatures ranging from 450 to 600 °C so that the mineral-free, liquid product (fast pyrolysis bio-oil) can be fractionated into separate, depolymerized fractions that maintain the key chemical functionalities (https://www.ieabioenergy.com/wp-content/uploads/2018/11/902-BTGBiomass-VandeBeld-smaller.pdf). The pyrolytic oil is then separated from the gaseous products of the pyrolysis using a condenser. The primary byproducts of the pyrolytic oil are extractives, sugars, and lignin. The biochar and biogas from pyrolysis, in addition to the pyrolytic oil, are burned to produce flue gas for energy generation. The system can handle any kind of biomass resource, so even waste biomass which is greener than agricultural crops can be used to make energy, sugar, and lignin in this system. This plant is said to be able to process 50 tons of biomass per hour and produce 3.2 tons of fast pyrolysis biofuel per hour, yielding 6.4%.

BTG claims that the fast pyrolysis lignin is suitable for further modification because it reacts more strongly than native and kraft lignin (https://bio4products.eu/wp-content/uploads/2019/11/4.-Hans-Heeres_Thermo-Chemical-Fractionation.pdf). Since no chemical is added during pyrolysis, it reacts more vigorously than kraft and native lignin. Fast pyrolysis lignin has the potential to partially replace fossil fuel with an additional treatment upgrade using commercial catalysts. Fast pyrolysis can produce pyrolytic sugars that can be used to make furan-based resins and extracts similar to tall oil liquids. These sugars can also be used as co-feed for making diesel-like products. In a char combustor, biochar from pyrolysis is burned to produce electricity-producing flue gas. Additionally, the condenser yields biogas for energy generation. Pyrolytic lignin cannot be completely separated from the produced mixed oil product due to the fact that the bio-oil is a mixture of lignin, sugar, and other extractives. Consequently, the biomass's carbohydrate may have contaminated pyrolytic lignin with sugar.

5.2.8 Sequential Liquid-lignin Recovery and Purification Technology

A pilot-scale sequential liquid-lignin recovery and purification(SLRP)-based lignin recovery technique was developed by the Liquid Lignin Company (South Carolina, USA) (Lake and Blackburn, 2014). Using this method, the kraft pulping process's lignin and black liquor are separated by gravity as a true liquid phase. This is not the same as the conventional methods, which precipitate lignin into smaller solid particles. Carbon dioxide and black liquor from the kraft pulping process are passed into the carbonation reactor, respectively, through the reactor's top and bottom injection ports. As a result of this design, the slurry contains more lignin than lignin particles. At a pH of 9.0–10.0, black liquor reacts with carbon dioxide to produce liquid lignin. After that, droplets and liquid lignin collide to form an easy-to-separate bulk dense slurry. The carbonation reactor maintains a pH of 2.0–3.0 and a temperature of 100–150 °C to prevent the formation of lignin particles. After that, the slurry is moved to an acidification reactor, where it will react with sulfuric acid to form a precipitate with a pH between 2.0 and 3.0. Following filtration, washing, and drying, solid lignin is produced. As the production of SLRP lignin needs a substantial amount of carbon dioxide, the gas capture system in this system contributes to the reduction of carbon dioxide loss as well as air pollution. In the black liquor, the SLRP procedure was able to recover 55% of the lignin (Lake and Blackburn, 2014).

Due to the small equipment that is a result of the continuous operation of SLRP, it is claimed that the capital and operating costs are significantly lower than those of conventional processes. This value in polymeric applications of this process is limited by the purity and molecular weight of the lignin it produces (Lake and Blackburn, 2014; Velez and Thies, 2016). Additionally, it has been reported that the SLRP process yields lignin with a high content of sulfur and ash (2–4%) (Kihlman, 2015). However, in the context of a biorefinery, due to its low production cost, SLRP lignin can still be utilized for biofuel applications with the right purification to cut down on sulfur and ash.

Notably, because of their protracted, energy-intensive protocols and lower lignin yields, the aforementioned methods are still primarily investigated for research purposes at pilot or laboratory scales. These lignin separation techniques may nevertheless lead to the commercialization of lignin because they could result in lignin that is less contaminated and more reactive and functional group-rich than conventional industrial lignin (Chakar and Ragauskas, 2004). These efficacy of the lignin separation technique has also improved as a result of modern technology. These lignin separation techniques could be scaled up for future commercialization with systematic optimization.

5.3 Commercial scale

Commercial lignin production can be broken down into four categories: kraft lignin derived from kraft pulping, lignosulfonate from sulfite pulping, organosolv from organosolv pulping, and enzymatic lignin from the enzymatic ethanol manufacturing process. Typically, industrial lignin is burned as a fuel to generate energy for biomass treatment (Korbag and Mohamed Saleh, 2016). Compared to laboratory-produced lignin, lignin from industrial pulping processes exhibits greater heterogeneity and more severe degradation (Brodin, 2009). Compared to lignin made in laboratories, industrial lignin has higher dispersity indices and lower molecular weights. For instance, MWL from spruce has a Mw of approximately 23 500 g mol^{-1} and a polydispersity index of 3.7; kraft lignin from spruce, on the other hand, has 4.5 and 4500 g mol^{-1} values (Anderson et al., 2006; Brodin, 2009). This is due to the possibility that the structure of lignin could be altered by the cooking liquor used in industrial processes. With the right modifications, industrial lignin is widely used for a variety of value-added applications because it is a cheap, sustainable, and abundantly available material (Chen and Li, 2000). Table 5.3 provides list of commercial-scale processes (Salaghi et al., 2021).

5.3.1 LignoForce™ Technology

The LignoForce™ technology was developed by FPInnovations and NORAM in Canada, which can be used to recover lignin from black liquor (Kouisni et al., 2014). West Fraser (Quebec, Canada) uses this technology on a commercial scale. Kraft lignin is purified through a series of steps in this system. Oxygen is injected into black liquor to initiate oxidation. The oxidized black liquor is then coagulated after being acidified with carbon dioxide at 70–75 °C to a pH of 9.5–10.0. The acidified black liquor is precipitated and filtered in the following step. Sulfuric acid and water are

Table 5.3 Lignin production on commercial scale.

LignoForce™ technology
LignoBoost™ technology
SunCarbon lignin
Organosolv and soda lignin
Thermo-mechanical pulp-bio lignin
Lignosulfonate production
Kraft lignin production

Adapted from Salaghi et al. (2021)

used to wash the liquor cake that has been enriched with lignin through this step. Pressing and air drying produce LignoForce™ lignin (Kouisni et al., 2012). With lignin particles ranging from 5 to 10 micrometers, LignoForce™ was able to recover 60–62% of the lignin in the black liquor. FPInnovations built a lignin demonstration plant to produce 12.5 kg h^{-1} of lignin of superior quality at the Resolute Thunder Bay mill (Diels and Browne, 2018).

LignoForce™ lignin has a number of advantages over standard kraft lignin. Without affecting lignin structure or purity, LignoForce™ increases lignin slurry filtration rates. In the steps of precipitation and washing, this process lessens the demand for acid (Kouisni et al., 2012). The ash content, which is in the range of 0.1–0.7%, is less than that of standard kraft lignin (0.2–15%). Because of these facts, LignoForce™ lignin is an excellent material for a variety of uses, including the production of biofuels and phenolic chemicals. Due to the need for oxygen, the oxidation step of LignoForce™ lignin may result in an increase in the cost of production.

5.3.2 LignoBoost™ Technology

The LignoBoost™ technology was developed by Innventia and Chalmers University of Technology in Sweden for the extraction of lignin from the black liquor produced during the kraft pulping process (https://bioplasticsnews.com/2014/05/19/domtar-starts-up-commercial-scale-lignoboost-lignin-separation-plant-based-on-metsos-technology; 2014). Domtar Inc. in North Carolina, USA, was the company that brought this technology to market (Durruty, 2017; Kong et al., 2015). In order to precipitate lignin, carbon dioxide is used to acidify the black liquor that results from kraft pulping to a pH of 9.0. The obtained lignin is redispersed and acidified once more with sulfuric acid following filtration. The lignin is then separated from the above mixture through a second filtration step. After being washed and dried with acid and water, the final product is LignoBoost™ lignin (Kong et al., 2015). Using LignoBoost™ lignin, a pilot plant can produce 75 tons of lignin per day.

A benefit of this LignoBoost™ lignin production is the reduction of the concentration gradient during production due to the nearly identical temperature and pH of the liquid media for redispersed lignin (Zainab et al., 2018). LignoBoost™ lignin has been developed for a wide range of markets and applications, including biothermoplastics, resins, and fuels.

5.3.3 SunCarbon Lignin

SunCarbon, based in Sweden, came up with a method for making lignin oil from biomass in order to enhance the quality of the lignin that was obtained from the black liquor of the kraft process (https://suncarbon.se/en/technology). SunCarbon intends to finish a plant for extracting lignin oil that can produce approximately 45 000 tonnes of lignin oil annually at the beginning of 2022 (https://bioenergyinternational.com/biofuels-oils/preem-and-sveaskog-take-stake-in-suncarbon-plans-for-a-lignin-oil-plant; 2019). Notably, black liquor from pulp mills serves as the system's feedstock. Membrane filtration is the first step in separating lignin from black liquor. The retentate stream rich in lignin is then processed using a hydrothermal treatment. A thermal catalytic cracker treats the lignin-rich retentate stream under subcritical conditions during the hydrothermal treatment. After that, additional purification steps are taken to get rid of its ash. After further desalting, the depolymerized lignin oil is ready for a variety of applications that add value (https://www.hulteberg.com/hulteberg-suncarbon). After a series of purifications, this process could recover up to 25% of the lignin in black liquor.

Compared to standard pulping lignin, SunCarbon lignin oil is a superior quality product that has been systematically filtered and purified to reduce sugar, ash, and sulfur contamination. As a result, it is better for the environment in downstream applications, particularly biofuel production. Through membrane filtration, the pulping liquor and catalysts in a SunCarbon process can be recycled and reused to cut production costs. However, pulp mill upstream black liquor production has a significant impact on SunCarbon lignin quality. SunCarbon lignin is also expensive to produce because the lignin's quality is improved through a series of purification steps. SunCarbon lignin's low production yield—between 10 and 25% of the pulp and paper mill's total lignin—is another drawback. Due to the aforementioned flaws, for value-added applications, SunCarbon lignin does not appear to be economically viable.

5.3.4 Production of Lignosulfonates

Sulfite pulping produces lignosulfonates, which are lignin products. Prior to the development of Kraft pulping in the 1940s, this was the most common method of pulping (Cohen, 1987). To dissolve lignin in biomass, a sulfite pulping procedure involves cooking biomass in fresh liquor containing base solvents and sulfur dioxide. To produce pulp, the residues that do not contain lignin are filtered, washed, and bleached. However, in order to precipitate lignin from the spent liquor, a basic oxide is added. After filtration and drying, lignosulfonate is collected (Aro and Fatehi, 2017). The system uses recycling units to recover sulphur dioxide and the base solvent. Base solvent cooking liquors and precipitation chemicals of all kinds can be used in the sulfite process to produce sodium, magnesium, and ammonium lignosulfonates in addition to calcium lignosulfonate. Sulfite pulping allows for the recovery of 60–70% of the lignin from corn cobs.

The application of lignosulfonate as a binder has seen widespread use. One of the largest suppliers of lignosulfonate products in the world, Borregaard, has successfully turned lignosulfonates into commercial binder products like Norlig A, Borrebond FP, Borresperse CA, (https://www.lignotech.com/Industrial Applications/Agriculture/Agricultural-Chemicals). These binders may also function in high-temperature conditions due to lignin's thermal reluctance. Bornstein (1978) produced the lignosulfonate-based binder. The board produced by this binder has superior stability, even in conditions of high temperature and humidity and is compatible with conventional binders utilized in the production of boards. In a sulfite pulping, sulfite (like $CaSO_3$ or Na_2SO_3) and ethanol can be recovered and reused alongside lignin to cut production costs. Lignosulfonate's high sulfur content, which ranges from 5 to 6 weight % (Korntner et al., 2018) may cause an odor when used.

5.3.5 Kraft Lignin Production

Kraft lignin is the primary by product of kraft pulping. It accounts for 85% of all lignin sold worldwide (Chen, 2015). White liquor is used in the kraft pulping process to break the link between fibers and lignin at temperatures between 150 and 170 °C (Chakar and Ragauskas, 2004). The residues free of lignin are used to make pulp following filtration. Lignin-enriched black liquor is produced from the spent liquor. Precipitation of kraft lignin is done by adding carbon dioxide or sulfuric acid to the black liquor. The LignoForce and LignoBoost technologies have enabled the commercialization of the acidification-based production of kraft lignin.

Kraft lignin is a suitable feedstock for producing fertilizers, plastics, binders, and carbon fibers among other biomaterials (García et al., 1996; Li and Sarkanen, 2003; Olivares et al., 1988; Sagues et al., 2019).

Ramirez et al. (1997) came up with a way to ammoniate kraft lignin to make sustainable, multipurpose fertilizers based on lignin.

Eckert and Abdullah (2010) proposed a process for producing carbon fibers from softwood lignin.

Araújo (2008) examined the viability of producing vanillin from kraft lignin and improved the manufacturing process. Due to its abundant availability and the rising worldwide demand for vanillin, kraft lignin has the potential to be used for producing vanillin. Kraft lignin is a sulfur-bearing material, just like lignosulfonate. Kraft lignin's sulfur content—1.5–3 weight % may hinder its subsequent applications (Svensson, 2008). Due to its degradation during pulping, kraft lignin has a relatively low molecular weight in comparison to other industrial lignins (Lange et al., 2013). This might be a problem in some cases, like when lignin-based flocculants need a lot of polymers to make flocculation easier.

5.3.6 Organosolv and Soda Lignin

Organosolv and soda pulping use cooking with pulping liquors that do not contain sulfur to break down biomass. Organic solvents like acetone and ethanol/water are used to cook biomass (https://www.e-education.psu.edu/egee439/node/658). The lignin-free residues are then obtained through filtration. After being washed with ethanol and water, the produced solid residues are turned into pulp. In contrast, after filtration, the spent liquor is collected for the production of organosolv lignin. Evaporation is used to recycle the organic solvent, which reduces the need for chemicals (Pan et al., 2006). The liquor is then diluted and precipitated by adding water. Following filtration, washing with water, and drying, Organosolv lignin is produced (da Rosa et al., 2017). Additionally, precipitation yields some water-soluble sugar. The dry original biomass feedstock yields approximately 20% organosolv lignin.

Typically, straw and bagasse are used as the feedstock for soda pulping (Doherty and Rainey, 2006). Anthracenedione and aluminum oxide are added to the NaOH solvent, which is used to cook the biomass between 150 and 170 °C. For the production of pulp, the lignin-free residue is obtained through a filtration process. Due to the high silica content of the feedstock used in soda pulping, silica gels are produced when flue gas blows the black liquor (Doherty and Rainey, 2006). The black liquor is then treated with sulfuric acid to precipitate lignin at 70–80 °C (Sameni et al., 2014). In order to reduce production costs, the filtrate is collected for chemical recovery. For producing soda lignin, which can be used for recovery of energy and other purposes, the filtered residues are rinsed with water and dried. Between 4 and 10% of the dry original biomass material is produced by recovered soda lignin.

Organosolv and soda lignin are odorless and environment friendly because they do not contain any sulfur (Mandlekar et al., 2018). The structure of lignin without sulphur is more like native lignin than that of lignin bearing sulfur because of less destructive processes. In addition, of all industrial lignins, organosolv lignin is the purest (Ahmad and Pant, 2018). Lignin without sulfur is a potential source for producing phenolic compounds because of its lower molecular weight. In addition, resin can be made from lignin that does not contain sulfur.

Cook and Hess (1991) came up with a way to make organosolv lignin-phenol aldehyde resin in 1991. This resin is superior as compared to conventional lignin resin because it is unaffected and insolubilized by water and does not emit offensive sulphur odours when the adhesive is hot-pressed or when the resin is prepared.

Feng et al. (2016) suggested using soda lignin to make wood adhesive, which has been shown to be good for the environment and cheap. Pulping liquors like methanol, ethanol, and acetone, can be recovered alongside lignin. These solvents can then be utilized in a variety of processes. Organosolv lignin's relatively high production cost is a major drawback because the pulping process requires a lot of organic solvent (Shrotri et al., 2017). The high levels of silica and ash make soda lignin less useful in other applications (Mousavioun, 2011).

5.3.7 Thermo-mechanical Pulp-bio Lignin

A proprietary thermo-mechanical pulp-bio lignin (TMP-Bio) has been developed by FPInnovations (Quebec, Canada) for the industrial conversion of cellulosic biomass into fermentable sugars and hydrolysis lignin (https://web.fpinnovations.ca/fpinnovations-and-resolute-inaugurate-thermomechanical-pulp-bio-refinery-in-thunder-bay). After being moved to a feed bin, biomass is cooked in a pre-treatment tower before being combined with steam and caustic by means of a mixing screw (https://ifbc2017.files.wordpress.com).

After being heated, the slurry is neutralized before being put through a screw press to remove most of the water. In the reactor, the biomass is hydrolyzed enzymatically. Enzymatic hydrolysis generates sugars that are subsequently used to create fermentable sugars like glucose and xylose by filtering them from solid biomass (H-lignin). Conversely, a ring dryer is used to dry hydrolysis lignin, which is separated during filtration. Estimates show that a TMP-Bio process could convert 1000 kg of biomass into 400 kg of lignin and 500 kg of sugar products.

Products with added value, like resins, activated carbon, and additives, can be made with TMP-Bio lignin (Mao et al., 2017). It is important to note that no sulphur is present in TMP-Bio lignin. This makes it easier to use in a number of different ways, including the production of biofuels and adhesives. Traditional pulping lignin typically has a dark brown or black color. TMP-Bio lignin, on the other hand, is attractive for construction adhesive because of its light yellow color. Another advantage is that because it is cooked at a relatively low temperature (90 °C), its properties are similar to those of native lignin. The TMP-Bio technology has the potential to produce fermentable sugars as well as value-added chemicals like, butanol, ethanol, lactic acid, succinic acid, resins, activated carbon, and additives. The production cost of TMP-Bio is high because variety of chemicals like sulfuric acid, sodium hydroxide, enzyme, buffer solution, etc. are required. Due to the method's numerous heating and cooling cycles, TMP-Bio lignin production may require a significant amount of energy.

5.4 Future Perspectives

Many novel lignin extraction technologies, including AFEX, PROESA®, BioFlex, steam explosion, SLRP, and German lignocellulose feedstock biorefinery processes, have recently been developed as a result of growing interest in lignin and its potential applications. These procedures could eventually be scaled up for industrial production with additional optimization. Additionally, the SunCarbon, LignoBoost, LignoForce, and TMP-Bio lignin processes are listed along with other commercial lignin production techniques. Although there are still many obstacles in the way of lignin production and value-added applications, lignin can be valorized in a variety of ways because of the increasing demand for sustainable products made of lignin like lignin flocculants, dispersants, binders, and biofuels (Salaghi et al., 2021).

Bibliography

Abbati de Assis C, Greca LG, Ago M, Balakshin MY, Jameel H, Gonzalez R, and Rojas OJ (2018). Techno-economic assessment, scalability, and applications of aerosol lignin micro-and nanoparticles. *ACS Sustain Chem Eng*, 6: 11853–11868.

Agrawal A, Kaushik N, and Biswas S (2014). Derivatives and applications of lignin–an insight. *Sci Tech J*, 1: 30–36.

Ahmad E and Pant KK (2018). Lignin conversion: a key to the concept of lignocellulosic biomass-based integrated biorefinery. *Waste Biorefinery*, 409–444.

Anderson G, Filpponen I, Lucia LA, Saquing C, Baumberger S, and Argyropoulos DS (2006). Toward a better understanding of the lignin isolation process from wood. *J Agric Food Chem*, 54: 5939–5947.

Araújo JDP (2008). *Production of vanillin from lignin present in the Kraft black liquor of the pulp and paper industry (Doctoral dissertation)*. University of Porto, Porto.

Aro T and Fatehi P (2017). Production and application of lignosulfonates and sulfonated lignin. *ChemSusChem*, 10: 1861–1877.

Bajwa DS, Pourhashem G, Ullah AH, and Bajwa SG (2019). A concise review of current lignin production, applications, products and their environmental impact. *Industrial Crops and Products*, 139: 111526.

Bornstein LF (1978). Lignin-based composition board binder comprising a copolymer of a lignosulfonate, melamine and an aldehyde. *US patent 4130515*. 1978 December 19.

Brodin I (2009). *Chemical properties and thermal behaviour of kraft lignins (Doctoral dissertation)*. Stockholm: KTH.

Bruijnincx P, Weckhuysen B, Gruter G-J, and Engelen-Smeets E (2016). *Lignin Valorisation: The Importance of a Full Value Chain Approach*. Utrecht University.

Chakar FS and Ragauskas AJ (2004). Review of current and future softwood kraft lignin process chemistry. *Ind Crops Prod*, 20: 131–141.

Chen F and Li J (2000). Aqueous gel permeation chromatographic methods for technical lignins. *J Wood Chem Technol*, 20: 265–276.

Chen H (2015). Lignocellulose biorefinery feedstock engineering. *Lignocellulose Biorefinery Engineering*, 1st ed., Woodhead Publishing, Cambridge, p. 37–86.

Cohen AJ (1987). Factor substitution and induced innovation in North American kraft pulping: 1914–1940. *Explorations in Economic History*, 24: 197–217.

Cook PM and Hess SL (1991). Organosolv lignin-modified phenolic resins and method for their preparation. *US Patent 5010156*. 1991 April 23.

da Rosa MP, Beck PH, Müller DG, Moreira JB, da Silva JS, and Durigon AMM (2017). Extraction of organosolv lignin from rice husk under reflux conditions. *Biol Chem Res*, 87–98.

Damen KJ, Smit AT, Huijgen WJJ, and Van HJW (2017). Fabiola: fractionation of biomass using low-temperature acetone. *13th International Conference on Renewable Resources & Biorefineries*.

Diels L and Browne T (2018). Lignin international conference. Edmonton: Alberta, Sep 18–20.

Doherty WOS and Rainey T (2006). Bagasse fractionation by the soda process. *Proceedings of the Australian Society of Sugar Cane Technologists*. 2–5 May 2006, Mackay, Queensland, Australia.

Durruty J (2017). *On the local filtration properties of lignoboost lignin, studies of the influence of xylan and ionic strength (Doctoral dissertation)*. Gothenburg: Chalmers University of Technology.

Eckert RC and Abdullah Z (2010). Carbon fibers from kraft softwood lignin. *US Patent 7678358*.

EPA (2018). *Inventory of U.S. Greenhouse Gas Emissions and Sinks*. United States Environmental Protection Energy.

Feng MW, He G, Zhang Y, Wang X, Kouisni L, and Paleologou M (2016). High residual content (HRC) kraft/soda lignin as an ingredient in wood adhesives. *US Patent Application 15/130,107*.

García MC, Diez JA, Vallejo A, Garcia L, and Cartagena MC (1996). Use of kraft pine lignin in controlled-release fertilizer formulations. *Ind Eng Chem Res*, 35: 245–249.

Holladay JE, White JF, Bozell JJ, and Johnson D, (2007). *Top Value-added Chemicals From Biomass-volume II—Results of Screening for Potential Candidates From Biorefinery Lignin*. Pacific Northwest National Lab. (PNNL), Richland, WA, USA.

Kihlman J (2015). Acid precipitation lignin removal processes integrated into a kraft mill. *NWBC 2015 The 6th Nordic Wood Biorefinery Conference*. p. 402–410.

Klinger KM., Liebner F, Fritz I, Potthast A, and Rosenau T (2013). Formation and ecotoxicity of N-heterocyclic compounds on ammoxidation of mono-and polysaccharides. *J Agric Food Chem*, 61: 9004–9014.

Kong F, Wang S, Price JT, Konduri MK, and Fatehi P (2015). Water soluble kraft lignin-acrylic acid copolymer: synthesis and characterization. *Green Chem*, 17: 4355–4366.

Korbag I and Mohamed Saleh S (2016). Studies on mechanical and biodegradability properties of PVA/lignin blend films. *Int J Environ Stud*, 73: 18–24.

Korntner P, Schedl A, Sumerskii I, Zweckmair T, Mahler AK, Rosenau T, and Potthast A (2018). Sulfonic acid group determination in lignosulfonates by headspace gas chromatography. *ACS Sustain Chem Eng*, 6: 6240–6246.

Kouisni L, Gagné A, Maki K, Holt-Hindle P, and Paleologou M (2014). LignoForce system for the recovery of lignin from black liquor: feedstock options, odor profile, and product characterization. *ACS Sustain Chem Eng*, 4: 5152–5159.

Kouisni L, Holt-Hindle P, Maki K, and Paleologou M (2012). The lignoforce system: a new process for the production of high-quality lignin from black liquor. *J Sci Technol For Prod Processes*, 2: 6–10.

Lake MA and Blackburn JC (2014). SLRP-an innovative lignin-recovery technology. *Cellul Chem Technol*, 48: 799–804.

Lange H, Decina S, and Crestini C (2013). Oxidative upgrade of lignin–recent routes reviewed. *Eur Polym J*, 49: 1151–1173.

Li Y and Sarkanen S (2003). Biodegradable Kraft lignin-based thermoplastics. *Biodegrad. Polym. Plastics*, 121–139.

Ľudmila H, Michal J, Andrea Š, and Aleš H (2015). Lignin, potential products and their market value. *Tech. Rep. CRDLR US Army Chem. Res. Dev. Lab.*, 60: 973–986. https://doi.org/10.13140/RG.2.2.10427.21289.

Luo H and Abu-Omar MM (2017). Chemicals from lignin. *Encyclopedia of Sustainable Technologies*, Abraham MA (ed.). Elsevier, pp. 573–585. ISBN: 9780128046777.

Mandlekar N, Cayla A, Rault F, Giraud S, Salaün F, Malucelli G, and Guan J-P (2018). *An overview on the use of lignin and its derivatives in fire retardant polymer systems, lignin-trends and applications*. InTech. Matheus Poletto, IntechOpen. https://doi.org/10.5772/intechopen.72963.

Mao C, Yuan Z, Fernando DRL, Wafa ADW, Wong D, and Browne T (2017). TMP-Bio for converting cellulosic biomass to 2nd generation sugar and near-native lignin. In: *Proceedings International Forest Biorefining Conference (IFBC)*.

Moniruzzaman M, Dale BE, Hespell RB, and Bothast RJ (1997). Enzymatic hydrolysis of high-moisture corn fiber pretreated by AFEX and recovery and recycling of the enzyme complex. *Appl Biochem Biotechnol*, 67: 113–126.

Mousavioun P (2011). *Chemical and thermal properties of bagasse soda lignin (dissertation)*. Queensland: Queensland University of Technology.

Olivares M, Guzman JA, Natho A, and Saavedra A (1988). Kraft lignin utilization in adhesives. *Wood Sci Technol*, 22: 157–165.

Palmisano P (2013). Lignin conversion into bio-based chemicals. *Engineering Conferences International, BioEnergy IV: Innovations in Biomass Conversion for Heat, Power, Fuels and Chemicals Proceedings*.

Pan X, Gilkes N, Kadla J, Pye K, Saka S, Gregg D, Ehara K, Xie D, Lam D, and Saddler J (2006). Bioconversion of hybrid poplar to ethanol and co-products using an organosolv fractionation process: optimization of process yields. *Biotechnol Bioeng*, 94: 851–861.

POET-DSM contracts with ANDRITZ for biomass pre-treatment process for cellulosic ethanol plant (2012a). https://www.greencarcongress.com/2012/10/andritz-20121025.html.

POET-DSM makes major technology, process purchase for commercial cellulosic bio-ethanol (2012b). https://poet.com/pr/poet-dsm-makes-major-technology-process-purchase-for-cellulosic-bio-ethanol.

Pucciariello R, Bonini C, D'Auria M, Villani V, Giammarino G, and Gorrasi G (2008). Polymer blends of steam-explosion lignin and poly (ε-caprolactone) by high-energy ball milling. *J Appl Polym Sci*, 109: 309–313.

Ragauskas AJ, Beckham GT, Biddy MJ, Chandra R, Chen F, Davis MF, Davison BH, Dixon RA, Gilna P, and Keller M (2014). Lignin valorization: improving lignin processing in the biorefinery. *Science*, 344: 1246843. https://doi.org/10.1126/science.1246843.

Ramirez F, González V, Crespo M, Meier D, Faix O, and Zúñiga V (1997). Ammoxidized kraft lignin as a slow-release fertilizer tested on Sorghum vulgare. *Bioresour Technol*, 61: 43–46.

Sagues WJ, Jain A, Brown D, Aggarwal S, Suarez A, Kollman M, Park S, and Argyropoulos DS (2019). Are lignin-derived carbon fibers graphitic enough? *Green Chem*, 21: 4253–4265.

Salaghi A, Zhou L, Saini P, Konduri M, Kong F, and Fatehi P (2021). Lignin production in plants and pilot and commercial processes, Chapter 20. *Biomass, Biofuels, Biochemicals, Biodegradable Polymers and Composites – Process Engineering to Commercialization*, Binod P, Raveendran S, and Pandey A (eds.). Elsevier, pp. 551–587.

Sameni J, Krigstin S, dos Santos Rosa D, Leao A, and Sain M (2014). Thermal characteristics of lignin residue from industrial processes. *BioResources*, 9: 725–737.

Shrotri A, Kobayashi H, and Fukuoka A (2017). Catalytic conversion of structural carbohydrates and lignin to chemicals. *Adv Catal*, 60: 59–123.

Sun R and Tomkinson J (2000). Fractionation and characterization of water-soluble hemicelluloses and lignin from steam-exploded birchwood. *Int J Polymer Mater*, 45: 1–19.

Svensson S (2008). *Minimizing the sulphur content in Kraft lignin (Thesis)*. Mälardalen University, Västerås and Eskilstuna.

Teymouri F, Laureano-Perez L, Alizadeh H, and Dale BE (2005). Optimization of the ammonia fiber explosion (AFEX) treatment parameters for enzymatic hydrolysis of corn stover. *Bioresour Technol*, 96: 2014–2018.

Velez J and Thies MC (2016). Liquid lignin from the SLRP™ Process: the effect of processing conditions and black-liquor properties. *J Wood Chem Technol*, 36: 27–41.

Zainab AK, Pradhan R, Thevathasan N, Arku P, Gordon A, and Dutta A (2018). Beneficiation of renewable industrial wastes from paper and pulp processing. *AIMS Energy*, 6: 880.

https://web.fpinnovations.ca/fpinnovations-and-resolute-inaugurate-thermomechanical-pulp-bio-refinery-in-thunder-bay.
https://suncarbon.se.
https://www.bio.org/sites/default/files/legacy/bioorg/docs/beta%20renewables%20proesa%20technology%20june%202013_bio_michele_rubino.pdf.
https://www.filtsep.com/chemicals/features/bioflex-turns-wood-into-green-bioproducts-10.
https://www.hulteberg.com/hulteberg-suncarbon.
https://www.lixea.co/#bioflex-process.
https://www.e-education.psu.edu/egee439/node/658.
https://bioenergyinternational.com/preem-and-sveaskog-take-stake-in-suncarbon-plans-for-a-lignin-oil-plant.
http://unravel-bbi.eu/successful-scale-up-of-the-fabiola-process-in-a-lignocellulose-biorefinery-pilot-plant.
https://www.borregaard.com.
https://www.yumpu.com/en/document/read/41220925/the-german-lignocellulose-feedstock-biorefinery-project.
https://bioeconomy.msu.edu/2015/09/25/a-patent-and-milestone-why-mbis-afex-biomass-pretreatment-is-of-global-importance.

https://ifbc2017.files.wordpress.com.

https://bio4products.eu/wp-content/uploads/2019/11/4.-Hans-Heeres_Thermo-Chemical-Fractionation.pdf.

https://www.ieabioenergy.com/wp-content/uploads/2018/11/902-BTGBiomass-VandeBeld-smaller.pdf.

https://bioplasticsnews.com/2014/05/19/domtar-starts-up-commercial-scale-lignoboost-lignin-separation-plant-based-on-metsos-technology.

6

Applications of Lignin

Abstract

Lignin is the second most abundant natural material on the earth. Commercially, it is generated as a waste product from the paper and ethanol production. The application of lignin and its derived aromatics in various sectors such as aromatics, phenolics and flavoring compounds, carbon materials, lignin-based nanomaterials, biomedical applications, lignin-based nanocomposites, urethanes and epoxy resins, controlled release fertilizers, biosensors and bioimaging, hydrogen production, battery materials for energy storage, dust control agents, bitumen modifiers in the road industry, cement additives and building materials, and many more have been presented.

Keywords   *Lignin applications; Lignin valorization; Aromatics; Phenolics and flavoring compounds; Carbon materials; Lignin-Based Nanomaterials; Biomedical application; Lignin based nanocomposites; Urethanes and epoxy resins; Controlled release fertilizer; Biosensor and bioimaging; Hydrogen production; Battery material for energy storage; Dust control agent; Bitumen modifier in road industry; Cement additives; Bioplastics; Dispersant; Food Additives; Sequestering agent; Bio-Oil*

6.1 Introduction

Lignin is the second-largest plant polymer in the world after cellulose. Global pulping and papermaking facilities produce approximately 50–70 million tons of kraft lignin annually. Organic solvent lignin, most commonly hydrolyzed lignin, is also produced during the production of second-generation biofuels. Organic products derived from lignin are sustainable alternatives to petroleum refinery-produced chemicals. Lignin's current and potential uses can be broken down into three groups: heat, power, green fuel, and syngas products; aromatic compounds with low molecular weights that result from lignin depolymerization; and macromolecules such as drugs and nutraceuticals (Holladay et al., 2007). The current and potential applications of technical lignin are shown in Figure 6.1 (Mandlekar et al., 2018).

Depolymerization of Lignin to Produce Value Added Chemicals, First Edition. Pratima Bajpai.
© 2024 John Wiley & Sons, Inc. Published 2024 by John Wiley & Sons, Inc.

6.2 Applications

6.2.1 Aromatics, Phenolics and Flavoring Compounds

Lignin is the only renewable substance that could be utilized for making aromatic compounds such as xylene, benzene, toluene, pyrogallol, p-hydroxybenzoic acid, cis-muconic acid as well as flavoring compounds like guaiacol, eugenol, and catechol, among other things. Even though lignin has a complicated structure, getting a lot of aromatic compounds of higher quality from it is hard in the long run, but it does seems to be possible. Various approaches, such as the thermo-chemical or engineered microbes approach, were utilized for the biotransformation and fermentation of lignin and its constituents into large quantities of chemicals with molecules that add value like vanillin, pyrogallol, muconic acid, guaiacol, and catechol.

Vanillin (4-hydroxy- 3-methoxybenzaldehyde) is widely used in the food, cosmetic, pharmaceutical, and other industries. It is typically synthesized through chemical synthesis or extraction from natural plants. Using lignin as a feedstock for the biosynthesis of vanillin is an efficient and promising process. Microbial catalysis can either produce vanillin from ferulic acid or depolymerize lignin during the lignin metabolic process. In order to depolymerize lignin and produce vanillin through an oxidative reaction mediated by hydrogen peroxide, a microbial fuel cell system was developed. Ferulic acid can be produced by *Bacillus subtilis*, *Streptomyces* sp., and the genus *Amycolatopsis*. To boost the amount of vanillin produced by microorganisms, metabolic engineering has been developed.

Vanillin, one of the aromatic monomers produced by lignin depolymerization, has been commercialized, producing approximately 3000 tons annually (Fache et al., 2016; Li and Takkellapati, 2018). Despite extensive research into the production of additional biobased aromatics from lignin depolymerization, the level of technology is still not very mature (Lange et al., 2013). Compared to vanilla beans and petroleum-based chemicals, lignin due to its economic viability and environmental sustainability is thought to be a promising source for the production of vanillin. According to Bajpai (2018), worldwide, food additives account for approximately 60% of industrial vanillin sales whereas approximately 33% of it is used in cosmetics and 7% in pharmaceuticals. In the food industry, vanillin is frequently found in beverages and sweet foods particularly chocolate, cakes, biscuits, and ice cream. In addition to its use in food, vanillin is utilized as a feed for pigs, cattle, and poultry (Mohammadi and Kim, 2018). However, other flavoring agents like zingiberene or jasmine oil are extensively produced using vanillin (Banerjee and Chattopadhyay, 2018). Additionally, vanillin is utilized as a food preservative due to its antioxidant and antimicrobial properties (Banerjee and Chattopadhyay, 2018; Rupasinghe et al., 2006). The fragrance industry uses vanillin a lot in shampoos, body lotions, soaps, shower gels, room sprays, scented candles, and air fresheners. Instead of being a flavoring and fragrance ingredient, vanillin is now used as an important intermediate or precursor in the synthesis of pharmaceuticals and fine chemicals. Vanillin and other lignin derivatives like syringaldehyde, guaiacol, ferulic acid, and 4-hydroxybenzoic acid are the focus of extensive research as biobased building blocks. Diverse polymer products have been prepared through numerous efforts (Fache et al., 2015, 2016; Llevot et al., 2016). Another interesting aromatic aldehyde produced by lignin depolymerization is syringaldehyde, which is 3,5-dimethoxy-4-hydroxybenzaldehyde. Syringaldehyde, like vanillin, is used in the flavor and fragrance industry as an ingredient. Despite the fact that syringaldehyde is not nearly as well-known as vanillin, it has emerged as a promising lignin-derived chemical, especially given that it is a necessary precursor for several antibiotics (Ibrahim et al., 2012; Tarabanko and Tarabanko, 2017).

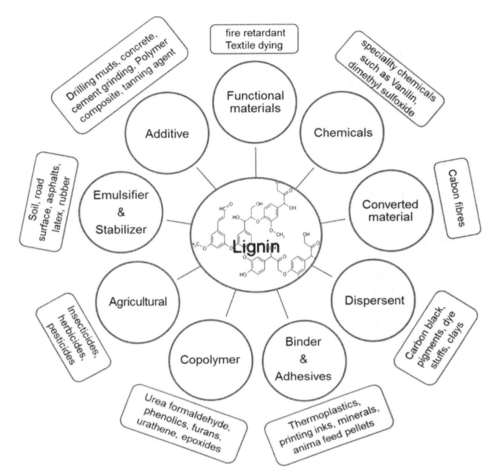

Figure 6.1 Current and potential applications of technical lignin. Mandlekar et al. (2018). Distributed under CC BY 3.0 license. doi: 10.5772/intechopen.72963.

The synthesis of trimethoprim also begins with vanillin, but syringaldehyde has the benefit of having two methoxyl groups (Mota et al., 2015). Up to this point, there have only been preliminary in vitro tests for ascertaining syringaldehyde's bioactive properties. Nonetheless, the expanding use of syringaldehyde benefits from these useful "cornerstones" and discovery of new areas of potential study. Additionally, syringaldehyde has antimicrobial properties.

González-Sarras et al. (2012) examined the anti-cancer properties of 51 different phenolic compounds purified from Canadian maple syrup extract. Seven other phenolic compounds, including syringaldehyde, had a greater potential for inhibiting the growth of colon cancer cells. Syringaldehyde's potential as a food additive, pharmaceutical, and natural health product, could be further studied in light of its bioactive qualities and positive health effects.

According to Kumar and Pruthi (2014), ferulic acid is also a significant step in the industrial production of vanillin. The, hydroxyl, carboxylic acid, and methoxy groups on the benzene ring of ferulic acid allow for the stabilization of the intermediate phenoxyl radical and, ultimately, the end of free radical chain reactions when they react with a free radical (Ghosh et al., 2017). These functional groups, according to Mancuso and Santangelo (2014), make ferulic acid an effective scavenger of reactive nitrogen and oxygen species. Because of its potent antioxidant qualities, ferulic acid can be utilized as a preservative in the food industry. The first use of ferulic acid as a food preservative was reported by Tsuchiya and Takasawa (1975). Additionally, ferulic acid helps to prevent bananas and green tea from changing color because of oxidation (Kumar and Pruthi, 2014). Another use for ferulic acid is as a cosmetic ingredient because at pHs between neutral and acidic, it easily penetrates the skin and is a potent UV absorber. According to Parmar et al. (2015), ferulic acid has been identified as a cardiovascular agent, an anti-cancer agent, a neuroprotective agent, and an anti-diabetic agent. In induced diabetic animals, ferulic acid reduced blood glucose levels effectively and alleviated diabetes-related physical symptoms (such as obesity) (Ramar et al., 2012; Song et al., 2014). Ferulic acid has the potential to be used as an adjuvant in cancer treatment and heart problems (Alam, 2019; Bumrungpert et al., 2018; Fahrioglu et al., 2016).

6.2.2 Carbon Materials

As a polymeric and polyelectrolytic material, all commercial lignins have potential uses. This could be greatly expanded to produce macro monomers of higher value for polymer applications. The fundamental understanding of lignin reactivity has advanced significantly, facilitating the creation of various macromolecules like: filler, polymer extender, carbon fiber, polyurethane, and polymer alloy. The value of the global market for carbon fiber was USD 2.23 billion in 2020, and it is anticipated to increase to USD 4.08 billion in 2028 from USD 2.33 billion in 2021. Around 48.7 billion USD will be spent on carbon-reinforced plastic worldwide. But, the cost of carbon fiber is the primary obstacle to its use. According to one study, the manufacturing cost of carbon fiber based on polyacrylonitrile (PAN) is 15 USD per kilogram in the United States, 18 USD per kilogram in Japan, and 20 USD per kilogram in Germany (Gill et al., 2016). The high cost of PAN is the reason for the higher prices. For the production of carbon fiber, a low-cost source of lignin, a natural precursor with a higher carbon content, may be ideal (Mainka et al., 2015; Ragauskas et al., 2014).

Carbon fiber is used extensively in the aerospace and sports industries. But, carbon fiber's potential to replace steel in the automotive industry may be promising. Carbon fiber should cost between 7 USD and 11 USD per kilogram when compared to the prices of regular steel, which is 4 USD per kilogram and high strength steel, which is 8 USD per kilogram. As a result, the plan is to use lignin instead of PAN because it is inexpensive and contains a lot of carbon. According to Baker and Rials (2013), the price of carbon fiber derived from lignin is anticipated to be approximately 7 USD per kilogram. Lignin-based high-quality carbon fiber is being developed by BASF & SGL (Germany), GrafTech (USA), Plasan, Ford Motor (USA), Oak Ridge National Lab (USA), DowAska (Netherlands). Kraft lignin from hardwoods and softwoods has been used to make carbon fibers based on lignin (Kadla et al., 2002; Norberg et al., 2013). Carbon materials like carbon black, activated carbon, graphitic carbon, and structural carbon fibers have been the focus of recent research. Advanced materials, absorbent materials, electrodes in electrochemical applications, energy storage, and the removal of organic pollutants or heavy metals from gas and aqueous media all make use of carbon-rich lignin-based materials (Puziy et al., 2018; Wang et al., 2019a). The abundance of lignin, energy efficiency, and reduced environmental impact are the driving factors of ongoing research to create carbon fibers and related products from lignin. However, the main obstacles to producing high-quality carbon-rich products have been the heterogeneity of the lignin molecule, the amount of ash, and the behavior of thermoplastic foaming.

6.2.3 Lignin-based Nanomaterials

Due to lignin's complex macromolecular structure and poor solubility in water, which limit its use on a large scale, the transformation of raw lignin into uniformly sized and shaped aqueous nanoparticle dispersions is considered a significant advancement (Chen et al., 2018; Lievonen et al., 2016). Various methods, including, anti-solvent precipitation, sonication, interfacial crosslinking and solvent exchange have been used to successfully produce a variety of nanomaterials from lignin, including nanoparticles, nanotubes, nanofibers, and nanogels (Figueiredo et al., 2018; Zhao et al., 2016). Because of their higher volume-to-surface area ratios, nanoparticles typically possess distinct or superior properties to their parent polymers or bulk materials (Yearla and Padmasree, 2016). The antioxidant and superoxide radical scavenging activity of LNPs made from organosolv lignin through the supercritical antisolvent process was higher than that of their non-nanoscale counterparts. The results of the experiment suggested that the pharmaceutical industry, animal husbandry, and food processing industries could further benefit from the use of LNPs (Lu et al., 2012). Using the nanoprecipitation method, dioxane lignin and alkali lignin were used to create LNPs with an average particle size of 104 nm. These LNPs had stronger antioxidant and UV-protective properties than dioxane lignin and alkali lignin. LNPs appear to be a promising option for use in the food, cosmetic, and pharmaceutical industries. Yearla and Padmasree (2016) have reported similar results. LNPs are suitable for numerous biomedical applications, including drug delivery, in addition to their potential use as an antioxidant. By incorporating pH-sensitive polymers, LNPs may also permit the loading of hydrophilic drugs and the pH-responsive release of drugs. Additionally, targeting moieties can be added to LNPs to enhance cellular interaction with particular cells for disease treatment because of their unique surface structure. LNPs containing anti-cancer drugs showed an anti-proliferation effect in a variety of cancer cells, including breast, colon cancer cells, and prostate cancer endothelial cells (Figueiredo et al., 2017). Additionally, LNPs can interact intimately with polymer matrices and be evenly distributed, enhancing their barrier and mechanical properties (Beisl et al., 2017; Nair et al., 2014). Due to this unique property, LNPs can be utilized as strengthening agents in polymer matrices and nanocomposites.

6.2.4 Biomedical Application

Products derived from lignin are becoming increasingly popular as materials in biomedical engineering. Most studies are still in the proof-of-concept stage. The primary focus is on the creation of smart bio-based nanocarriers that can be used to physically attach or encase active bio-agents. Medical diagonistic and therapeutic uses of lignin nanoparticles are promising. Hydrogels made from lignin and nanotubes made from lignin play a crucial role in DNA delivery and tissue engineering. Lignin-based hydrogels are used in in vitro diagnostics, cancer and stem cell research, tissue engineering, immunomodulation, and cell therapy (Muir, 1996) According to Ten and Vermerris (2015), the production of lignin nanowires and lignin nanotubes is thought to offer a low-cost means of delivering DNA and therapeutic agents. Additionally, lignin has numerous pharmacological uses. Numerous diseases, including cancer, are protected by lignin, according to epidemiological research. According to Aro and Fatehi (2017) and Martinez et al. (2012) drugs that boost the body's immune system and reduce oxidative activity could also be made with the help of lignosulfonates. Drug delivery, healing, and purification of water are all possible applications for lignin-based hydrogels (Thakur and Thakur, 2015; Vashist et al., 2014). Sustainable methods of producing natural fire retardants rather than synthetic chemicals have been investigated for meeting the constant demand for fire protection. Xanathan gum and lignin are capable of producing biodegradable hydrogels that offer enhanced thermo-oxidative properties and increased thermal chemical reactivity. Natural flame retardants can also be used with lignin (Cayla et al., 2016). Additionally, application of lignin in polymer systems that are resistant to fire has garnered attention. Polymers like polypropylene, PBS, ABS, and PET can effectively reduce their flammability if lignin is present (Mandlekar et al., 2018). Biopolymers and synthetic polymers based on lignin are less flammable and produce less smoke. Lignin is able to act as an antioxidant because it contains methoxy and hydroxyl rings, giving it a complex aromatic character. Through their free radical scavenging activities, the oxidative propagation reaction is put to an end by the functional groups. When it comes to antioxidant and UV protection, lignin nanoparticles perform better than bulk lignin. It is anticipated that the food, pharmaceutical, and cosmetics industries will make use of the lignin nanoparticles (Yearla and Padmasree, 2016).

6.2.5 Lignin-based Nanocomposites

Recently, there has been a lot of interest in using nanocomposites made of lignin for use in 3D printing. The best material for 3D printing is capable of maintaining its structure after printing and has good extrudability, making 3D printing simpler. Lignin's structure, which makes it ideal for 3D printing, includes oxygenated aromatic bonds, β-O-4 connections, and aliphatic ether groups (Bhagia et al., 2021; Nguyen et al., 2018).

By adding lignin to common plastics for 3D printing, environment friendly and affordable nanocomposites can be produced. Energy-sacrificial bonds, such as lignin, have been suggested as a technique for creating strong biomaterials. These sacrificial links break and stretch to reform and release energy. Huang et al. (2019) made use of coordination bonds based on zinc between lignin nanoparticles and an elastomer matrix. The composite became stronger as a result of the role of these connections in spreading lignin throughout the matrix. According to the findings of this study, the ductility, toughness, and strength of these thermoplastic elastomers could be enhanced by adding up to 30% lignin, which could be helpful for 3D printing. It has been demonstrated that the structural qualities of a material for 3D printing are enhanced by lignin. Kraft lignin and acrylonitrile butadiene styrene were mixed in a study to create a composite that was stiffer but had lower tensile strength (Bhagia et al., 2021). When 10% acrylonitrile butadiene rubber (NBR41) was added, the mechanical properties of composite were better than those of petroleum-based thermoplastics because of the physical and chemical crosslinks between the NBR41 and lignin. With 40% lignin, 10% NBR41, and half ABS, this composite was extremely 3D-printable.

In an experiment with nylon, it was discovered that organosolv hardwood lignin enhanced 3D printability by strengthening the thermoplastic framework and increasing stiffness (Nguyen et al., 2018). A study also found that lignin was used as a nucleating agent that helped Poly lactic acid crystallize in a 3D printing matrix with 20–40% lignin and PLA. The lignin did not clump in this composite material, which flowed and extruded easily (Tanase-Opedal et al., 2019).

Acetylating alkali and organosolv lignin with PLA improved their compliance in another study. However, lignin improved PLA thermal stability while also reducing PLA crystallization and hydrolysis (Gordobil et al., 2014).

Lignin can be successfully combined with a wide range of different polymers to make materials for 3D printing, according to other studies. To make photoactive acrylate resins for 3D printing, researchers used resin bases, a reactive diluent, 15% acylated organosolv lignin, and other compounds. Although the resulting lignin resin had properties suitable for 3D printing, it had lower thermal stability compared to commercial resin. The resin made the material more ductile and made durable, high-resolution, evenly fused 3D prints (Sutton et al., 2018).

When compared to PHB on its own, poly-hydroxy-butyrate (PHB) and 20% biorefinery lignin from *Pseudomonas radiata* reduced material shrinkage and improved surface texture during 3D printing (Vaidya et al., 2019). When comparison was made with PHB alone, which served as the sole printing material, the addition of lignin reduced wrapping by 38–78%. The complex structure of the lignin and the lack of electrostatic repulsion between the lignin and the polymer matrices may have contributed to reduced shrinkage.

6.2.6 Urethanes and Epoxy Resins

Additionally, eco-friendly thermoset resin polyurethane can be made from lignin as a starting material. Comfort, energy absorption, and a high strength-to-weight ratio are all characteristics of polyurethane. According to Banik and Sain (2008), it is utilized extensively in the shipbuilding, automotive, furniture insulation, and packaging industries. A diisocyanate and a diol are used in the processing of conventional polyurethane, which is not good for the environment.

According to Lee and Deng (2015), the biosynthesis of lignin can make polyurethane biodegradable, making it better for the environment. A urethane linkage is formed when hydroxyl groups and isocyante groups react in the urethanization reaction. Polyurethanes based on lignin are made by chemically altering lignin with polyols or diols. There have typically been two evaluations: the first method involves adding diisocyanate or diol as a comonomer in a single step. The two steps in the second approach are polymerization with lignin and prepolymer production with polyol and isocyanate (Ahvazi et al., 2011). According to Mahmood et al. (2016), the properties of lignin-based polyurethane foam are comparable to or superior to those of conventional polyurethanes. Polyesters made from lignin are made by functionalizing hydroxyl groups by esterifying or etherifying them (Figueiredo et al., 2018). The properties of composites are influenced by the density of hydrogen-bonding groups, which is linked to the kind of lignin used and how it is extracted. Lignin-epoxy resins can improve the anti-aging properties of asphalt (Xin et al., 2016).

6.2.7 Controlled Release Fertilizer

There are several active groups in lignin that can be broken down into humus by soil microorganisms. The soil's urease enzyme is negatively affected by humus, which also has a great release-controlling effect and doubles as a soil and fertilizer enhancer (Chen et al., 2020). In order to make nutrients like potassium, phosphorus, and others available to plants, lignin may fix them and reduce the capacity of the soil to adsorb them. A fertilizer that contains all of these elements can also be made by combining it with other trace elements like manganese, zinc, molybdenum, copper, iron, and boron. Lignin is commonly used in agriculture, as a biological control agent, animal feed, soil conditioner, soil water holding regulator, fruit preserves, and fertilizer additive among other things (Lin et al., 1998; Tingda 2007). Lignin fertilizers include nitrogen, composite, control-release phosphate, chelating fertilizers, and others. Lignin urea, lignosulfonate, and ammonia-oxidized lignin are examples of controlled-release nitrogen fertilizers that can be applied to lignin. Non-volatilization, hard draining, delayed release, and excessive consumption are all characteristics of N-fertilizer.

In a lab experiment, Meier et al. (1994) demonstrated that in terms of crop production, controlled-release nitrogen fertilizer performed 82% better than ammonia. Urea could also be altered using the sodium lignosulfonate from industrialized straw pulp as a feedstock. By converting some of the nitrogen in urea, ammonia can be produced, which can bond directly to lignin. When the lignin breaks down, the nitrogen is gradually released, which helps the plant use more nitrogen. A temperature of 70 °C, a pH of 4, and a reaction time of four hours are the best process conditions for attaining better controlled-release properties (Yiqin et al., 2009).

An interesting development in the process of producing controlled-release nitrogen fertilizer was provided by making use of the lignin that is produced through the production of pulp and paper (Behin and Sadeghi, 2016). Using acetic acid and sodium metabisulfite, an acetylation procedure was used for chemically modifying extracted lignin for making it more hydrophobic. A fluidized-bed method was used in another study to apply a thin layer of acetylated synthetic lignin (kraft and sulfite) to granular urea. After seven days, about 36.3 and 45.3%, respectively, of the nitrogen were released into the soil.

6.2.8 Biosensor and Bioimaging

A biosensor is a device with a bioreceptor that recognizes the sample and interacts with it to generate a physiological response (Sawant, 2017). After that, this signal is converted into an electrical signal by a component of the transducer. The same amount of biological material that was present in the test samples is represented by an analog waveform that is produced. Lignin is highly compatible with carbon-based substances and has excellent sorption onto sp2-hybridized carbon interfaces because of its availability of aromatic units (Jędrzak et al., 2019a). Silver nanoparticles can also be stabilized by lignin (Tai et al., 2021). Lignin is an enticing source for biomedical imaging and biological recognition due to its biocompatibility. Using glucose's oxidation reaction mechanism, Jedrzak et al. (2019a, 2019b) looked into biosensors for glucose detection based on oxides and lignin at concentrations between 0.5 and 9.0 mM. Redox mediators like Ga2O3/lignin ZrO2/lignin with carbon paste electrode, magnetite/lignin/polydopamine (Fe3O4/Lig/PDA) and ferrocene were used to immobilize glucose oxidase in their experiments.

Beaucamp et al. (2021) developed a third-generation blood glucose sensor using glucose oxidase immobilized on highly porous nanomaterials made of 50% polylactic acid and 50% alcell lignin, which is ideal for glucose oxidase immobilization and has a density of 2–4 nm. This has a larger surface area, which enables it to have a comparatively higher coverage of 3.010 10 mol cm^2 and a straightforward electron transfer speed of 1.3 s^{-1} between the glucose oxidase's active redox groups and highly porous carbon. In ideal conditions, the sensor system exhibits a linear response between 0.15 and 2.7 mM, with a LoD of 89 mol L^{-1} and a high specificity of 50 A mM1 cm^2.

Zhang et al. (2018) created a novel graphene film composite sensor with 2.3 ng L^{-1} microcystin-LR sensing as an alternative to the traditional analysis methods used to monitor and evaluate groundwater resources in a variety of water sources using the temporary recommendation limit established by the World Health Organization (WHO), which are time-consuming, costly, unable to be transported, and frequently necessitate specialized knowledge. An eco-friendly graphene nanofiber laser biosensor decked with artificial metal nanoparticles (AgNPs) obtained from oil palm lignin was developed by Tai et al. (2021).

Using preferential interbreeding and incompatibility testing, the binding affinity of a preferential DNA sample recorded on AgNPs with the *Mycobacterium tuberculosis* DNA template was evaluated to validate the detecting performance. Studies using cyclic voltammetry demonstrated a sensitivity of up to 1 fM. This technique provides a low-potential sensing system for determining *Mycobacterium tuberculosis* biomarkers, opening up a new way to diagnose diseases.

Nishan et al. (2021) discovered that, at pH 7.5 and ambient temperature, they were able to rapidly synthesize lignin-stabilized silver nanoparticles coated with ionic liquids for the quantitative measurement of hydrogen peroxide in blood samples of high blood pressure patients. Carbon quantum dots (CQDs) appear to be biomaterial and non-toxic carbon core particles with a limit of 10 nm. They are widely used for in vitro and in vivo biomedical imaging and can produce radiant energy in the blue to red range.

O-aminobenzenesulfonic acid (A-acid) was used by (Wang et al. 2019b) in a mixture for dispersing lignin into lignin nanoparticles and hydrothermally fusioning them into lignin-based GQDs for hydrogen peroxide sensing. Long-term photo-stability, excellent fluorescence, solubility in water, and cytocompatibility are all characteristics of LGQDs. In principle, a growing area of medical science research could be the lignin combination of various substances and systematic approaches for effective biosensing. For even more precise and trustworthy medical diagnostics, these biological recognition techniques may be an excellent choice. Nevertheless, additional biofluid testing using all of these biosensors is required. The in vivo function is still under investigation, in addition to the in vitro application fields. A comprehensive overview of lignin-based biosensors was recently published (Sugiarto et al., 2022). The most important results from lignin-related biosensors and bioimaging are summarized in Table 6.1.

Table 6.1 Important results from lignin-based biosensors and use in bioimaging.

Lignin	Matrix component	From	In vitro research	Significant discoveries
Lignin from wood	–	Bioimaging	Human dermal fibroblasts	Fluorescent capability is excellent, and it is photostable for 30 days. Approximately 10–30% of radical scavenging behavior. A drop in fibroblasts of less than 20% is deemed non-cytotoxic
Alkali lignin	—	Biosensor	Mouse L929 fibroblasts Mouse 3 T3 fibroblasts	High sensitivities in the measurement of hydrogen peroxide at small concentrations as 0.13 nM low cytotoxicity (cell viability of 90.32%)
Lignin	—	Bioimaging	Human lung A549 adenocarcinoma cells HeLa cells	Biocompatible and low-cytotoxic (cell viability of 80%). HeLa cells have a good luminescence properties for imaging
Alkali lignin	Silver	Biosensor Breast cancer MCF-7 cells Melanoma A375 cells	Alkali lignin Silver Biosensor Breast cancer MCF-7 cells Melanoma A375 cells	Having good sensitivity and a significant correlation, hydrogen peroxide may be detected in concentrations ranging from 10^{-1} – 10^{-6} M. Properties of cytotoxicity (30% cell viability)
Organosolv lignin	Concanavalin A Horseradish peroxidase Glucose oxidase	Biosensor	–	Glucose measurement by chromogenic reagents is possible. At 0.85 M, the limit of detection is higher or equivalent.
Lignosulfonate	Nitrogen MXene/ Prussian blue Biosensor	Biosensor	–	Detecting behaviour for hydrogen peroxide (0–10 mM), glucose (10M 5.3 mM), lactate (0–20 mM), and alcohol (0–50 mM) over a range of concentration was effective

Table 6.1 (Continued)

Lignin	Matrix component	From	In vitro research	Significant discoveries
Corn stover lignin	Magnetite Anti-prion protein aptamer	Biosensor	–	Improved sensitivity by tenfold and effective identification of prion protein at concentrations ranging from 0.1 to 200 ng mL^{-1}
Organosolv lignin	Antigenic p17-1 peptide sequence	Biosensor	–	Specific anti-p17 human immunodeficiency virus antibodies as low as 0.1 ng mL^{-1} were successfully detected.
Kraft lignin	Silica Glucose oxidase	Biosensor	-	Immobilization of glucose oxidase (25.28 mg g^{-1}) improved by twice. Measurement of glucose in a regression was used to test (0.5–9 mM) with such a limit of detection of 145 mM and a high accuracy of 0.78 A mM^{-1}
Kraft lignin	Gallium oxide Zirconium (IV) oxide Glucose oxidase	Biosensor		Immobilization of glucose oxidase at 24.7 and 27.1 mg g^{-1} sensing glucose successfully
Kraft lignin	Magnetite Polydopamine Glucose oxidase	Biosensor		Increased glucose oxidase dosage (29.44 ± 2.39 mg g^{-1}). Professional biosensors have equivalent glucose detection precision and specificity

Based on Chauhan et al., 2022; Sugiarto et al., 2022; Chen et al., 2016; Aadil et al., 2016; Capecchi et al., 2020; Lei et al., 2020; Yuan et al., 2019; Cerrutti et al., 2015; Jędrzak et al., 2019a,b; Wang et al., 2019b.

6.2.9 Hydrogen Production

One suitable and cost-effective option for adding value is to turn lignin into hydrogen. In the not-too-distant future, hydrogen is likely to play a significant role in power generation as a well-known energy transporter. A hydrogen–oxygen redox reaction's chemical potential can be used to generate clean, sustainable electricity. For instance, since 2015, major automakers have been offering H2-fuel cell vehicles for sale. At an onset voltage of 0.25 V, direct lignin electrolysis was demonstrated with efficiencies of nearly 100% for producing hydrogen at the cathode and 85% for producing carbon dioxide at the anode (Hibino et al., 2017).

Caravaca et al. (2019) created pure hydrogen by "electrolysis of lignin solutions in continuous-flow mode in a polymer electrolyte membrane reactor" with an "OH conductor/anode/anion exchange membrane/cathode" electrochemical cell. They discovered that, in comparison to water, which can be electrolyzed at 1.2 V (from 0.45 V), lignin can be electrolyzed at much lower potentials.

Using supercritical water gasification, Sato et al. (2006) produced hydrogen from lignin with the help of nickel–magnesium oxide catalysts. Supercritical water gasification is one potential thermochemical method for conversion of biomass with a high moisture content into gas. At a temperature of 374.12 °C and pressure 221.2 bar above its critical point, water reaches supercritical conditions in which the gas and liquid phases do not separate. In addition, it functions well as a reactant (a source of hydrogen and free radicals) and excellent reaction medium, producing higher concentrations of H+ and OH–. It was discovered that an increase in gases was caused by an increase in the nickel catalyst. Through lignin gasification, the nickel probably made the formation of gases easier. At 673 K and a water density of 0.3 g cm^{-3}, about 80% of the gas was produced. In a similar fashion, Furusawa et al. (2007) produced hydrogen by calcining Ni/MgO catalysts made by impregnation at 773–1173 K for eight hours in air. Under test conditions, 10 weight % Ni/MgO produced the best catalytic performance. The results indicate that the Ni/MgO catalyst could be used to gasify lignin in supercritical water.

Kang et al. (2016) improved the method for biomass supercritical water gasification using cellulose and lignin as biomass and Ni/Ce/Al2O3 as the best catalysts. It was discovered that a catalyst loading of 100% and high temperatures (650 °C) were beneficial for increasing hydrogen production.

Kadam et al. (2014) came up with a green method for breaking down lignin and producing hydrogen by the use of nanostructured carbon, nitrogen, sulphur-doped zinc oxide under solar light and water splitting. This sample calcined at 500 °C demonstrated the greatest lignin degradation as compared to the sample calcined at 600 °C. 1-phenyl-3-buten-1-ol, a valuable fine chemical, is produced as a byproduct of lignin breakdown. The photo-catalytic activity found is substantially higher than that of visible light active oxide and sulphide photo-catalysts that have previously been documented. Hydrogen and electrons were produced through oxidation of water in a different method, as shown in the following equation:

$$2H2O \rightarrow O2 + 4H++ 4e-$$

The reaction's electrode potential was 1.23 V, but the four-electron transfer mechanism's slow kinetics make efficient water oxidation—also known as the oxygen evolution reaction—still a challenging endeavor. Even though OER catalysts have come a long way, there are still some drawbacks, like having to use expensive materials like platinum, ruthenium, and iridium and having poor electrode and catalyst stability, especially at acidic pHs. As a result, Oh et al. (2020) investigated alternative electron sources by oxidative depolymerizing biomasses containing phosphomolybdic acids at 1.23 V. Phosphomolybdic acids serve as an electron mediator and a catalyst for depolymerization.

A novel method for producing hydrogen from biomass using polyoxometalates was developed by Liu et al. (2016). Protons from biomass diffused to the cathode and were transformed into hydrogen there. At 0.2 A cm^2, the amount of electricity consumed per normal cubic meter of hydrogen may be as low as 0.69 kWh, or just 16.7% of the energy required for water electrolysis.

6.2.10 Battery Material for Energy Storage

Due to its low impact on the environment and higher carbon content (approximately 60 weight %), lignin has been extensively researched over the past ten years for its potential application in the production of batteries and supercapacitors (Yu and Kim 2020). Lignin is a cheap chemical that can be utilized for making binders, gel electrolytes, anodes, Pb-acid, and Na batteries for lithium batteries (Zhu et al., 2020). In supercapacitor applications, specific functional groups of lignin, such as the benzyl and phenolic groups, serve as active reaction sites for the storage of ions. Moreover, redox reactions and adsorption of ions from electrolytes by supercapacitors depend on lignin's numerous oxygen atoms. Supercapacitors can use lignin for porous carbon structures because of its cross-linked structure (Wu et al., 2020). It has been reported that electrospinning procedures to produce non-materials can benefit from alkali lignin whereas other forms of lignin, such as lignosulfonate, can be utilized in batteries or supercapacitors (Zhu et al., 2020). Usually, carbon from lignin could be extracted in two ways for use in energy materials. In one process, a precursor carbonization and activation were used, while in another, a chemical enhancer was used before carbonization and activation (Wu et al., 2020). Lignocellulosic materials are inert and useful for preventing active elements in battery recharge from corroding (Orsino and Harmon, 1945). In contrast, their potential application in batteries has been hindered by their high solubility. The acid insolubility of lignosulfonates can be improved by lowering their sulphur concentration, making them suitable for use in batteries in the future (Harmon, 1945). The generation of acidic bubbles was reduced by adding reduced-sulfur lignosulfonates to the lead compound at 0.1–0.2 weight %. This allows for a longer battery life and less harm to occur. It has been demonstrated that batteries with lignosulfonates can last for several years, whereas batteries without them can only operate for a few days due to lead plate corrosion. After 6000 charging and discharging cycles with electrospun carbon nanofibers produced from alkali lignin/PVA solutions, it was found that the super capacitor's capacitance reduced by 10%. Additionally, a power density of 91 kWh kg^{-1} and an energy value of 42 Wh kg^{-1} were reported. In addition, an experiment suggested that a higher concentration of lignin in the precursor nanofibers increased pore volume and specific area and reduced pore size (Lai et al., 2014).

Alkali lignin electrodes were found to have one of the highest specific capacitances (205 Fg^{-1}) (Ago et al., 2016). According to the literature, the mesopore class of carbon fibers had a wide hole dispersion, which contributed to their high electrochemical efficiency. Moreover, utilizing the carbonization-actuation technique, progressive permeable carbons can be obtained from steam blast lignin. Lignin-based hierarchical porous carbons had a capacitance of 286.7 F g^{-1} at 0.2 A g^{-1} (Zhang et al., 2015). Because it had a large surface area and easily accessible ion transport routes, the structure of Lignin made electrochemical performance easier. Compounds derived from lignin produced outcomes comparable to those of commercial graphite, which is frequently utilized in industrial batteries. Ball-milled hydrolysis lignin can be used to produce low-rate energy sources. Gnedenkov et al. (2014) looked into how well a lithium battery that used hydrolyzed lignin as the cathode performed. The cathode material is composed of 13 weight % of carbon black, 76 weight % of hydrolysis lignin, and 11 weight % of PTFE-based binder. The discharge capacity of lithium batteries that use hydrolysis lignin as an electrode material is 450 mAh g^{-1} (Chang et al., 2015).

Culebras et al. (2019, 2020) created a piezoelectric material out of macro carbon nanotube yarns (CNTYs) treated with 23% gaseous lignin. The Seebeck coefficient and electrical properties are significantly higher than in clean CNTY specimens. The result is a power factor of 132.2 W m1 K2, which is significantly greater than six times that of pure CNTY. At a temperature gradient of 30 K, a thermoelectric generator made of 20 CNTY/lignin nanocomposite yarns generates a high voltage gain of 3.8 W.

Dalton et al. (2019) also made bio-based carbon nanofibres using electrospinning from PAN and lignin combinations. Lignin increases cross fusion by up to 70%, improves specimen mobility, and reduces the size of CNFs by 450 to 250 nm. The lignin content of the carbonized and progenitor counterparts was used to measure the Seebeck coefficient and electrical resistance. In the end, a hydrazine vapour process was used to change the semi-conductive properties of p-type materials into n-type ones. The maximum p-type output power of CNFs carbonized at 900 °C with 70% lignin is 34.5 fold higher than that of CNFs with no lignin.

6.2.11 Dust Control Agent

Chemical suppressants were frequently used to reduce the environmental impact of road dust, which significantly pollutes the environment. Due to their low cost, eco-friendliness, and biodegradability, the suppressant based on lignin has scientific and practical significance. For instance, lignosulfonates preferentially bind to dust molecules that are polar and non-polar. The lignin and its derivatives absorb the tiny dust molecules, creating a heavier complex that settles the dust. On gravel roads with a lot of sand, lignosulfonate treatments were better at getting rid of dust than chlorides (calcium or magnesium) (Aro and Fatehi, 2017). The water that was sprayed onto filthy roads caused the lignosulfonates to thicken, trapping dust and preventing pollution (Calvo-Flores et al., 2015). Because they work better, they are better than other chemicals that prevent dust, like CaCl2. By improving surface drainage, they improve road occupancy and reduce maintenance requirements (Brown and Elton, 1994). However, the only drawback of using lignosulfonates is that they are only solubilized in water. Lignosulfonates may escape from the road surface during heavy rain.

A lignin-based liquid dust suppressant for gluing and shaping the surface of soil with a light texture was developed by Shulga and Betkers (2011). The binder that was made was easy to make and did not need any special tools or high temperatures or pressures. Its effectiveness as a dust suppressant was demonstrated by testing its resistance to water, wind, and mechanical forces in a variety of climates.

Katra (2019) compared several dust management products for wind-induced dust discharge from unsurfaced roadways. According to the control sample, unpaved roadway dust emissions contribute considerably to air pollution and mass transfer. Some of the products tested for controlling dust included lignin, resin, bitumen, PVA, and brine. The initial assessment of the findings under controlled laboratory conditions, made it possible to quantitatively evaluate the effectiveness of the product against wind erosion without the use of vehicles. After that, the products were tested on plots parallel to the road in an active quarry. The field demonstration continued for a number of days and weeks after the quarry-haul trucks were installed. According to the findings, the amount of dust released typically rises with wind speed. To determine how well dust control products worked in each area, the PM10 (particulate matter 10 m) fluxes from the road surface were used. The results demonstrated that lignin, brine, and PVA significantly reduced dust and that less brine was used with less emission.

6.2.12 Bitumen Modifier in Road Industry

Bitumen is a black, amorphous thermoplastic with a temperature-dependent stiffness. Construction of motorways and highways frequently makes use of it. Bitumen is used to pave between 90% and 95% of all roads worldwide. 18 million tons of asphalt are needed annually for a smooth and safe highway, and they need to work at both high and low temperatures (Xie et al., 2017). Asphalt production necessitates expensive petroleum and releases harmful greenhouse gases. In addition, the availability of bitumen is negatively impacted by the petrochemical industry's increased capacity to break down longer-chain hydrocarbons into shorter ones with more added value than bitumen. As a result, alternatives that emit less CO_2 are being looked into by the asphalt industry. Lignin has been investigated as a "partial" alternative to bitumen and as a means of lowering CO_2 emissions (Van Vliet et al., 2016; Wang et al., 2015). Bitumen blends and a variety of native and modified lignins have been investigated. These changes were made for improving or restoring the characteristics of the original binder (bitumen). Kraft lignin, organosolv, klason, and SHS, can be mixed with used bitumen fraction up to 25% of the time. Visco-elastic behavior is influenced by the type and quantity of lignin, as shown by DSR. The impacts of lignin content in the mix turned out to be more articulated, causing significant varieties in firmness and stage point. The visco-elastic behavior of the lignin–bitumen mixture is ultimately altered as a result of this impact, which was influenced by lignin content and bitumen type (Van Vliet et al., 2016).

Perez et al. (2019a, 2019b) looked into including waste that contained lignin in asphalt binders. The conventional asphalt was mixed directly with the lignin-rich material that was undesirable to the hardboard industry at concentrations of 0%, 5%, 10%, 20%, and 40%. The waste made fatigue and resistance to rutting stronger. Utilizing asphalt binder with 20% waste is preferable as an extender and enhancer for asphalt pavements. The lignin-containing by-products used as bitumen antioxidants led to an enhancement of high-temperature properties but detriment of low-temperature properties of bitumen (Bourzac, 2015). A test bike trail was constructed in Iowa in 2010 using bio-asphalt to replace 3% of the bitumen. The bike path produced consistent results. The use of lignin as an antioxidant for bitumen was demonstrated by Pan (2012). Temperature control is required to prevent lignin from aging. The 70-meter test piece was made of low-temperature bio-asphalt, and after two years, the performance of the road was found to be reasonable. Lignin-modified bitumen improves the asphalt mixture's durability and rutting capacity at high temperatures (Xu et al., 2017). According to Boomika et al. (2017), 15% lignin and 20% plastic were efficient alternatives to bitumen. Additionally, the Netherlands Organization for Applied Scientific Research demonstrated that using chemically altered hydrophobic lignin in bitumen modified with lignin increases asphalt mixes' resistance to thermal cracking at 12 °C (Bourzac, 2015; Slaghek et al., 2017; Van Vliet et al., 2016).

The modified lignin was combined with bitumen (up to 25% weight-to-weight). The resulting bitumen lasts a long time and is resistant to a wide range of climates. However, the price of this bitumen is higher than that of conventional bitumen.

6.2.13 Cement Additives and Building Material

Modern concrete mixtures require plasticizers to increase the operability of composites without increasing the water–cement ratio. In a cement mortar, these might save water while maintaining mobility. The concrete mixture is more durable and effective because the water–cement ratio is lower (Jędrzejczak et al., 2021). They aid in the formation of a lubricant layer between the binder and hydrophobic interactions by producing electrostatic activity. Water-reducing chemicals come in two varieties: superplasticizers and plasticizers. The former results in a loss of water that is greater than 12%, while the latter does not. Lignosulfonates have been widely used for more than eight decades, despite the fact that there appear to be a number of different plasticizers on the market. These could be made, for example, through alkaline lignin sulfonation, or sulfite liquor (Jędrzejczak et al., 2021).

Brick ceramics and plasterboard components also contain these materials as additives (0.5–2.0%). Demand for building materials, particularly concrete, has significantly increased as a result of civilization's incredible growth. One of the possible technical advancements in contemporary concrete is the use of functional mineral admixtures like limestone, alumina, silica, titanium dioxide slag, and fly ash (Elgalhud et al., 2016; Klapiszewski et al., 2019; Norhasri et al., 2017; Staub de Melo and Triches, 2018).

Admixtures improve the properties of concrete by replacing some cement. They all improve water resistance, compressive strength, and settling time control. Klapiszewski et al. (2019) followed this trend and made hybrid materials for cement mortars with lignosulfonate, also known as lignin, and Al2O3. In addition, it was discovered that concrete with less or modified lignin may offer some high-performance concrete strengths, a stronger effect of dispersion in the cement matrix, and less harm to the exterior wall from moisture and acid rain. The plasticizing capacity of concrete increased from 161 to 185 mm when the dose of lignosulfonates was increased by functionalizing lignosulfonates through oxidation and sulfomethylation at a weight-to-weight ratio of 0.3 (He and Fatehi, 2015; Yu et al., 2013). Concrete admixtures' ability to plasticize was found to be affected by the process of lignosulfonate nitration. Nitration decreased the quantity of water required to be added to concrete mixtures while maintaining 25 mN strength to almost 0.6% weight-to-weight (Aro and Fatehi, 2017). The fluidity increased from 161 to 185 mm (Staub de Melo and Trichês, 2018) as a result of the oxidation and sulfomethylation of lignosulfonates, which improved the concentration of sulphur from 0.65 to 1.45 mmol g^{-1}. The material's capacity for plasticization was enhanced as a result. The charge density increased to 4.6 meq g^{-1} and the kraft lignin sulfonation rate increased to 2.04 meq g^{-1} as a result of the same conversion, increasing the cement paste fluid. In contrast to commercial lignosulfonates and lignosulfonate acid, however, a high molecular mass or high sulphur content may hinder lignosulfonate diffusion (He and Fatehi, 2015). Therefore, it is reasonable to draw the conclusion that dissolution and dispersal require the S-quantity of sulfonated lignin.

A composite made from waste biomass and unseparated biomass that is stronger, more durable, and reusable than Portland cement was developed by Lauer et al. (2020). They did this by employing a 100% atom economical polymerization process. Mechanical strength was superior to that of Portland cement in the composite carbon-negative cement (APS95) that was produced as a result. APS95 is also a product that absorbs carbon, whereas the production of Portland cement is the primary source of anthropogenic CO_2. In addition, when modified lignosulfate is doped into concrete, its strength increases slowly in the beginning and decreases when the quantity is higher. Modified lignosulfate has a higher plasticizing capacity than lignosulfate with a content of 0.25–0.11% (Chen et al., 2011). The reduction rate of water can be anywhere from 1.5–19.6%. In fact, costs could be reduced by combining lignosulfate with polycarboxylate superplasticizers, naphthalene, aliphatic acid salts, sulfamic acid, and water loss correction and consolidation as dehydrating agents. The efficiency of water-forming can be improved when amino-based superplasticizers and lignosulfonate are combined (Zhenyang et al., 2011). The correlations between the properties and composition of lignin, as well as their effects on the use of lignin in the construction industry and the life cycle assessment of that use, have been discussed by Jędrzejczak et al. (2021).

6.2.14 Bioplastics

While the global production of bioplastics is less than 200 000 tons annually, the production of oil-based plastics exceeds 30 million tons annually. When compared to conventional plastics, bioplastics are better for the environment because their manufacturing results in the emission of fewer greenhouse gases like carbon dioxide which is one of the primary causes of air pollution and contributes to environmental problems like climate change and global warming. Lignin is combined with natural fibers like cellulose, hemp and flax, as well as additives, to create bioplastics that can be processed at high temperatures. Strength, rigidity, dimensional stability, and other mechanical properties of bioplastics can be changed by changing how they are made. Musical instruments, furniture, automobile interiors, garden supplies, and other products all use bioplastics (Bogomolova, 2013). Global bioplastics industry is expected to expand significantly in the next few years (*Bioplastics News*, 2014). It reached USD 7.02 billion in 2018.

6.2.15 Use of Lignin as a Binder

It has been discovered that lignosulfonates are an extremely cost-effective adhesive that can be used as a binding agent or "glue" for compressed materials or pellets. When applied to unpaved roads, lignosulfonates stabilize the road surface and lessen the impact of airborne dust on the environment. Due to its ability to bind to other substances, it serves as a useful component in a variety of goods, including biodegradable plastics, carbon black, coal briquettes, particle board and plywood, ceramics, animal feed pellets, fertilisers and herbicides, fibre glass insulation, linoleum paste, soil stabilizers, and others. Lignin holds together glass wool insulation for buildings. It is used to bind glass fibres when the fibre pads are formed and is applied to hot glass as recovered ammonium salt solid from the kraft paper manufacturing process. The binder creates composite materials with a respectable wet strength that are also reasonably priced. A lignin-based modifier is added to formaldehyde-based binder systems, such as urea formaldehyde, phenol formaldehyde, resorcinol formaldehyde, and/or tannin formaldehyde resins, to create panel boards like hardboard plywood, medium density fiberboard, or particle boards.

6.2.16 Lignin as Dispersant

For coatings, paints, and other products, chemically modified lignin has been utilized as an auxiliary agent, thickener, flocculent, or dispersing agent. Undissolved particles in suspensions are prevented from clumping and settling by lignosulfonate. According to the TIFAC report, a substance obtained from lignin has been utilized as a dispersant in soils, compounds for cleaning and/or laundry detergent, oil drilling muds, leather tanning, cement mixes, insecticides, and pesticides Aluminum plates have been cleaned with a mixture of lignin sulfonic acid and polycarboxylic acid to prevent calcium scaling. For emulsion or dispersion polymerization, lignosulfonates are utilized as biodegradable and non-toxic emulsifiers or dispersants.

6.2.17 Lignin as Food Additives

The food industry is always looking for new, natural ingredients that make food better for health. According to VTT, lignin, fibrillated cellulose, and xylan have characteristics that set them apart from other ingredients that are commonly used. The surface-active properties of lignin could be utilized for producing emulsions and foams with better texture. Additionally, lignin can be used to stop food products from oxidizing. Lignin in muffin production was examined by VTT. Lignin proved to be a surprisingly effective alternative to whole eggs and egg yolks, giving muffins a fluffier texture. Lignin was found to be an emulsifier in mayonnaise and to support juiciness in meat products. Potential applications of lignin include including alkali lignins as a roughage or fiber source in human and pet food. According to extensive research conducted in this field, consuming a diet high in dietary fiber is associated with lower rates of colon cancer (Meister, 2007).

6.2.18 Lignin as Sequestering Agent

Heavy metals present a serious risk to both the environment and human health because of their high toxicity. Metal plating, mining, tanning, chloralkali processing, alloying and smelting, plants, as well as the storage battery industry, all produce waste that is contaminated with heavy metals. Even though the use of active carbon is the most common method for removing heavy metals from water, it is very expensive (Hegazi, 2013). Toxic heavy metal ions such as lead, cadmium, chromium, arsenic, zinc, copper, and nickel must be eliminated in a cost-effective manner. It has been demonstrated that lignin's ability to absorb heavy metals can be enhanced by altering its structure (Ge et al., 2018; Guo et al., 2008). Lignosulphonates are strong sequestering agents which are able to bind metal ions and form complexes. Lignosulphonates can be used to improve soil quality, condition soil, or remove harmful metals thanks to this property. In addition, lignin's potential as a control release agent in fertilizers, pesticides, and herbicides has been investigated (Chowdhury, 2014). In agricultural and forestry operations, testing has been done on lignin hydrogels as a soil conditioner that can hold water. The ability of superabsorbent lignosulphonate-g-acrylic acid hydrogels to remove dye (methylene blue) and contaminants from waste water has been demonstrated (Yu et al., 2016). According to Flores-Céspedes et al. (2015), pesticides containing lignin have a better resistance to photodegradation.

6.2.19 Lignin Bio-oil

Bio-oil is a fluid produced by the thermochemical transformation of biomass, primarily through rapid pyrolysis. It holds tremendous promise as a sustainable source of chemicals for the production of adhesives, polymers, and resins. The majority of the compounds in the pyrolysis biooil are coniferyl alcohol, sinapyl alcohol, isoeugenol, vanillin, catechol, guaiacol, vinyl guaiacol, methyl guaiacol, and many other compounds. The various products produced by bio-oil are influenced by the pyrolysis time and/or temperature; for instance, bio-oils used to produce fuel require longer pyrolysis times, whereas optimal yields of valuable compounds require shorter times (Collard and Blin, 2014; Triantafyllidis et al., 2013). Bio-oil can be used to make fuel, isolate various chemicals, and make products based on oil. The TRL for the isolation of lignin bio-oil in 2018 was six, indicating that there is a significant growth potential (Fraunhofer, 2018).

Bibliography

Aadil KR, Barapatre A, Meena AS, and Jha H (2016). Hydrogen peroxide sensing and cytotoxicity activity of Acacia lignin stabilized silver nanoparticles. *Int J Biol Macromol*, 82: 39–47.

Ago M, Borghei M, Haataja JS, and Rojas OJ (2016). Mesoporous carbon soft-templated from lignin nanofiber networks: microphase separation boosts supercapacitance in conductive electrodes. *RSC Adv*, 6(89): 85802–85810.

Ahvazi B, Wojciechowicz O, Ton-That TM, and Hawari J (2011). Preparation of lignopolyols from wheat straw soda lignin. *J Agri Food Chem*, 59: 10505–10516.

Alam MA (2019). Anti-hypertensive effect of cereal antioxidant ferulic acid and its mechanism of action. *Front Nutr*, 6: 121.

Aro T and Fatehi P (2017). Production and application of lignosulfonates and sulfonated lignin. *Chem Sus Chem*, 10(9): 1861–1877.

Bajpai P (2018). Value-added products from lignin. *Biotechnology for Pulp and Paper Processing*, Springer, Singapore, pp. 561–571.

Baker DA and Rials TG (2013). Recent advances in low cost carbon fiber manufacture from lignin. *J Appl Polym Sci Symp*, 130: 713–728.

Banerjee G and Chattopadhyay P (2018). Vanillin biotechnology: the perspectives and future. *J Sci Food Agric*, 99(2): 499–506.

Banik I and Sain MM (2008). Water blown soy polyol-based polyurethane foams of different rigidities. *J Reinf Plast Compos*, 27: 357–373.

Beaucamp A, Culebras M, and Collins MN (2021). Sustainable mesoporous carbon nanostructures derived from lignin for early detection of glucose. *Green Chem*, 23(15): 5696–5705.

Behin J and Sadeghi N (2016). Utilization of waste lignin to prepare controlled-slow release urea. *Int J Recycl Org Waste Agric*, 5(4): 289–299.

Beisl S, Friedl A, and Miltner A (2017). Lignin from micro- to nanosize: applications. *Int J Mol Sci*, 18(11): 2367.

Bhagia S, Bornani K, Agarwal R, Satlewal A, Ďurkovič J, Lagaňa R, Bhagia M, Yoo CG, Zhao X, Kunc V, Pu Y, Ozcan S, and Ragauskas AJ (2021). Critical review of FDM 3D printing of PLA biocomposites filled with biomass resources, characterization, biodegradability, upcycling and opportunities for biorefineries. *Appl Mater Today*, 24: 101078.

Bogomolova A (2013). *Bioplastic advances innovative, green architecture*. Stuttgart University's, Institute of Building Structures and Structural Design (ITKE).

Boomika A, Naveen MA, Daniel Richard J, Mythili A, and Vetturayasudharsanan R (2017). Experimental study on partial replacement of bitumen with lignin and plastic. *SSRG Int J Civ Eng*, 9–14.

Bourzac K (2015). Inner workings: paving with plants. *Proc Natl Acad Sci U.S.A.*, 112(38): 11743–11744.

Brown DA and Elton DJ (1994). *Guidelines for dust control on unsurfaced roads in Alabama*. Final report No. IR-94-02.

Bumrungpert A, Lilitchan S, Tuntipopipat S, Tirawanchai N, and Komindr S (2018). Ferulic acid supplementation improves lipid profiles, oxidative stress, and inflammatory status in hyperlipidemic subjects: a randomized, double- blind, placebo-controlled clinical trial. *Nutrients*, 10(6): 713.

Calvo-Flores FG, Dobado JA, Isac-García J, and Martín-Martínez FJ (2015). *Lignin and Lignans as Renewable Raw Materials: Chemistry, Technology and Applications*. John Wiley & Sons.

Capecchi E, Piccinino D, Tomaino E, Bizzarri BM, Polli F, Antiochia R, Mazzei F, and Saladino R (2020). Lignin Nanoparticles are Renewable and Functional Platforms for the Concanavalin a Oriented Immobilization of Glucose Oxidase–peroxidase in Cascade Bio-sensing. *RSC Adv*, 10(48): 29031–29042.

Caravaca A, Garcia-Lorefice WE, Gil S, de Lucas-Consuegra A, and Vernoux P (2019). Towards a sustainable technology for H2 production: direct lignin electrolysis in a continuous-flow polymer electrolyte membrane reactor. *Electrochem Commun*, 100: 43–47.

Cayla A, Rault F, Giraud S, Salaün F, Fierro V, and Celzard AJP (2016). PLA with intumescent system containing lignin and ammonium polyphosphate for flame retardant textile. *Polymers*, 8(9): 331.

Cerrutti BM, Moraes ML, Pulcinelli SH, and Santilli CV (2015). Lignin as immobilization matrix for HIV p17 peptide used in immunosensing. *Biosens Bioelectron* Sep 15, 71: 420–426.

Chang ZZ, Yu BJ, and Wang CY (2015). Influence of H2 reduction on lignin-based hard carbon performance in lithium ion batteries. *Electrochimica Acta*, 176: 1352–1357.

Chauhan PS, Agrawal R, Satlewal A, Kumar R, Gupta RP, and Ramakumar SSV (2022). Next generation applications of lignin derived commodity products, their life cycle, techno-economics and societal analysis. *Int J Biol Macromol*, 197: 179–200.

Chen G, Gao J, Chen W, Song S, and Zhenhua PENG (2011). Method for preparing concrete water reducer by grafting of lignosulfonate with carbonyl aliphatics. *U.S. Patent Application No. 12/674,645*.

Chen J, Fan X, Zhang L, Chen X, Sun S, and Sun R (2020). Research progress in lignin-based slow/controlled release fertilizer. *ChemSusChem*, 13(17): 4356–4366.

Chen L, Zhou X, Shi Y, Gao B, Wu J, Kirk TB, Xu J, and Xue W (2018). Green synthesis of lignin nanoparticle in aqueous hydrotropic solution toward broadening the window for its processing and application. *Chem Eng J*, 346: 217–225.

Chen W, Hu C, Yang Y, Cui J, and Liu Y (2016). Rapid synthesis of carbon dots by hydrothermal treatment of lignin. *Materials (Basel)* Mar 9, 9(3): 184.

Chowdhury MA (2014). The controlled release of bioactive compounds from lignin and lignin-based biopolymer matrices. *Int J Biological Macrom*, 65: 136–147.

Collard F-X and Blin J (2014). A review on pyrolysis of biomass constituents: mechanisms and composition of the products obtained from the conversion of cellulose, hemicelluloses and lignin. *Renew Sustain Energy Rev*, 38: 594–608.

Culebras M, Geaney H, Beaucamp A, Upadhyaya P, Dalton E, Ryan KM., and Collins MN (2019). Bio-derived carbon nanofibers from lignin as high performance Li-ion anode materials. *ChemSusChem*, 8: 4516–4521.

Culebras M, Ren G, O'Connell S, Vilatela JJ, and Collins MN (2020). Lignin doped carbon nanotube yarns for improved thermoelectric efficiency. *Adv Sustain Syst*, 4(11): 2000147.

Dalton N, Lynch RP, Collins MN, and Culebras M (2019). Thermoelectric properties of electrospun carbon nanofibres derived from lignin. *Int J Biol Macromol*, 121: 472–479.

Elgalhud AA, Dhir RK, and Ghataora G (2016). Limestone addition effects on concrete porosity. *Cem Concr Compos*, 72: 222–234.

Fache M, Boutevin B, and Caillol S (2015). Vanillin, a key-intermediate of biobased polymers. *Eur Polym J*, 68(C): 488–502.

Fache M, Boutevin B, and Caillol S (2016). Vanillin production from lignin and its use as a renewable chemical. *ACS Sustain Chem Eng*, 4(1): 35–46.

Fahrioğlu U, Dodurga Y, Elmas L, and Seçme M (2016). Ferulic acid decreases cell viability and colony formation while inhibiting migration of MIA PaCa-2 human pancreatic cancer cells in vitro. *Gene*, 576(1): 476–482.

Figueiredo P, Lintinen K, Hirvonen JT, Kostiainen MA, and Santos HA (2018). Properties and chemical modifications of lignin: towards lignin-based nanomaterials for biomedical applications. *Prog Mater Sci*, 93: 233–269.

Figueiredo P, Lintinen K, Kiriazis A, Hynninen V, Liu Z, Bauleth-Ramos T, Rahikkala A, Correia A, Kohout T, Sarmento B, Yli-Kauhaluoma J, Hirvonen J, Ikkala O, Kostiainen MA, and Santos HA (2017). In vitro evaluation of biodegradable lignin-based nanoparticles for drug delivery and enhanced antiproliferation effect in cancer cells. *Biomaterials*, 121: 97–108.

Flores-Céspedes F, Martínez-Domínguez GP, Villafranca-Sánchez M, and Fernández-Pérez M (2015). Preparation and characterization of azadirachtin alginate-biosorbent based formulations: water release kinetics and photodegradation study. *J Agric Food Chem*, 63: 8391–8398.

Forecast Bioplastics Industry Growth (2014). *Bioplastics news*. https://bioplasticsnews.com/2014/01/28/forecasted-bioplastics-industry-growth (accessed December 20, 2022).

Fraunhofer ISI (2018). *Directorate-general for research and innovation(European Comission)*. Detailed Case Studies on the Top 20 Innovative Bio-Based Products. In Top 20 Innovative Bio-Based Products EU Publications, University of Bologna, Luxembourg, pp. 53–242.

Furusawa T, Sato T, Sugito H, Miura Y, Ishiyama Y, Sato M, Itoh N, and Suzuki N (2007). Hydrogen production from the gasification of lignin with nickel catalysts in supercritical water. *Int J Hydro Energy*, 32(6): 699–704.

Ge Y and Li ZJASC(2018). Application of lignin and its derivatives in adsorption of heavy metal ions in water: a review. *ACS Sustain Chem Eng*, 2018: 657181–657192.

Ghosh S, Basak P, Dutta S, Chowdhury S, and Sil PC (2017). New insights into the ameliorative effects of ferulic acid in pathophysiological conditions. *Food Chem Toxicol*, 103: 41–55. https://doi.org/10.1016/j.fct.2017.02.028.

Gill AS, Visotsky D, Mears L, and Summers JD, (2016). Cost estimation model for PAN based carbon fiber manufacturing process. *ASME 2016 11th International Manufacturing Science and Engineering Conference*, American Society of Mechanical Engineers.

Gnedenkov SV, Opra DP, Sinebryukhov SL, Tsvetnikov AK, Ustinov AY, and Sergienko VI (2014). Hydrolysis lignin: electrochemical properties of the organic cathode material for primary lithium battery. *J Ind Eng Chem*, 220(3): 903–910.

González-Sarrías A, Li L, and Seeram NP (2012). Anticancer effects of maple syrup phenolics and extracts on proliferation, apoptosis, and cell cycle arrest of human colon cells. *J Funct Foods*, 4(1): 185–196.

Gordobil O, Egüés I, Llano-Ponte R, and Labidi J (2014). Physicochemical properties of PLA lignin blends. *Polym Degrad Stab*, 108: 330–338.

Guo X, Zhang S, and Shan X-q (2008). Adsorption of metal ions on lignin. *J Hazard Mater*, 151: 134–142.

Harmon C (1945). *Lignin compounds and process classification*. U.S. Patent 2371136 A.

He W and Fatehi P (2015). Preparation of sulfomethylated softwood kraft lignin as a dispersant for cement admixture. *RSC Adv*, 5(58): 47031–47039.

Hegazi HA (2013). Removal of heavy metals from wastewater using agricultural and industrial wastes as adsorbents. *HBRC J*, 9: 276–282.

Hibino T, Kobayashi K, Nagao M, and Teranishi S (2017). Hydrogen production by direct lignin electrolysis at intermediate temperatures. *Chem Electro Chem*, 4(12): 3032–3036.

Holladay JE, White JF, Bozell JJ, and Johnson D (2007). *Top Value-added Chemicals From Biomass-Volume II—Results of Screening for Potential Candidates From Biorefinery Lignin*. Pacific Northwest National Lab. (PNNL), Richland, WA, USA.

Huang J, Liu W, and Qiu X (2019). High performance thermoplastic elastomers with biomass lignin as plastic phase. *ACS Sustain Chem Eng*, 7(7): 6550–6560.

Ibrahim MNM, Sriprasanthi RB, Shamsudeen S, Adam F, and Bhawani SA (2012). A concise review of the natural existance, synthesis, properties, and applications of syringaldehyde. *BioResources*, 7(3): 4377–4399.

Jędrzak A, Rębi's T, Kuznowicz M, Kołodziejczak-Radzimska A, Zdarta J, Piasecki A, and Jesionowski T (2019a). Advanced Ga2O3/Lignin and ZrO2/Lignin hybrid microplatforms for glucose oxidase immobilization: evaluation of biosensing properties by catalytic glucose oxidation. *Catalysts*, 9(12): 1044.

Jędrzak A, Rębiś T, Kuznowicz M, and Jesionowski T (2019b). Bio-inspired magnetite/lignin/polydopamine-glucose oxidase biosensing nanoplatform. From synthesis, via sensing assays to comparison with others glucose testing techniques. *Int J Biol Macromol* Apr 15, 127: 677–682.

Jędrzejczak P, Collins MN, Jesionowski T, and Klapiszewski L (2021). The role of lignin and lignin-based materials in sustainable construction–a comprehensive review. *Int J Biol Macromol*, 187(2021): 624–650.

Kadam SR, Mate VR, Panmand RP, Nikam LK, Kulkarni MV, Sonawane RS, and Kale BB (2014). A green process for efficient lignin (biomass) degradation and hydrogen production via water splitting using nanostructured C, N, S-doped ZnO under solar light. *RSC Adv*, 4(105): 60626–60635.

Kadla J, Kubo S, Venditti R, Gilbert R, Compere A, and Griffith W (2002). Lignin-based carbon fibers for composite fiber applications. *Carbon*, 40: 2913–2920.

Kang K, Azargohar R, Dalai AK, and Wang H (2016). Hydrogen production from lignin, cellulose and waste biomass via supercritical water gasification: catalyst activity and process optimization study. *Energy Convers Manag*, 117: 528–537.

Katra I (2019). Comparison of diverse dust control products in wind-induced dust emission from unpaved roads. *Appl Sci*, 9(23): 5204.

Klapiszewski Ł, Klapiszewska I, Slosarczyk A, and Jesionowski T (2019). Lignin-based hybrid admixtures and their role in cement composite fabrication. *Molecules*, 24(19): 3544.

Kumar N and Pruthi V (2014). Potential applications of ferulic acid from natural sources. *Biotechnol Rep*, 4(1): 86–93.

Lai C, Zhou Z, Zhang L, Wang X, Zhou Q, Zhao Y, Wang Y, Wu X, Zhu Z, and Fong H (2014). Free-standing and mechanically flexible mats consisting of electrospun carbon nanofibers made from a natural product of alkali lignin as binder-free electrodes for high-performance supercapacitors. *J. Power Sources*, 247: 134–141.

Lange H, Decina S, and Crestini C (2013). Oxidative upgrade of lignin – recent routes reviewed. *Eur Polym J*, 49(6): 1151–1173.

Lauer MK, Karunarathna MS, Tennyson AG, and Smith RC (2020). Recyclable, sustainable, and stronger than portland cement: a composite from unseparated biomass and fossil fuel waste. *Materials Adv*, 1(4): 590–594.

Lee A and Deng Y (2015). Green polyurethane from lignin and soybean oil through nonisocyanate reactions. *Eur Polym J*, 63: 67–73.

Lei Y, Alshareef AH, Zhao W, and Inal S (2020). Laser-Laser-scribed graphene electrodes derived from lignin for biochemical sensing. *ACS Appl Nano Mater*, 3(2): 1166–1174.

Li T and Takkellapati S (2018). The current and emerging sources of technical lignins and their applications. *Biofuels, Bioprod Bioref*, 12(5): 756–787.

Lievonen M, Valle-Delgado JJ, Mattinen ML, Hult EL, Lintinen K, Kostiainen MA, Paananen A, Szilvay GR, Setälä H, and Österberg M (2016). A simple process for lignin nanoparticle preparation. *Green Chem*, 18(5): 1416–1422.

Lin C, Jim Y, and Zhongzheng L (1998). The application of lignin in fertilizers. *China Paper*, 10(2): 68–70.

Liu W, Cui Y, Du X, Zhang Z, Chao Z, and Deng Y (2016). High efficiency hydrogen evolution from native biomass electrolysis. *Energy Environ Sci*, 9(2): 467–472.

Llevot A, Grau E, Carlotti S, Grelier S, and Cramail H (2016). From lignin- derived aromatic compounds to novel biobased polymers. *Macromol Rapid Commun*, 37(1): 9–28.

Lu Q, Zhu M, Zu Y, Liu W, Yang L, Zhang Y, Zhao X, Zhang X, Zhang X, and Li W (2012). Comparative antioxidant activity of nanoscale lignin prepared by a supercritical antisolvent (SAS) process with non-nanoscale lignin. *Food Chemistry*, 135: 63–67.

Mahmood N, Yuan Z, Schmidt J, and Xu C (2013). Production of polyols via direct hydrolysis of kraft lignin: effect of process parameters. *Bioresour Technol*, 139(C): 13–20.

Mahmood N, Yuan Z, Schmidt J, and Xu CC (2016). Depolymerization of lignins and their applications for the preparation of polyols and rigid polyurethane foams: a review. *Renew Sustain Energy Rev*, 60: 317–329.

Mainka H, Täger O, Körner E, Hilfert L, Busse S, Edelmann FT, and Herrmann AS (2015). Lignin–an alternative precursor for sustainable and cost-effective automotive carbon fiber. *J Mater Res Technol*, 4: 283–296.

Mancuso C and Santangelo R (2014). Ferulic acid: pharmacological and toxicological aspects. *Food Chem Toxicol*, 65: 185–195.

Mandlekar N, Cayla A, Rault F, Giraud S, Salaün F, Malucelli G, and Guan JP (2018). *An overview on the use of lignin and its derivatives in fire retardant polymer systems, lignin-trends and applications.* InTech. Matheus Poletto, IntechOpen. https://doi.org/10.5772/intechopen.72963.

Martinez V, Mitjans M, and Pilar Vinardell M (2012). Pharmacological applications of lignins and lignins related compounds: an overview. *Curr Org Chem*, 16: 1863–1870.

Meier D, Zúñiga-Partida V, Ramírez-Cano F, Hahn NC, and Faix O (1994). Conversion of technical lignins into slow-release nitrogenous fertilizers by ammoxidation in liquid phase. *Bioresource Technol*, 49(2): 121–128.

Meister JJ (2007). Modification of lignin. *J Macromol Sci Part C*, 42(2): 235–289.

Mohammadi Gheisar M and Kim IH (2018). Phytobiotics in poultry and swine nutrition - a review. *Ital J Anim Sci*, 17(1): 92–99.

Mota MIF, Rodrigues Pinto PC, Loureiro JM, and Rodrigues AE (2015). Recovery of vanillin and syringaldehyde from lignin oxidation: a review of separation and purification processes. *Separat Purif Rev*, 45(3): 227–259.

Muir M (1996). DMSO: many uses, much controversy. *Alt and Compl Ther*, 2: 230–235.

Nair SS, Sharma S, Pu Y, Sun Q, Pan S, Zhu JY, and Ragauskas AJ (2014). High shear homogenization of lignin to nanolignin and thermal stability of nanolignin- polyvinyl alcohol blends. *ChemSusChem*, 7(12): 3513–3520.

Nguyen NA, Barnes SH, Bowland CC, Meek KM., Littrell KC, Keum JK, and Naskar AK (2018). A path for lignin valorization via additive manufacturing of high-performance sustainable composites with enhanced 3D printability. *Sci Adv* Dec 14, 4(12).

Nishan U, Niaz A, Muhammad N, Asad M, Khan N, Khan M, Shujah S, and Rahim A (2021). Non-enzymatic colorimetric biosensor for hydrogen peroxide using lignin-based silver nanoparticles tuned with ionic liquid as a peroxidase mimic. *Arab J Chem*, 14(6): 103164.

Norberg I, Nordström Y, Drougge R, Gellerstedt G, and Sjöholm E (2013). A new method for stabilizing softwood kraft lignin fibers for carbon fiber production. *J Appl Polym Sci*, 128: 3824–3830.

Norhasri MM, Hamidah MS, and Fadzil AM (2017). Applications of using nano material in concrete: a review. *Constr Build Mater*, 133: 91–97.

Oh H, Choi Y, Shin C, Nguyen TVT, Han Y, Kim H, and Ryu J (2020). Phosphomolybdic acid as a catalyst for oxidative valorization of biomass and its application as an alternative electron source. *ACS Catal*, 10(3): 2060–2068.

Orsino JA and Harmon C (1945). *US 2371137 A*.

Pan (2012). A first-principles based chemophysical environment for studying lignins as an asphalt antioxidant. *Constr Build Mater*, 36: 654–664.

Parmar I, Bhullar KS, and Rupasinghe HPV (2015). Anti-diabetic effect of ferulic acid and derivatives: an update. *Ferulic Acid: Antioxidant Properties, Uses and Potential Health Benefits*, Warren B. (ed.). Nova Science Publishers, Inc., Hauppauge, NY, USA, pp. 93–116. 978-1-63463-299-7.

Pérez I, Pasandín AR, Pais JC, and Pereira PA (2019a). Feasibility of using a lignin-containing waste in asphalt binders. *Waste Biomass Valori*, 1: 1–14.

Pérez IP, Pasandín AMR, Pais JC, and Pereira PAA (2019b). Use of lignin biopolymer from industrial waste as bitumen extender for asphalt mixtures. *J Clean Prod*, 220: 87–98.

Puziy AM, Poddubnaya OI, and Sevastyanova O (2018). Carbon materials from technical lignins: recent advances. *Top Curr Chem*, 376: 33.

Ragauskas AJ, Beckham GT, Biddy MJ, Chandra R, Chen F, Davis MF, Davison BH, Dixon RA, Gilna P, and Keller M (2014). Lignin valorization: improving lignin processing in the biorefinery. *Science*, 344: 1246843.

Ramar M, Manikandan B, Raman T, Priyadarsini A, Palanisamy S, Velayudam M, Munusamy A, Marimuthu Prabhu N, and Vaseeharan B (2012). Protective effect of ferulic acid and resveratrol against alloxan-induced diabetes in mice. *European Journal of Pharmacology*, 690(1–3): 226–235.

Rupasinghe HPV, Boulter-Bitzer J, Ahn T, and Odumeru JA (2006). Vanillin inhibits pathogenic and spoilage microorganisms in vitro and aerobic microbial growth in fresh-cut apples. *Food Res Int*, 39: 575–580.

Sato T, Furusawa T, Ishiyama Y, Sugito H, Miura Y, Sato M, and Itoh N (2006). Hydrogen production from lignin with supported nickel catalysts through supercritical water gasification. *Int Ass for Hydrogen Energy - IAHE* 38 43 2006 RN: 38099900.

Sawant SN (2017). Development of biosensors from biopolymer composites. *Biopolymer Composites in Electronics*, 353–383.

Shulga G and Betkers T (2011). Lignin-based dust suppressant and its effect on the properties of light soil. *Environmental Engineering. Proceedings of the International Conference on Environmental Engineering*. ICEE, 8, 1210. Vilnius Gediminas Technical University, Department of Construction Economics & Property.

Slaghek TM, Van Vliet D, Giezen C, and Haaksman IK (2017). U.S. Patent Application No. 15/125, 268.

Song Y, Wu T, Yang Q, Chen X, Wang M, Wang Y, Peng X, and Ou S (2014). Ferulic acid alleviates the symptoms of diabetes in obese rats. *J Funct Foods*, 9: 141–147.

Staub de Melo JV and Trichês G (2018). Study of the influence of nano-TiO2 on the properties of Portland cement concrete for application on road surfaces. *Road Mater Pavement Des*, 19(5): 1011–1026.

Sugiarto S, Leow Y, Tan CL, Wang G, and Kai D (2022). How far is lignin from being a biomedical material? *Bioactive Mater*, 8: 71–94.

Sutton JT, Rajan K, Harper DP, and Chmely SC (2018). Lignin-containing photoactive resins for 3D printing by stereolithography. *ACS Appl Mat Interfaces*, 10(42): 36456–36463.

Tai MJY, Perumal V, Gopinath SCB, Raja PB, Ibrahim MNM, Jantan IN, Suhaimi NSH, and Liu WW (2021). Laser-scribed graphene nanofiber decorated with oil palm lignin capped silver nanoparticles: a green biosensor. *Sci Rep* Mar 9, 11(1): 5475.

Tanase-Opedal M, Espinosa E, Rodríguez A, and Chinga-Carrasco G (2019). Lignin: a biopolymer from forestry biomass for biocomposites and 3D printing. *Materials*, 12(18): 3006.

Tarabanko VE and Tarabanko N (2017). Catalytic oxidation of lignins into the aromatic aldehydes: general process trends and development prospects. *Int J Mol Sci*, 18(11). https://doi.org.10.3390/ijms18112421.

Ten E and Vermerris W (2015). Recent developments in polymers derived from industrial lignin. *J Appl Polym Sci*, 132.

Thakur VK and Thakur MK (2015). Recent advances in green hydrogels from lignin: a review. *Int J Biol Macromol*, 72: 834–847.

Tingda J (2007). *Lignin*, 2nd ed., Chemical Industry Press, Beijing.

Triantafyllidis KS, Lappas AA, and Stöcker M (eds.) (2013). *The Role of Catalysis for the Sustainable Production of Bio-Fuels and Bio-Chemicals*, 1st ed., Elsevier, Amsterdam, The Netherlands; Boston, MA, USA, ISBN 978-0-444-56330-9.

Tsuchiya T and Takasawa M (1975). Oryzanol, ferulic acid, and their derivatives as preservatives. *Jpn Kokai*, 07: 518–521.

Vaidya AA, Collet C, Gaugler M, and Lloyd-Jones G (2019). Integrating softwood biorefinery lignin into polyhydroxybutyrate composites and application in 3D printing. *Mater Today Commun*, 19: 286–296.

Van Vliet D, Slaghek T, Giezen C, and Haaksman I (2016). Lignin as a green alternative for bitumen. *Proceedings of the 6th Euroasphalt & Eurobitume Congress*, pp. 1–3.

Vashist A, Vashist A, Gupta Y, and Ahmad SJ (2014). Recent advances in hydrogel based drug delivery systems for the human body. *J Mater Chem B Mater Biol Med*, 2: 147–166.

Wang H, Pu Y, Ragauskas A, and Yang B (2019a). From lignin to valuable products–strategies, challenges, and prospects. *Bioresour Technol*, 271: 449–461.

Wang P, Dong Z, Tan Y, and Liu Z (2015). Investigating the interactions of the saturate, aromatic, resin, and asphaltene four fractions in asphalt binders by molecular simulations. *Energy & Fuels*, 29: 112–121.

Wang R, Xia G, Zhong W, Chen L, Chen L, Wang Y, Min Y, and Li K (2019b). Direct transformation of lignin into fluorescence-switchable graphene quantum dots and their application in ultrasensitive profiling of a physiological oxidant. *Green Chem*, 21(12): 3343–3352.

Wu X, Jiang J, Wang C, Liu J, Pu Y, Ragauskas A, Li S, and Yang B (2020). Lignin-derived electrochemical energy materials and systems. *Biofuels, Bioprod Bioref*, 14: 650–672.

Xie S, Li Q, Karki P, Zhou F, and Yuan JS (2017). Lignin as renewable and superior asphalt binder modifier. *ACS Sustain Chem Eng*, 5(4): 2817–2823.

Xin J, Li M, Li R, Wolcott MP, and Zhang J (2016). Green epoxy resin system based on lignin and tung oil and its application in epoxy asphalt. *ACS Sustain Chem Eng*, 4: 2754–2761.

Xu G, Wang H, and Zhu H (2017). Rheological properties and anti-aging performance of asphalt binder modified with wood lignin. *Constr Build Mater*, 151: 801–808.

Yearla SR and Padmasree K (2016). Preparation and characterisation of lignin nanoparticles: evaluation of their potential as antioxidants and UV protectants. *J Exp Nanosci*, 11: 289–302.

Yiqin Y, Baoyu L, and Yunfeng C (2009). Study on preparation of slow-release nitrogen fertilizer by sodium lignin lignosulfonate. *Zhonghua Paper*, 29(13): 55–58.

Yu C, Wang F, Zhang C, Fu S, and Lucia LA (2016). The synthesis and absorption dynamics of a lignin-based hydrogel for remediation of cationic dye-contaminated effluent. *React Funct Polym*, 106: 137–142.

Yu G, Li B, Wang H, Liu C, and Mu X (2013). Preparation of concrete superplasticizer by oxidation-sulfomethylation of sodium lignosulfonate. *Bioresources*, 8(1): 1055–1063.

Yu O and Kim KH (2020). Lignin to materials: a focused review on recent novel lignin applications. *Appl Sci*, 10(13): 4626.

Yuan C, Lou Z, Wang W, Yang L, and Li Y (2019). Synthesis of $Fe_3C@C$ from pyrolysis of Fe_3O_4-Lignin clusters and its application for quick and sensitive detection of PrP^{Sc} through a Sandwich SPR detection assay. *Int J Mol Sci*, 20(3): 741.

Zhang W, Jia B, and Furumai H (2018). Fabrication of graphene film composite electrochemical biosensor as a pre-screening algal toxin detection tool in the event of water contamination. *Sci Rep*, 8(1): 1–10.

Zhang W, Zhao M, Liu R, Wang X, and Lin H (2015). Hierarchical porous carbon derived from lignin for high performance supercapacitor. *Colloids Surf A Physicochem Eng Asp*, 484: 518–527.

Zhao W, Simmons B, Singh S, Ragauskas A, and Cheng G (2016). From lignin association to nano-/micro-particle preparation: extracting higher value of lignin. *Green Chem*, 18(21): 5693–5700.

Zhenyang L, Jie C, Ming H et al. (2011). Properties of lignin modified amino-based superplasticizer. *New Build Mater*, 1: 5–8.

Zhu J, Yan C, Zhang X, Yang C, Jiang M, and Zhang X (2020). A sustainable platform of lignin: from bioresources to materials and their applications in rechargeable batteries and supercapacitors. *Prog Energy Combust Sci*, 76: 100788.

7

Lignin – Business and Market Scenario

Abstract

The second most prevalent substance in typical biomass is lignin, which has the peculiar property of being one of the few sources of aromatic compounds that is not petroleum. The primary byproduct of lignocellosic bio-refineries and a vital renewable resource for the chemical industry is lignin. There are many opportunities for turning plant lignin into value-added products that could significantly increase bio-profitability. The demand for lignin is rising significantly across a wide range of end-use industries, which is helping the market. For a variety of factors, such as quick urbanization, ongoing infrastructure development, industrial expansion, and technological advancement, market growth is accelerated in developing economies. But it is anticipated that a lack of awareness and technological limitations will stifle industry expansion. The lignin business and market scenario is presented in this chapter.

Keywords   Lignin; Renewable resource; Lignin market; Lignin market share; Kraft lignin; Lignosulphonates; Soda lignin; Organosolv lignin; High purity lignin; Low purity lignin

7.1 Introduction

Industrially, more than 70 million tons of lignin is produced annually, but only a small amount—between 1% and 2%—is utilized. Lignin has been shown to be a reliable source of numerous chemicals with added value because it is polyaromatic. As a result, it is anticipated that the lignin market will expand worldwide as a result of industry adoption of lignin.

The market for lignin is extremely concentrated. The market was worth USD 976.6 million. From 2022 to 2029, the total revenue is expected to increase by 2.37%, reaching nearly USD 1177.8 million.

Lignin is one of the few non-petroleum sources of aromatic compounds (Glasser, 1981; Glasser et al., 2000; Glasser and Sarkanen, 1989). Lignin is the primary byproduct of lignocellosic bio-refineries and is a crucial renewable resource for the chemical industry. There are numerous opportunities to transform plant lignin into products with added value that have the potential to significantly increase bio-profitability. In the coming years, it is anticipated that the growth of the market will be driven by an increase in the demand for lignin as an organic additive. The market for lignin is being pushed by ongoing efforts in research and development. The lignin market is expanding in a positive direction as a result of the rising demand from numerous end-use industries. Market expansion is accelerated in developing economies for a variety of factors, including

Depolymerization of Lignin to Produce Value Added Chemicals, First Edition. Pratima Bajpai.
© 2024 John Wiley & Sons, Inc. Published 2024 by John Wiley & Sons, Inc.

rapid urbanization, ongoing infrastructure development, industrial expansion, and technological advancement. However, it is anticipated that technological constraints and a lack of awareness will stifle industry expansion.

7.2 Lignin Market

Table 7.1 lists the major companies in the lignin market https://www.databridgemarketresearch.com/reports/global-lignin-market. The major players in the industry are Borregaard AS, Domtar Corporation, Aditya Birla Group, Nippon Paper Industries Co., Ltd., The Dallas Group of America, and The Lenzing Group, among others.

In order to encourage the industrial use of lignin, the industry in Europe and North America has been heavily involved in research. Some of the most important industry developments are shown in Table 7.2 (Bajwa et al., 2019).

Prices per metric ton (MT) of different types of lignin vary significantly (Table 7.3). Lignin with low purity has a price range of 50–280 USD/MT, and lignin with high purity has a price range of up to 750 USD/MT. Other types of lignins, such as lignin from the kraft process, which has a market value of between 260 and 500 USD/MT, and lignosulphonates, which have a price range of between 180 and 500 USD/MT, are in between. Soda lignin, which costs 200–300 USD/MT, is more affordable when made from sulphur-free lignins (Hodásová et al., 2015). Organosolv lignin is more expensive than other lignin types available on the market, with prices ranging from 280 USD/MT to 520 USD/MT (Gosselink, 2011; Hodásová et al., 2015).

Table 7.1 Lignin market key players.

Borregaard LignoTech (US)
Bu Burgo Group Spa (Italy)
Chengzhou Shanfeng Chemical Industry Corporation (China)
Domsjo Fabriker (Sweden)
Fibria Cellulose (Brazil)
Green Value SA (US)
Ingevity (US)
Innventia (Sweden)
Lenzing AG (Austria)
Lignin Lignin Company LLC (US)
Metsa Group (Finland)
Nippon Paper Industries Co Ltd. (Japan)
Northway Lignin Chemical (Canada)
Rayonier Advanced Materials (US)
Sigma-Aldrich Co. (US)
Stora Enoo Oyj (Finland)
Tembec (Canada)
The Dallas Group of America (US)
West Fraser (Canada)
West Rock Company (US)

Adapted from https://www.databridgemarketresearch.com/reports/global-lignin-market / last accessed 4 May 2023.

Table 7.2 Current industrial applications of lignin.

Alberta-pacific (Alpac), Canada
Chemicals, Materials
Kraft pulping process
Innventia, Sweden
Carbon Fiber (Innventia, 2018)
Kraft pulping
Domtar Corporation, Canada
Adhesives, agricultural films and chemicals, carbon products (e.g., carbon fiber, graphene, graphite, activated carbon, etc.), coatings, dispersants, fuels and fuel additives, natural binders, plastics, resins (Domtar, April 2018)
Technaro GmbH, Germany
Thermo plastics, carbon fiber (TECNARO - The Biopolymer Company https://www.tecnaro.de/en/2010/05)
Borregaard LignoTech, USA
Dispersing agents in concrete, textile dyes, pesticides, batteries and ceramic products—or as binding agents in animal feeds, briquetting and various dust suppression applications. (Borregaard - the sustainable biorefinery https://www.borregaard.com)
MeadWestvaco, USA
Propoxylated lignin polypols, animal feeds, particle board, wax emulsion, dyes, lead acid batteries, ceramics, concrete, and refractories (Agrawal et al., 2014)
Non-sulfonated kraft lignin
Northway Lignin Chemicals, Canada
Emulsifiers, organic binders, dispersants, and liquid/powder agglomeration (Agrawal et al., 2014)
Sulfur free kraft lignin
KMT lignin. UK
Concrete mixture, dust abatement, leather tanning, ceramics, insecticides sprays etc. (Agrawal et al., 2014)
Lennox Polymers Ltd., USA
Formaldehyde-free resins, adhesives (Holladay et al., 2007)

Reproduced from Bajwa et al (2019) / with permission of ELSEVIER.

Table 7.3 Lignin market value.

Low purity lignin
50–280 USD/MT
High purity lignin
750 USD/MT
Lignin from kraft process
260–500 USD/MT
Lignosulphonates
180 USD/MT- 500 USD/MT
From sulphur-free lignins
200–300 USD/MT
Organosolv lignin
280USD/MT- 520 USD/MT

Adapted from Hodásová et al. (2015).

Table 7.4 Lignin market share (by product) 2022.

Low purity lignin	77.9%
Lignosulfonates	14.7%
Kraft lignin	4.9%
Organosolv lignin	0.9%
Others	1.7%
Total	100%

Adapted from Gmisights / https://www.gminsights.com/industry-analysis/lignin-market / last accessed 4 May 2023.

Table 7.4 shows lignin market share (by product) in 2022. Increased lignin use as an energy source, the development of new lignin extraction technologies, and increased lignin use in biomass production are all helping the market. The creation of novel separation techniques and the utilization of lignin in the creation of aromatic monomers are anticipated to drive the industry. Europe, Asia Pacific, North America, Latin America, the Middle East, and Africa, are the major geographical areas in this sector. The market's two main application categories are macromolecules and aromatics (https://www.expertmarketresearch.com/reports/lignin-market).

According to a new report from Global Market Insights Inc., the high purity lignin market is expected to reach a value of USD 46 million by 2030 (https://www.globenewswire.com/en/news-release/2022/10/05/2528401/0/en/High-Purity-Lignin-Market-to-Surpass-USD-45-Million-by-2030-says-Global-Market-Insights-Inc.html). Among the primary factors that have contributed to the growth of the high purity lignin industry are the rising production of automobiles and the subsequent rise in carbon fiber consumption.

High-purity lignin manufacturers will benefit greatly from rising vehicle production, particularly in Asia-Pacific, due to carbon fiber's status as a key composite material. For instance, vehicle production in China reached over 26 million units in 2021. Automobile sales will rise in tandem with increased consumption of high purity lignin as disposable income rises. Lignins recovered from chemical pulp mills already have significant commercial markets. In most of these markets, the value of lignin is significantly higher than the value of fuel. According to Nadányi et al. (2022), chemical pulping alone could produce anywhere from 44 to 66 million tons of lignin in 2025.

Rather than being sold for its fuel value, a lot of lignin might be sold for its chemical value since biorefineries that process lignocellulosic feedstocks are expected to be built soon (Pye, 2005). In addition, lignin derived from particular kinds of biorefineries is more likely to have superior performance characteristics and a higher commercial value for its chemical properties than the lignin that is currently derived from chemical pulping. With improved chemical and physical properties, these "new" lignins will undoubtedly open up significant new markets for this resource that is renewable. But it is highly probable that these materials will initially target the same market niches as the lignin products that are already on the market.

Currently, the kraft, sulfite, and soda pulping processes are primarily used by the chemical pulping industry for separating lignin from woody plant materials and some annual plants like baggase, flax, and straw (Bajpai, 2018; Smook, 2003). The majority of this lignin is burned in chemical recovery boilers as parts of the concentrated black liquor feed rather than being isolated and recovered. As a result, it supplies the pulp mill with low-cost fuel for the production of steam and electricity (Bajpai, 2018; Smook, 2003). On the other hand, very few established businesses sell systems that partially recover and purify lignin for a wide range of applications to numerous industries (Gargulak and Lebo, 2000).

In the future, biorefineries that process lignocellulosic materials in order to primarily produce fermentable sugars from hemicelluloses and cellulose will result in extremely high levels of lignin production. These procedures will almost certainly result in lignin that is chemically distinct from the industrial lignins that are currently available and will almost certainly be closer to the native material's chemical structure. As a result, these more adaptable products will find new uses, but they will also be able to compete with the lignin products that are currently available in the markets for the lignin. The lignins that are presently produced and sold come from the pulping liquors used in the chemical pulping industry. However, these are affected in some way by the chemistry of the particular chemical pulping process from which they are obtained (Alén, 2000). Lignin that has been partially hydrolyzed in the wood receives sulfonic acid groups during the sulfite pulping process, transforming it into a lignosulfonate that is completely water-soluble. Together with the hemicellulose sugars, this becomes a significant component of the spent sulfite liquors. The action of sodium sulfide, which introduces thiol groups, transforms the native lignin into thiolignin fragments with lower molecular weights during the kraft pulping process. The cooking liquor's strong alkali dissolves this thiolignin. However, the kraft black liquor's acidification can precipitate and recover it (Alén, 2000). The lignin is hydrolyzed into smaller fragments following soda pulping, which dissolve in the highly alkaline pulping liquors. However, the lignin that is produced is chemically relatively unaltered.

As a result, the types of lignin produced by each process determine the applications and markets for the variety of lignin forms currently recovered from commercial chemical pulping facilities. As was mentioned earlier, the chemical pulping industry extracts a lot of wood and woody fibers, but only a small amount of each type of lignin is recovered and sold. These amounts only make up a small fraction of what the industry extracts. Due to the necessity of recovering and recycling inorganic cooking chemicals, chemical recovery furnaces burn nearly all of the lignin extracted from woody raw materials by the pulping industry worldwide. In some less developed nations, however, chemical recovery boilers are out of reach for a few smaller soda pulp mills that use annual fibers as their raw material so they either directly or with minimal treatment release their pulping liquors into the environment. Almost all governments actively discourage this latter activity at the moment, which has led to the closure of numerous smaller chemical pulp mills, particularly in China, over the past decades (Zhong and Gan 1997). Lignin is a wet, solid material that is left after the saccharification and/or fermentation stages, mostly in a form that contains contamination from cellulose. It is produced by a number of lignocellulosic-based biorefineries that are as of now in the improvement stage. The lignin is a component of a low-value fuel in bigger chemical pulp mills, where recovery boilers are a financial requirement. This fuel supplies the pulp mill with steam and power. It is therefore most practical to utilize this lignin residue as a cheap boiler fuel for producing steam and electricity for these specific biorefinery processes (Zimbardi et al., 2002).

Alternately, pure lignin, which is simple to produce using biorefinery technologies like organosolv process, may be a very good source of industrial specialty and commodity chemicals and replace crude oil and natural gas-based materials (Lora et al., 1989). Lignin is a possible new wellspring of sweet-smelling synthetic compounds and different items that are much of the time utilized in industry as a result of its dominating fragrant substance structure. This adaptable material would be most cost-effectively utilized as a fuel. Using established technology, purified lignin of higher value could theoretically be recovered from either the initial feedstock of woody biomass before saccharification or the residues of the saccharification or fermentation stages. But, it is now widely accepted that the presence of hemicellulose and lignin can, to a certain extent, appreciably slow down the rate and effectiveness of acid hydrolysis or cellulolytic enzyme-mediated cellulose saccharification (Pan et al., 2004). As a result, almost all of the biorefinery processes that have been proposed include mechanical or chemical feedstock pretreatments that partially remove or modify these two materials. The majority of these pre-treatments, like, ammonia fiber explosion, dilute acid hydrolysis, and steam explosion disrupt the highly integrated physical and possibly chemical relationships of hemicelluloses, cellulose, and lignin in native woody biomass (Bajpai, 2016; Boussaid et al., 2000; Teymouri et al., 2004; van Walsum, 2001).

Lignin's chemical structure and physical properties can be altered by such pre-treatments, which can increase or decrease its value in potential commercial applications. According to Alén (2000), the majority of the biorefinery pre-treatments that have been proposed will result in partially depolymerized hydrolyzed lignin, primarily at the aryl–alkyl ether linkages, hydrolyzed lignin. In contrast to the majority of lignins currently available, this will produce a lignin product that does not contain sulfur-bearing substitutes. The average number and weight of these lignin products should be smaller (1000–2000) and the polydispersity should be lower if the pre-treatment and lignin recovery conditions do not result in significant recondensation of the lignin fragments (Lora et al., 1989), which produced a very uniform product. With the commercial lignins that are currently available, this is not the case.

On the topic of the development of specialty and common applications for pure, unmodified lignin, much research has been published and numerous conferences have been held over the past 50 years (Gargulak and Lebo, 2000; Glasser and Sarkanen, 1989; Hu, 2002), but few of these exciting applications have been made commercially available as of yet. It is easy to see why. Until very recently, the only commercially available lignins were thiolignins from the sulfate cooking process and lignosulfonates from the sulfite cooking process. According to Gargulak and Lebo (2000), the Borregaard Lignotech division of Borregaard Industries and the Specialty Chemicals division of MeadWestvaco Corporation are currently the two primary suppliers of these chemically modified lignins. Chemically nonderivatized lignins, like those that biorefineries are likely to produce and that are used in numerous studies, are not available in sufficient amounts to sustain a substantial commercial market. This is about to change with the expected commercial deployment of biorefineries that process lignocelluloses. These biorefineries may be able to produce significant amounts of lignin in a form that is relatively uniform and purified. Low-cost fuel can be made with these lignins or for industrial chemical applications of much higher value, which would significantly boost the profitability of the biorefineries that produce them. The lignin of these upcoming biorefineries could serve as the foundation for a significant new industry that creates products that either replace or are novel to the market for chemicals produced from crude oil or natural gas.

Bibliography

Agrawal A, Kaushik N, and Biswas S (2014). Derivatives and applications of lignin—an insight. *Scitech J*, 1: 30–36.

Alén R (2000). Basic chemistry of wood delignification. *Papermaking Science and Technology, Book 3- Forest Products Chemistry*, Stenius P (ed.). Fapet Oy, Helsinki, Finland, pp. 43–52.

Bajpai P (2016). Pretreatment of lignocellulosic biomass. *Pretreatment of Lignocellulosic Biomass for Biofuel Production*. Springer Briefs in Molecular Science. Springer, Singapore, https://doi.org/10.1007/978-981-10-0687-6_4.

Bajpai P (2018). *Biermann's Handbook of Pulp and Paper: Volume 1: Raw Material and Pulp Making*. Elsevier, USA.

Bajwa DS, Pourhashem G, Ullah AH, and Bajwa SG (2019). A concise review of current lignin production, applications, products and their environmental impact. *Ind Crops Prod*, 139: 111526.

Boussaid AL, Esteghlalian AR, Gregg DJ, Lee KH, and Saddler JN (2000). Steam pretreatment of Douglas-fir wood chips. Can conditions for optimum hemicellulose recovery still provide adequate access for efficient enzymatic hydrolysis? *Appl Biochem Biotechnol* 2000 Spring, 84-86: 693–705.

Gargulak JD and Lebo SE (2000). In lignin: historical, biological and materials perspectives, *ACS Symposium Series 742*, American Chemical Society, Washington, DC, p. 304.

Glasser WG (1981). Potential role of lignin in tomorrow's wood utilization technologies. *Forest Prod J*, 31: 24–29.

Glasser WG, Northey RA, and Schultz TP (2000). Lignin: historical, biological and materials perspectives. *ACS Symposium Series 742*, American Chemical Society, Washington, DC.

Glasser WG and Sarkanen S (Eds.) (1989). *Lignin: Properties and Materials*. American Chemical Society, Washington, DC, p. 545.

Gosselink RJA (2011). *Lignin as a renewable aromatic resource for the chemical industry*. PhD Thesis, Wageningen University, Wageningen, NL, p. 45.

Hodásová L, Jablonský M, Škulcová A, and Ház A (2015). Lignin, potential products and their market value. *Wood Res*, 60 (6): 973–986.

Holladay JE, Bozell JJ, White JF, and Johnson D (2007). *Top Value-Added Chemicals from Biomass*.

Hu TQ (2002). *Chemical Modification, Properties, and Usage of Lignin*. Kluwer Academic/Plenum Publishers, New York.

Lora JH, Wu CF, Pye EK, and Balatinecz JJ (1989). *Lignin: Properties and Materials*. Glasser W.G. and Sarkanen S. (eds.). American Chemical Society Symposium Series 397, Washington, DC, USA, pp. 312–324.

Nadányi R, Ház A, Lisý A, Jablonský M, Šurina I, Majová V, and Baco A (2022). Lignin modifications, applications, and possible market prices. *Energies*, 15: 6520. https://doi.org/10.3390/en15186520.

Pan X, Zhang X, Gregg DJ, and Saddler JN (2004). Enhanced enzymatic hydrolysis of steam-exploded Douglas fir wood by alkali-oxygen post-treatment. *Appl Biochem Biotechnol*, 115(1–3): 1103–1114.

Pye EK (2005). Industrial lignin production and applications. *Biorefineries-Industrial Processes and Products*, Kamm B., Gruber P.R., and Kamm M. (ed.). https://doi.org/10.1002/9783527619849.ch22.

Smook GA (2003). *Handbook for Pulp and Paper Technologists*, 2nd ed., Joint Textbook Committee of the Paper Industry of the United States and Canada, Atlanta, USA, and CPPA, Montreal, Canada. 425.

Teymouri F, Laureano-Perez L, Alizadeh H, and Dale B (2004). Ammonia fiber explosion treatment of corn stover. *Appl Biochem Biotechnol*, 113–116: 951–963.

van Walsum GP (2001). Severity function describing the hydrolysis of xylan using carbonic acid. *Applied Biochem Biotechnol*, 91/93: 317–328.

Zhong XJ and Gan A (1997). *Profit Through Innovation, Pira International*. London, UK, p. 222–224.

Zimbardi F, Ricci E, and Braccio G (2002). Technoeconomic study on steam explosion application in biomass processing. *Applied Biochem Biotechnol*, 18/20: 89–99.

https://www.databridgemarketresearch.com/reports/global-lignin-market.

Lignin Market - By Raw Material (Hardwood, Softwood, Straw, Sugarcane Bagasse, Corn Stover), By Product (Kraft Lignin, Lignosulphonates, Organosolv Lignin), By Application (Aromatics, Macromolecules), By Downstream Potential & Forecast, 2023-2032. https://www.gminsights.com/industry-analysis/lignin-market.

High-Purity-Lignin-Market-to-Surpass-USD-45-Million-by-2030-says-Global-Market-Insights. https://www.globenewswire.com/en/news-release/2022/10/05/2528401/0/en/High-Purity-Lignin-Market-to-Surpass-USD-45-Million-by-2030-says-Global-Market-Insights-Inc.html.

8

Challenges and Perspectives on Lignin Valorization

Abstract

Lignin is the largest aromatic source on earth and the non-carbohydrate component of plant cell walls with the highest abundance. Due to its potential for numerous industrial applications, it is receiving much attention. Because of its inherent rigidity, heterogeneity, and complexity, the application of lignin is challenging in comparison to that of carbohydrate components like hemicelluloses and cellulose. On the other hand, the range of renewable resources that can be used to make chemicals, fuels, and materials can be expanded by successfully utilizing lignin. Additionally, many biofuels and biochemicals primarily made from hemicellulose and cellulose may benefit from an increase in the economic competitiveness of value-added lignin products. Combining the first step of thermochemical depolymerization with the biological transformation of lignin-derived monomers into products with added value would be the most practical approach to lignin valorization. This chapter discusses lignin valorization challenges and perspectives.

Keywords   Lignin; Lignin valorization; Depolymerization; Thermochemical depolymerization; Biological conversion; Biofuels; Biochemicals; Biorefinery; Low-value macromolecules; Lignin-first biorefining

8.1 Introduction

Despite being regarded as the aromatic resource with the greatest abundance, less than 2% of lignin is utilized in the production of value-added products because of the limitations of the available conversion technology. The fact that it is anticipated that more biofuels will be produced from lignocellulosic biomass will also significantly increase the availability of lignin residue, requiring the development of suitable chemical or biological conversion technologies.

The evaluation of lignin is an important step in developing biomass utilization strategy. It can help ensure the long-term viability of biorefinery processes by increasing revenue and lowering waste stream costs.

In the 1990s, the idea of a biorefinery emerged to maximize the value of biomass by combining cutting-edge machinery with processes for converting biomass into biofuel, power, syngas, and value-added chemicals. The lignocellulosic biorefinery industry can operate more efficiently thanks to lignin's abundance, biodegradability, biocompatibility, chemical versatility (functional groups), capacity for reinforcement, antioxidant properties, and potential pharmacological activities (Li and Takkellapati, 2018; Luo et al., 2023; Paone et al., 2020; Rajesh Banu et al., 2021; Rinaldi et al., 2016; Singhania et al., 2022; Yuan et al., 2013).

8.2 Challenges and Perspectives on Lignin Utilization

The biorefinery industry has an intermediate upcoming opportunity to utilize lignin to generate green fuel, syngas, heat, and power. Lignin's polymer and polyelectrolyte properties are utilized in almost all current commercial applications, with the exception of the immediate opportunity. The production of low-value macromolecules like binders, dispersants, emulsifiers, and sequestrants based on this medium-term opportunity is one of the primary objectives of using lignin in the biorefinery industry (Zhou et al., 2022).

With the assistance of "tailored" catalysts and solvents, the production of polymers of higher value and macro-monomers like carbon fibers and thermosetting resins extensively increases the number of industrial uses for lignin. This is made possible by developing selective conversion processes. At this point, it is considered difficult but doable to efficiently extract aromatic chemicals of high value from lignin through depolymerization, such as monomer molecules and BTX chemicals. Current research on lignin depolymerization focuses primarily on producing specific end products and improving the efficiency of biomass. Utilizing novel catalysts in traditional and lignin-first biorefining processes is of primary interest for improving the proportion of intended products in the mixture of depolymerized products and increase the efficiency of lignin depolymerization (Huang et al., 2018; Shu et al., 2018).

The efficiency of lignin degradation can be drastically enhanced by using nanotechnology for lignin depolymerization (Du et al., 2020). By mimicking peroxidase activity, a novel nanoparticle catalyst made of iron oxide doped with cerium was able to directly convert corn cob lignin into a mixture of low-molecular-weight products (Rajak et al., 2021). This lignin degradation process only required low operating temperatures (25 °C) to reach 44 weight % degradation within 30 hours. To improve the efficiency of lignin degradation, catalyst-based systems that depolymerize lignin continuously can be utilized (Nandiwale et al., 2020). But, the lower ratio of solvent to biomass (w/v) may impede the economic commercialization of these techniques. The proportions of the final products are significantly influenced by the lignin depolymerization metal catalysts (Shu et al., 2018). The requirement for additional steps to purify can be eliminated and production costs reduced by continuing research into the development of catalysts that are able to provide the desired final products.

The utilization of lignin in the biorefinery sector presents a number of challenges in addition to significant opportunities. Lignin's higher molecular weight, ambiguous chemical properties, and complex structure strictly limit its use in high-tech applications. In addition, lignin's physicochemical properties—such as reactivity, solubility, distribution of molecular weight, and a number of functional groups—variate frequently depending on where it comes from (biomass source) and how it is recovered. Due to the variability of lignin, it will be difficult and more expensive to isolate and purify each individual compound to be used in subsequent projects. Naturally, this will result in product mixtures.

To determine the most effective methods for depolymerizing isolated lignin into valuable products and removing lignin from raw biomass, it is essential to have a comprehensive comprehension of each conversion process in a lignin biorefinery as well as the life cycle of products obtained from lignin from upstream to downstream. These difficulties have not yet been completely resolved. Because of its multifunctionality, depolymerized lignin produces a variety of product streams that necessitate expensive procedures for purification and separation. Depolymerized products like hydroxybenzoates and their derivatives, for instance, can be utilized in a variety of industries, including the food, natural health products, cosmetics, and pharmaceutical sectors. But, evaluation of their bio-efficacy and safety is needed. The challenges that are unique to each product stream should be the focus of future research efforts. Plant feedstock engineering for fine-tuning lignin monomer and functionality may also assist in mitigating these issues. The extent to which the lignin structure of plants can be altered for yet-unknown new human applications may also be the subject of future research.

Lignin's diverse application strategies have been the subject of extensive research, ranging from its decomposition into fuels and chemicals with added value to the creation of cutting-edge materials like nanoparticles, hydrogels, and carbon fiber.

There are two steps that must be taken into account when developing technology for lignin valorization:

- depolymerization of lignin
- improvement of lignin degradation products

One of the requirements for the depolymerization of lignin is identifying the precise structures of native biomass lignin. In order to take into account the fact that lignin structures and component ratios would differ between biomasses, it is important to first investigate the lignin structure of the biomass that is utilized the most.

Lignin can be depolymerized by biological or thermochemical processes, both of which have their own benefits and drawbacks, as this book explains. Because it takes a long time and is associated with low productivity and yield, biological degradation of lignin does not offer any industrial advantages at the current stage of technology development. Even though well-known lignin-degrading enzymes like dye-decolorizing peroxidase, laccase, lignin peroxidase, manganese peroxidase, and versatile peroxidase have been used to degrade lignin model compounds in vitro, these enzymes can only break the particular linkage of the model compounds for lignin and rarely degrade the actual complex lignin residues (Chen and Wan, 2017). However, thermochemical depolymerization can rapidly degrade actual lignin residue into products with low molecular weights (Nguyen et al., 2021). It should be noted here that lignin undergoes structural modification during pre-treatment of lignocellulosic biomass. According to Cao et al. (2018), the low bond dissociation energy of fifty to seventy kcal mol^{-1} is due to the soft bond known as the β-O-4-aryl ether linkage, which is anticipated to be present in half of native lignin. Chemical or biological decomposition of the β-O-4-aryl ether linkage yields the aromatic compounds that make up lignin. However, for separating lignin from lignocellulosic biomass during the pretreatment process, the use of a catalyst like a strong acid and the generally high temperature cause the β-O-4 aryl ether linkages of lignin to decompose and/or transform. Although a "lignin-first" strategy is under consideration for preventing the lignin decomposition, which would separate lignin instead of cellulose for obtaining biosugar, the β-O-4-aryl ether linkages would unavoidably decompose and become denatured. Given these issues, using appropriate enzymes like laccase and peroxidase to depolymerize such denatured lignin is practically impossible. From this angle, chemical depolymerization of lignin might be more practical than the biological approach (Nguyen et al., 2021).

Genetic modification of lignocellulosic plants has recently been considered for the purpose of regulating lignin biosynthesis and altering the types or ratios of lignin linkages for altering lignin content and structure in order to facilitate lignin depolymerization. For thermochemical degradation and biological conversion, a suitable substrate could be an engineered plant with total or enriched G, H, or S unit monomers.

Pazhany and Henry (2019) state that introducing the novel lignin monomers like sinapyl ferulate or coniferyl ferulate—whose weak ester linkages can increase lignin extractability and reduce the need for a pre-treatment process—may be a potential strategy for modifying lignin. Hereditary designing of lignocellulosic biomass is in early phase, and further specialized issues to conquer the adverse consequences related with adjusting lignin, for example, issues with water conductivity, development imperfections, and plant pressures should be tended to. In addition, further improvement of lignin degradation products is required for lignin valorization. Prior to determining the catalytic reaction, separating the products of lignin depolymerization ought to be taken into consideration for chemical upgrading (Nguyen et al., 2021).

High-priced depolymerization products must first be separated for direct use without upgrading when a reasonably priced separation process is available. Chemo-catalytic conversion can be used to transform other products into products with a high value added. However, there are actually a lot of products of lignin depolymerization that are hard to separate using the current separation techniques. Vanillin and other phenolics like, syringaldehyde, coumaric acid, ferulic acid, and acetovanillone are found in varying amounts in thermochemical conversion (Fache et al., 2015). It is essential to use a method that efficiently separates the final product. However, at this time, only limited performance has been observed. From a practical standpoint, the development of a cost-effective method for separation may not be possible due to the small number of target products and numerous unidentified degradation products. As a result, upgrading lignin, such as HDO of products of lignin depolymerization, has been conducted primarily to enhance fuel properties because the technology used for chemical upgrading does not take separation into account. The HDO reaction mostly produces phenols, aromatics, cycloalkanes, and cyclohexanol/cyclohexanone as upgrading products.

According to Ponnusamy et al. (2019), cost competitiveness necessitates the conversion of cyclohexanol/cyclohexanone into value-added products like caprolactone, caprolactam, and adipic acid. Biological lignin upgradation with genetically engineered microorganisma is one option when lignin upgradation without a separation process is considered. Microbial cells can be programmed to use lignin degradation products sequentially through a variety of metabolic pathways and can mix a specific substrate. The aromatic compounds derived from lignin can go through the G-, S-, or H-lignin degradation pathways; metabolized into intermediates with aromatic cores; and then transformed into a variety of products with added value, including lipid, vanillin, PHA, and muconic acid. This is known as a "funneling" tactic.

One of the most promising approaches is metabolic engineering-based biological upgradation of the lignin degradation mixture in light of these benefits. In order to accomplish this, it is necessary to engineer microbial cells for expanding their lignin assimilation pathway in order to maximize the use of a wide variety of lignin-derived monomers. It is necessary to acquire a deeper comprehension of the metabolic lignin degradation pathway, like the non-oxidative decarboxylase or side chain reduction pathways. The yield and productivity of biological upgrading processes can be significantly improved by comprehending the control and feedback inhibition of aromatic compounds that are particularly pertinent to the shikimate pathway and are found to significantly hinder the production of related products. The method of fermentation ought to be taken into consideration because a number of aromatic monomers are known to be harmful to microorganisms.

It is necessary to optimize fed-batch fermentation because for maintaining a concentration below the toxic threshold, aromatic monomers can be added gradually (Kohlstedt et al., 2018). Additionally, adaptive evolution, also known as the initial stage of microorganism adaptation, needs to be performed to lessen the negative effects of aromatic monomers as a substrate for microorganisms (Torres et al., 2009). The yield and productivity of biological upgrading with engineered microbial cells, both of which are currently rather low, can be easily increased by a methodical approach to metabolic engineering that considers the discovery of novel enzymes and synthetic pathways.

Malonyl-CoA and coumaric acid can be used to make resveratrol. However, only a small portion of lignin is coumaric acid, the model compound of the H-lignin unit. A brand-new enzyme can break the ether linkage that exists between the methoxy groups of G-lignin or S-lignin—two of the main components of lignin. It is possible to convert every unit of lignin into resveratrol thanks to this enzyme's ability to directly convert ferulic and syringic acids to coumaric acid.

The thermochemical depolymerization and the first steps in the practical approach to lignin valorization would be the biological transformation of lignin-derived monomers into products with added value. Combining biological upgrading and thermochemical depolymerization to get the most out of both processes and selecting the appropriate target product can dispel the prevalent belief that "you can make anything from lignin except money."

During the process of lignin valorization, the primary obstacles that need to be taken into consideration are (Ullah et al., 2022):

- delection of lignin source and the depolymerization capacity
- how to choose the best microorganisms
- enzyme and microbe synergy; improvements to catabolic pathways and the production of a wanted item

In addition to the aforementioned methods of lignin valorization, important downstream processes include the separation and purification of lignin and its byproducts. Numerous lignins are regarded as byproducts or coproducts of cellulose-based biorefinery processes. As a result, extractives and inorganic components as well as process solvent and chemicals can be found in the fractionated lignin stream. A few drying strategies are utilized in laboratory studies, however, their reasonable application in industry is still questionable. Additionally, further purification is required to avoid the devaluation of lignin caused by residual carbohydrates bound to it. Furthermore, the decomposition of lignin, in particular, results in the production of numerous products and unknown components. Consequently, it is essential to separate the targeted products. Sadly, contrasted with the use procedures of lignin, its downstream methodologies are not researched a lot. As a result, future lignin valorization strategies will require additional research on the pre- and post-processing of lignin and its products (Yoo and Ragauskas, 2021).

Bibliography

Cao L, Yu IKM, Liu Y, Ruan X, Tsang DCW, Hunt AJ et al. (2018). Lignin valorization for the production of renewable chemicals: state-of-the-art review and future prospects. *Bioresour Technol*, 269: 465–475.

Chen Z and Wan C (2017). Biological valorization strategies for converting lignin into fuels and chemicals. *Renew Sustain Energ Rev*, 73: 610–621.

Du B, Liu C, Wang X, Han Y, Guo Y, Li H et al. (2020). Renewable lignin-based carbon nanofiber as Ni catalyst support for depolymerization of lignin to phenols in supercritical ethanol/water. *Renew Energ*, 147: 1331–1339.

Fache M, Boutevin B, and Caillol S (2015). Vanillin, a key-intermediate of biobased polymers. *Eur Polym J*, 68 (C): 488–502.

Huang Y, Duan Y, Qiu S, Wang M, Ju C, Cao H et al. (2018). Lignin-first biorefinery: a reusable catalyst for lignin depolymerization and application of lignin oil to jet fuel aromatics and polyurethane feedstock. *Sustain Energ Fuels*, 2 (3): 637–647.

Kohlstedt M, Starck S, Barton N, Stolzenberger J, Selzer M, Mehlmann K, Schneider R, Pleissner D, Rinkel J, Dickschat JS, Venus J, van Duuren JBJH, and Wittmann C (2018). From lignin to nylon: cascaded chemical and biochemical conversion using metabolically engineered *Pseudomonas putida*. *Metab Eng*, 47: 279–293.

Li T and Takkellapati S (2018). The current and emerging sources of technical lignins and their applications. *Biofuels, Bioprod Bioref*, 12 (5): 756–787.

Luo Z, Qian Q, Sun H, Wei Q, Zhou J, and Wang K (2023). Lignin-first biorefinery for converting lignocellulosic biomass into fuels and chemicals. *Energies*, 16: 25.

Nandiwale KY, Danby AM, Ramanathan A, Chaudhari RV, Motagamwala AH, Dumesic JA et al. (2020). Enhanced acid-catalyzed lignin depolymerization in a continuous reactor with stable activity. *ACS Sustain Chem Eng*, 8 (10): 4096–4106.

Nguyen LT, Phan D, Sarwar A, Tran MH, Lee OK, and Lee EY (2021). Valorization of industrial lignin to value-added chemicals by chemical depolymerization and biological conversion. *Ind Crops Prod*, 161: 113219.

Paone E, Tabanelli T, and Mauriello F (2020). The rise of lignin biorefinery. *Green Sustain Chem*, 24: 1–6.

Pazhany AS and Henry RJ (2019). Genetic modification of biomass to alter lignin content and structure. *Ind Eng Chem Res*, 58: 16190–16203.

Ponnusamy VK, Nguyen DD, Dharmaraja J, Shobana S, Banu JR, Saratale RG, Chang SW, and Kumar G (2019). A review on lignin structure, pretreatments, fermentation reactions and biorefinery potential. *Bioresour Technol*, 271: 462–472.

Rajak RC, Saha P, Singhvi MS, Kwak D, Kim DA, Lee H, Deshmukh AR, Bu Y, and Kim BS (2021). An eco-friendly biomass pretreatment strategy utilizing reusable enzyme mimicking nanoparticles for lignin depolymerization and biofuel production. *Green Chem*, 23: 5584–5599.

Rajesh Banu J, Preethi Kavitha S, Tyagi VK, Gunasekaran M, Karthikeyan OP, and Kumar G (2021). Lignocellulosic biomass based biorefinery: a successful platform towards circular bioeconomy. *Fuel*, 302: 121086.

Rinaldi R, Jastrzebski R, Clough MT, Ralph J, Kennema M, Bruijnincx PCA, and Weckhuysen BM (2016). Paving the way for lignin valorisation: recent advances in bioengineering, biorefining and catalysis. *Angew Chem Int Ed*, 55: 8164–8215.

Shu R, Xu Y, Ma L, Zhang Q, Wang C, and Chen Y (2018). Controllable production of guaiacols and phenols from lignin depolymerization using Pd/C catalyst cooperated with metal chloride. *Chem Eng J*, 338: 457–464.

Singhania RR, Patel AK, Raj T, Chen CW, Ponnusamy VK, Tahir N, Kim S-H, and Dong CD (2022). Lignin valorisation via enzymes: a sustainable approach. *Fuel*, 311: 122608.

Torres BR, Aliakbarian B, Torre P, Perego P, Domínguez JM, Zilli M, and Converti A (2009). Vanillin bioproduction from alkaline hydrolyzate of corn cob by *Escherichia coli* JM109/pBB1. *Enzyme Microb Tech*, 44: 154–158.

Ullah M, Liu P, Xie S, and Sun S (2022). Recent advancements and challenges in lignin valorization: green routes towards sustainable bioproducts. *Molecules*, 27 (18): 6055.

Velvizhi G, Balakumar K, Shetti NP, Ahmad E, Kishore Pant K, and Aminabhavi TM (2022). Integrated biorefinery processes for conversion of lignocellulosic biomass to value added materials: paving a path towards circular economy. *Bioresour Technol*, 343: 126151.

Yoo CG and Ragauskas AJ (2021). Opportunities and challenges of lignin utilization. *Lignin Utilization Strategies: From Processing to Applications*, Yoo CG and Ragauskas AJ (eds.). ACS Publications, Washington, DC, http://doi.org/10.1021/bk-2021-1377.ch001.

Yuan T-Q, Xu F, and Sun R-C (2013). Role of lignin in a biorefinery: separation characterization and valorization. *J Chem Technol Biotechnol*, 88 (3): 346–352.

Zhou N, Thilakarathna WPDW, He QS, and Rupasinghe HPV (2022). A review: depolymerization of lignin to generate high-value bio-products: opportunities, challenges, and prospects. *Front Energy Res*, 9: 758744. https://doi.org/10.3389/fenrg.2021.758744.

Index

"metal-free" lignin first process   153, 159
1-(4-sulfobutyl)-3-methyl imidazolium hydrosulfate   43t
1-Butyl-3-methylimidazolium tetrafluoroborate   146
1-butylimidazolium hydrogen-sulfate   99t
1-ethyl-3-methylimidazolium acetate   43t
1-ethyl-3-methylimidazolium chloride   99t
1-ethyl-3-methylimidazolium trifluoromethanesulfonate   97, 146
2,6-dimethoxy-4-propylphenol   162
2-benzazepine   165
2-methoxy-4-vinylphenol   100t, 126t
2-methoxyphenol   100t
2-naphthol   94
2-phenoxy-1-phenylethanol   137
2-propanol   151
3,5-dimethoxy-4-hydroxybenzaldehyde   189
3D printing   193–194
3-methoxyphenol   45
4-ethylguaiacol   106, 107t, 110t
4-ethylphenol   89t, 99t, 105–106, 107t, 110t
4-hydroxy-3-methoxybenzaldehyde   189
4-hydroxybenzaldehyde   126t, 134t, 139
4-hydroxybenzoic acid   59t, 189
4-hydroxybenzyl alcohol   137
4-propylcyclohexanol   151
4-propylguaiacol   104t, 107t, 111t, 152, 165
4-vinylphenol   99t
5-hydroxymethylfurfural   159

a

Absorbent materials   191
Acetals   88, 157, 159–160
Acetic acid   8t, 16t, 21–22, 31, 98, 100t, 128, 175, 195
Acetic acid lignin   21–22
Acetic acid pulping   21–22
Acetophenone   152
Acetosolv lignin   21
Acetovanillone   99t, 100t, 126t, 222
Acetylation   21, 195
Acid catalysis   36, 87–88
Acid- catalyzed depolymerization   87–90, 89t, 106, 159
Acidolysis   88, 90, 153, 159
Acrylonitrile butadiene styrene   193
Activated carbon   109, 138, 157, 183, 191, 214t
Activation energy   87, 106–107, 135–136
Active pharmaceutical ingredients   164f, 165
Adhesives   18, 173, 176, 182–183, 203–204, 214t
Adipic acid   222
AFEX   23t, 174–175, 183
Agricultural crop   12, 14, 178
Air fresheners   189
Alcell lignin   21–22, 106, 108, 110t
Alcell process   21
Alcohol   90, 94, 102, 104, 157–158, 176
Aldehyde-assisted fractionation   159, 160f
Aliphatic acid   202
Alkaline metal chloride   137
Alkyl imidazolium chloride   43
Alkylation   39, 109
Alkyl-O-aryl bond   146
Alkylsulfonate anions   97
Alloy   105
Alloying   203

Depolymerization of Lignin to Produce Value Added Chemicals, First Edition. Pratima Bajpai.
© 2024 John Wiley & Sons, Inc. Published 2024 by John Wiley & Sons, Inc.

Allylic rearrangement 88
Alternative fuel 29
Alumina 108, 201
Aluminium chloride 137
Aluminium oxide 108
Aluminosilicates 36
Aluminum zeolites 35
Amino-based superplasticizers 202
Ammonia 35, 100, 175, 194–195
Ammonia fiber explosion (AFEX) 174–175, 217
Ammonium formate 151
Ammonium lignosulfonates 181
Amycolatopsis 60t, 61, 189
Angiosperm 153
Animal feed 194, 214t
Animal feed pellets 203
Animal husbandry 192
Anion exchange resin 146
Annual crops 20
Annual plants 15t, 215
Anthracenedione 182
Antibiotics 189
Anti-cancer agent 191
Anti-cancer drugs 192
Anti-cancer properties 190
Anti-depressant 165
Anti-diabetic agent 191
Antimicrobial properties 189–190
Antioxidant 45, 189, 191–193, 201, 219
Anti-solvent precipitation 192
Apple tree pruning 96t
Aprotic 146
Aromatic aldehydes 127–128, 189
Aromatic compounds 1, 7, 31, 34–36, 40t, 53, 57–58, 63, 70, 72, 88, 90, 92t, 95, 105, 127, 137, 139, 151, 165, 188–189, 212, 221–222
Aromatic hydrocarbon 33–35, 36f, 113, 138, 153
Aromatic monomers 7, 35, 88, 95, 136, 138, 151–153, 162, 189, 215, 222
Aromatic ring hydroxylation 125
Aromatic substitution 31, 32f
Aromaticity 3
Aromatics 34, 57, 63, 110t, 111t, 139, 154f, 165, 177, 188–191, 215, 222
Arsenic 204
Aryl ether linkage 22, 221
Ash 19–20, 22, 177, 179–180, 182

Asphalt 194, 200–201
Asphalt pavements 201
Automobile interiors 202
Automotive industry 191
Auxiliary agent 203

b

Bacillus subtilis 189
Bacteria 53, 56–64, 59t, 60t, 61t, 65, 65t, 69, 73
Bagasse 96t, 177, 182
Ball milling 95
Bamboo 161t
Base-catalyzed depolymerization 87, 90–95, 98, 104, 127
Battery material 198–199
Beech 162, 176
Benzene 34, 110t, 113, 165
Benzyl radical 34
Benzylic acid oxidation 125
Benzylic alcohol 127, 134t, 138
Benzylic cations 159
Benzylic ketones 44, 127
Bimetallic alloy 105
Bimetallic catalyst 105, 162
Binder 181, 198–201, 203
Bio-btx aromatics 165
Biochar 95, 109, 138, 178
Biochemicals 7, 44, 176, 219
Biocompatibility 42, 195, 219
Biodegradability 200, 203, 219
Biodegradable plastics 203
Bioethanol 4, 159, 163, 175–177
Bio-Flex 174t, 175–176, 183
Biofuels 7, 164, 164f, 173, 176–177, 179–181, 183, 188, 219
Bioimaging 195–197, 196t
Biological control agent 194
Biological conversion 219, 221
Biological depolymerization 8, 53–73
Biological lignification 14
Biological lignin upgradation 222
Biological pretreatment 42t, 64, 156, 174, 177
Biomarker 195
Biomass 1, 2f, 7, 15, 20–22, 37–39, 53, 55t, 61t, 63–64, 97, 109, 113, 133, 136, 147, 151, 159–160, 163, 173, 175–183, 198, 215–216, 219–222

Biomass refineries   7, 12, 42
Biomaterial   165, 176, 181, 193, 196
Biomedical application   192–193
Bio-oil   7, 30, 33–36, 94, 108–109, 110t, 113, 137–139, 151, 178, 204
Bioplastics   177, 202
Biopolymer   1, 104, 165, 193
Biopolyols   135
Biorefinery   7–8, 29, 56, 127, 156, 158f, 164f, 165–166, 174, 176, 178–179, 215–217, 219–220
Biorefining   8, 156–172, 176
Biosensor   195–197, 196t, 197t
Biosphere   12
Biphenyl ether   2
Birch   22, 105–106, 108, 152, 153, 161t, 162–163
Bitumen   200–201
Bitumen modifier   200–201
Black liquor   18–19, 42, 178–182, 215
Body lotion   189
Brick ceramic   201
Brine   200
Bronsted acid   35–36, 90, 113
Building material   201–202
Butane   100
Butyl alcohol   136

C

Cadmium   203
Calcium hydroxide   90–91
Calcium scaling   203
Canadian maple syrup   190
Candlenut shells   151
Caprolactam   222
Caprolactone   222
Carbenium ion   139
Carbohydrates   2, 4, 18, 20, 158, 162, 175, 223
Carbon black   191, 203
Carbon dioxide   19, 44, 55, 57, 63, 101, 138–139, 146, 178, 180–181, 197, 202
Carbon fiber   19–20, 22, 174, 181–182, 191, 199, 214t, 215, 220–221
Carbon material   108, 191
Carbon nanotube yarns   199
Carbon quantum dots   196
Carbonium ion   20
Carbonization   19, 22, 36f, 42, 199
Carbonyl derivatives   125

Cascade process   158
Catalytic depolymerization   20, 108, 139, 140f, 147, 157
Catalytic hydrogenolysis   162
Catalytic hydrotreatment   108
Catalytic upgradation   139, 159
Catechol   32–33, 41t, 44, 45f, 46f, 58, 60t, 101, 106, 135, 139, 189
Cattle   189
Cell therapy   192
Cellulose   4, 8, 14, 18, 21, 43t, 55, 127, 153, 162–163, 175–176, 188, 198, 202–203, 221
Cellulosic bioethanol   4
Cellulosic ethanol   12
Cement additives   201–202
Cement mixes   203
Ceramics   203, 214t
Char   19, 30–31, 33–35, 91, 98, 100–101, 111t, 112t, 135, 138, 152
ChCl: Urea   99t
ChCl:Ethylene glycol   99t
Chelating fertilizers   194
Chemical depolymerization   30, 87–114, 221
Chemical industry   176, 212
Chemical oxidation   7, 128
Chemical pre-treatment method   7
Chemical pulping   12, 215–216
Chemical recovery boiler   215–216
Chemical recovery furnace   216
Chemical versatility   219
Chemicals   4, 7, 29, 58, 88, 96, 100–101, 106, 113, 135, 145, 147, 156–159, 173, 177, 182, 189, 200, 204, 212, 214t, 217, 221, 223
Chloralkali processing   203
Chlorine   125
Chromium   204
Chromium chloride   137
CIMV process   22
Cis-muconic acid   189
Club mosses   12
Coal briquettes   203
Coal tar   45
Coke   32–36, 36f, 39, 138
Colon cancer   203
Colon cancer cells   190, 192
Composite   193–195, 201–203
Compressive strength   201
Concrete   201–202

Concrete additive   18
Condensation   21, 90, 100, 138
Condensation reaction   20, 90, 98, 100, 104
Coniferyl alcohol   3f, 4f, 6–7, 6f, 13–14, 13f, 58, 93, 204
Construction waste   175
Controlled release fertilizer   194–195
Control-release phosphate   194
Cooking liquor   4, 21, 176, 179, 181, 216
Copper oxide   125, 136
Corn   181, 220
Corn stalk lignin   105–106
Corn stover   58, 63, 72, 96t
Corn stover lignin   63, 71, 92t, 126t, 197t
Cosmetic   189, 191–193, 221
Coumaric acid   222–223
Cyclic voltammetry   98, 195
Cycloalkanes   111t, 222
Cyclohexanol/cyclohexanone   222

d

Deep eutectic solvents   44, 98, 99t, 146
Degradation   7, 20–21, 39, 42, 43t, 57–58, 125–126, 136–137, 145, 163, 175, 179, 182, 220, 222
Dehydrating agent   202
Dehydrogenation   3, 6f, 138
Dehydroxylation   21, 55, 147, 159, 163, 177
Delignification rate   177
Demethoxylation   33–34, 57, 104, 139, 147
Demethylation   33, 125
Demethylation reaction   104
Deoxygenation   35, 108–109, 113, 151
Depolymerization   7, 30, 33, 36f, 39, 53, 58, 63, 72, 88, 90, 100, 102, 104–105, 107–108, 125, 145, 152, 156, 163
Destructive distillation   45
Dibenzodioxocin   2
Dicarboxylic acid   44
Dietary fiber   203
Diethyl phthalate   106, 107t, 134t
Dihydroconiferyl alcohol   100t
Dihydroeugenol   162
Diisocyanate   194
Diisopropylethylamine   98
Dilignol   3
Dimensional stability   202

Dimethyl formamide   136
Dimethyl sulfoxide   136
Diol-assisted fractionation   159, 160f
Diols   88, 194
Dioxane   88, 192
Dispersant   18, 173, 183, 203, 220
Dispersing agent   203
Divanillin   165
Double Enzymatic Lignin   99t
Drug delivery   192
Drugs   157, 188, 192
Dust control agent   200
Dust emission   200
Dust suppressant   200
Dye   204
Dye-decolorizing peroxidase   53, 221

e

Electrical properties   199
Electrochemical catalysis   98
Electrochemical degradation   146
Electrochemical depolymerization   99t, 145–147
Electrochemical processing   146
Electrodes   145–146, 191
Electromagnetic radiation   133, 136
Emulsifier   203, 220
Emulsion   203
Energy crops   175–177
Energy efficiency   136, 147, 191
Energy storage   191, 198–199
Enol ether   90
Environmental impact   98, 191, 200
Enzymatic ethanol manufacturing process   179
Enzymatic hydrolysis   106, 183
Enzymatic lignin   179
Enzymes   13, 53, 55–58, 59t, 60t, 63, 64–73, 164, 177, 183, 194, 221
Lignin-first biorefinery   158, 158f, 163
Epichlorohydrin   165
Epoxy resins   165, 194
Ethanediol   136
Ethanol   18, 21, 39, 63, 88, 90, 101, 103t, 104, 104t, 109, 161t, 164, 175, 177, 179, 182
Ethanol-water extraction   7
Ethanolysis   113
Ether bond   135
Ethyl alcohol   136

Index

Ethyl glycerol 105
Ethylene glycol 136, 157, 159, 162–163
Eucalyptus 161t
Eugenol 103t, 108, 135, 189
Extractives 2, 176, 178, 223

f

FABIOLA™ fractionation process 177
FABIOLA™ Lignin 177
Fabiola™ process 177
Fast pyrolysis 30, 33
Fast pyrolysis lignin 174t, 178
Feed 183, 189, 194, 203, 215
Fermentation 56, 70–71, 159, 177, 189, 216, 222
Ferns 12
Ferric chloride 137
Ferric sulfide 135
Ferrocene 195
Fertilizer 175, 194–195
Fertilizer additive 194
Fertilizer enhancer 194
Ferulic acid 54, 57, 189–191, 222
Fibre glass insulation 203
Fibrillated cellulose 203
Fixed bed 33–34
Flavoring agent 125, 189
Flavoring compounds 189–191
Flax 202, 215
Flax shive 161t
Flax straw 215
Flocculant 183
Fluidized bed 33–34, 195
Fly ash 201
Foam 203
Food 29, 189–193, 203, 220
Food additive 189–190, 203
Food preservative 189, 191
Formacell method 153
Formaldehyde 21, 88, 94, 104, 160
Formaldehyde resins 203
Formic acid 16t, 18, 21–22, 41t, 87–88, 94, 104, 106, 108, 137, 151
Fragrance industry 189
Fuel 1, 7, 29–30, 58, 147, 151, 156–157, 164, 173, 179–180, 204, 221
Fungi 53–56, 54t, 55t, 61t, 65, 65t, 67
Furfural 44, 159

g

Gallium oxide 197t
Gasification 7, 30, 37, 156
German lignocellulose feedstock biorefinery process 183
Glucose 183, 196t
Glucose oxidase 195, 196t, 197t
Glycerin 44
Graphene film 195
Graphitic carbon 191
Grass lignin 7, 14
Green fuel 188, 220
Greenhouse gas 200, 202
Guaiacol 34, 41t, 42t, 44–45, 45f, 87f, 88, 89t, 90, 92t, 93, 97, 99t, 101, 103t, 104t, 106, 107t, 110t, 113, 127, 135–136, 138–139, 189, 204
Guaiacylacetone 99t
Guaiacylglycerol-β-guaiacyl ether 90, 94
Gymnosperm 153

h

H2-fuel cell vehicles 197
Hardboard plywood 203
Hardwood 15t, 16t, 18, 20–22, 57, 61, 176
Hardwood lignin 7, 14, 14f, 19, 35, 60t, 128, 193
Health effects 190
Heat 7, 19, 34, 133, 188, 220
Heavy metal 176, 191, 203–204
Hemicellulose 1–2, 4, 8, 21, 29, 55, 127, 159, 162–164, 175–176, 216–217
Hemp 202
Herbaceous Lignin 100t
Herbicide 203–204
Heterogeneous catalyst 41t, 88, 108, 125, 157
Hibbert's ketones 87f, 88
High purity lignin 214t, 215
High-pressure steam 175
Homolysis 32–34
Homolytic cleavage 21
Horseradish peroxidase 196t
HUSY catalyst 138
Hydrocarbons 100, 138, 157
Hydrodeoxygenation 34, 108–109, 157, 162
Hydrogels 192–193, 221
Hydrogen 98, 102, 107, 113, 138, 153, 175, 197–198
Hydrogen donor 32, 104, 108, 139, 152–153, 166

Hydrogen donor solvent   108–109, 151
Hydrogen gas   151, 153
Hydrogen peroxide   56, 64, 70–72, 125–126, 126t, 128, 136–137, 139, 146, 189, 196, 196t
Hydrogen production   197–198
Hydrogenation   8t, 20, 108–109, 113–114, 137, 147, 152
Hydrogen-donating solvent   138
Hydrogenolysis   7, 44, 108–109, 114, 138, 151–153, 161–162
Hydrolysis   8, 21, 23t, 88, 102, 162–164, 193, 216
Hydrolytic cleavage   87, 91
Hydroperoxyl radical   146
Hydrophilicity   18
Hydrophobic   147, 195, 201
Hydrophobicity   18, 42
Hydrosilanes   152–153
Hydrothermal liquefaction   30, 37–46, 40f, 40t
Hydrotropic agents   22
Hydrotropic pulping   22
Hydroxyl groups   16t, 18, 22, 44, 58, 73, 90, 98, 125, 127, 138, 136, 162, 194
Hypochlorite   125

i

Imidazolium salt   43, 96, 136
Immobilization   195, 197
Immunomodulation   192
Insecticides   203, 214t
Interfacial crosslinking   192
Ionic liquid lignin   7
Ionic liquids   8, 22, 42, 43t, 90, 95–98, 95f, 136, 146–147, 176, 195
Ionosolv biorefining process   176
Isoeugenol   112t, 135t, 204
Isopropanol   109, 110t, 111t
Isopropyl alcohol   136–137

j

Jasmine oil   189

k

Ketocarbinol   88
Kraft lignin   7–8, 12, 15t, 16t, 18–21, 41t, 42t, 57, 59t, 61t, 61, 63–64, 109, 110t, 113, 128, 134t, 176, 181–182, 188, 193

l

Laccase   53, 54t, 56, 59t, 65–70, 221
Lactate   196t
Lead   4, 36, 174, 179, 199, 203, 214t
Leather tanning   203, 214t
Levulinic acid   147, 148f, 159
Lewis acid   89t, 153
Life cycle assessment   202
Lignans   165
Lignin   1–9, 12–23, 29–46, 53–73, 87–114, 125–129, 133–141, 145–148, 151–154, 156–166, 175–183, 188–204, 212–217, 219–223
Lignin applications   188
Lignin based nano-composites   188, 193–194
Lignin degradation   31f, 42–44, 57, 63, 65–67, 71–73, 94, 109, 136–137, 145, 157–158, 198, 220, 222
Lignin depolymerization   7–8, 29–30, 31f, 37, 39, 43t, 53, 56–64, 97, 101, 106–107, 109, 125–126, 137, 145–147, 151, 189, 220–221
Lignin extraction   97, 157, 159, 162–163, 173–174, 215
Lignin fertilizers   194
Lignin first-biorefining process   156–166
Lignin gasification   198
Lignin market   174, 212–217
Lignin market share   212, 215t, 215
Lignin Production   173, 174t, 175–176, 179–181, 179t, 183, 216
Lignin pyrolysis   32–38, 138–139
Lignin sulfonic acid   203
Lignin valorization   7, 173, 188, 219, 221–223
Lignin-based biosensors   196, 196t
Lignin-based nanomaterials   188, 192
Lignin-epoxy resins   194
Lignin-first biorefining   156, 219–220
Lignin-first method   156
LignoBoost   19, 173, 179t, 181, 183
LignoBoost technologies   19, 181
LignoBoost^{TM} technology   179t, 180
Lignocellosic bio-refineries   212
Lignocellulose polymer   2
Lignocelluloses   8, 29, 55, 69, 165, 217
Lignocellulosic biomass   21, 33, 147, 156–157, 160, 175–177, 219, 221–222

Lignocellulosic catalytic valorization   156
LignoForce   19, 173, 181, 183
Lignoforce®   19
LignoForce™ lignin   180
LignoForce™ technology   180
Lignosulfonate   7, 18, 106, 107t, 179, 181–182, 194, 196t, 199–203, 216
Lignosulfonate production   179
Lignosulfonate-based binder   181
Lignosulfonates   7, 12, 15t, 18, 174t, 181, 192, 199–203, 215t, 217
Limestone   201
Linoleum paste   203
Lithium hydroxide   91
Loblolly pine wood   96
Low purity lignin   212, 214t, 215t
Low-value macromolecules   219, 220
Lubricant layer   201

m

Macro-monomers   220
Magnesium hydroxide   91
Magnetite   195
Manganese chloride   137
Mechanical strength   202
Mechanochemical strategy   95
Medium density fiberboard   203
Melt-spinnable lignin   12
Mesoporous catalyst   138
Metal chloride   137
Metal nanoparticles   106, 137–138, 195
Metal plating   203
Metallic catalysts   104–108, 128
Metallic enhancer   106
Methacrylates   165
Methanol   15, 39, 100t, 100–101, 103t, 104, 161t, 163, 182
Methoxyl groups   34, 190
Methoxylated phenylpropane   1–2, 156
Methoxy-phenols   139
Methyl alcohol   136–137, 152
Methyl guaiacol   110t, 204
Methyl phydroxycinnamate   100
Methyl syringate   136
Methylene blue   204
Microalgae   138
Microbial catalysis   189

Microwave   44, 96t, 98, 106, 133, 135–138
Microwave assisted methylation   44
Microwave irradiation   98, 136–139
Microwave reactor   99t, 139
Microwave-aided depolymerization   133–141
Microwave-aided solvolysis   136, 140f
Milled wood lignin   99
Mineral acid catalyst   90
Mining   203
Miscanthus   16t, 152, 161
Miscanthus giganteus   152
Mixed oxide catalysts   42
Molecular oxygen   68, 126t, 127
Molecular weight distribution   136
Monocarboxylic acid   44
Monolignols   1–3, 7, 12, 29, 35–36, 152, 165
Mordenite   35
Muconic acid   189, 222
Mycobacterium tuberculosis   195

n

Nafion sac-13, 88, 106, 107t
N-allylpyridinium chloride   99t
Nanocomposites   188, 192–193
Nanofibers   192, 199
Nanofiltration system   177
Nanogels   192
Nanomaterials   188, 192, 195
Nanoparticles   99t, 106, 109, 137–138, 192–193, 195–196, 204, 221
Nanoprecipitation   192
Nanotubes   192
Naphthalene   202
Natural health product   190
Neuroprotective agent   191
Nickel   105–106, 108–109, 138, 146, 151, 153, 162, 164, 197, 204
Nickel catalyst   106, 109, 111t, 198
Nickel nanoparticles   109, 138
Nickel-bimetallic catalyst   105
Nitrobenzene   125, 126t, 128
Nitrogen-doped carbon   109
Noble metal   105–106, 108–109, 110t
Non-condensable gas   32
Non-oxidative decarboxylase   222

Nucleophilic attack   3, 98
Nucleophilic center   21
Nutraceuticals   188

O

o-aminobenzenesulfonic acid   196
o-cresol   101
Octanoic acid   98
Oganosolv pulping   12
Oil drilling muds   203
Oil palm   96t, 195
Optical activity   6
Organic biopolymer   1
Organic pollutants   191
Organic polymer   1–2, 12
Organic solvent lignin   188
Organic solvents   8, 18, 21, 94, 97, 101, 133, 136, 182
Organosolv lignin   7, 15t, 22, 41t, 61, 89t, 100–101, 112t, 128, 136–137, 151–153, 182, 192–193, 196t, 197t, 212–215
Organosolv pulping   21, 179
Oxalic acid   99t
Oxidation   3, 68–71, 113, 125, 127–128, 137–138, 175, 180, 191, 198, 202
Oxidative cracking   125, 127
Oxidative depolymerization   96, 125–129, 139, 152
Oxidative propagation   193
Oxirane ring   94
Oxygen   8, 12, 30, 34, 56, 70, 108, 113–114, 125, 138, 146, 153, 179–180, 191, 198
Oxygenated compounds   33, 35
Ozone   125

P

Palladium   106, 108, 138, 157, 162–163
Palladium/carbon catalyst   138, 162
Panel boards   203
Particle board   203, 214t
p-benzyloxyl phenol   146
p-coumaryl alcohol   1–2, 3f, 6, 12, 13f, 29
p-cresol   94, 101
Percarbonate   125
Permanganate   125, 128
Peroxidase   6, 53, 57, 68f, 70, 221
Peroxidase activity   71, 220

Peroxyacids   125
Pesticides   203–204, 214t
PFA resin   176
PHA   58, 59t, 62f, 63, 222
Pharmaceutical industry   192
Pharmaceuticals   156, 164, 166, 189
Phenol formaldehyde   92t, 203
Phenol oxidation   70, 125
Phenolic chemicals   105, 175–177, 180
Phenolic hydroxyl group   136–137
Phenolic monomers   90–91, 97–98, 109, 127, 133, 137, 139, 152, 165
Phenols   18–19, 32–33, 44, 58, 67, 88, 89t, 94, 106, 138–140, 152–153, 165, 176–177, 222
Phenoxy radical   34, 71
Phenylcoumaran   2–3
Phenylpropane   1–2, 12–13, 18, 156
Phenylpropanoid   3f, 6
Phenylpropanol   165
Phosphomolybdic acids   198
Photodegradation   204
p-hydroxybenzaldehyde   126t, 128
p-hydroxybenzoic acid   126t, 178, 189
p-hydroxyphenyl   3f, 15
Physicochemical properties   17, 220
Pine lignin   92t, 152
Plant polymer   188
Plasticizer   19, 22
Platinium   157, 162
Plywood   203
Poly (ethylene oxide)   19
Poly(ethyleneterephtalate)   19
Polyacetals   165
Polyaldimines   165
Polyaromatic   212
Polybenzoxazines   165
Polycaprolactone   175
Polycarbonates   165
Polycarboxylate superplasticizers   202
Polycarboxylic acid   203
Polydispersity   22, 97, 175, 179, 217
Polydopamine   195
Polyethylene   22, 138–139
Poly-hydroxy-butyrate   194
Polylactic acid   176, 195

Polymer   1–2, 12–13, 15, 29, 63, 69, 71, 90, 125, 146, 156, 163, 165, 188–189, 191–194, 197, 220
Polymer matrices   192, 194
Polyols   163–164, 194
Polyoxometalates   198
Polypropylene   138, 193
Polysaccharides   12, 56, 176–177
Poplar   22, 64, 125, 161t, 162, 164, 174
Populus tomentosa   96t
Portland cement   202
Potassium hydroxide   90–91, 101
Poultry   189
Power   145, 197, 199, 216, 219–220
Pre-impregnation   21
Pretreatment   42t, 106, 174, 177, 221
Primary pyrolysis   30, 33
Proesa®   173, 174t, 176, 183
PROESA® Lignin   174t, 176–177
Propane   65, 100
Propanylguaicol   106, 107t
Propenylsyringol   106–107
Propylguaiacol   104–105, 107t, 111t, 152, 165
Propylphenols   110t, 162
Propylsyringol   105, 111t
Protic   146, 159
Protolignin   2, 21
Protonation   19, 43
Pseudomonas radiata   194
Pulping   12, 18, 21–22, 181, 216
Pulping process   12, 18, 176, 178, 180–182, 214t, 216
Pyrocatechol   92, 106
Pyrogallol   33, 189
Pyrolysis   8, 30–36, 106, 108, 135–136, 138–139, 156, 178, 204

q
Quinone methide   3, 93, 94

r
Raney nickel   41t, 152
Reactor configuration   161, 164f
Redox-active   157–163, 165
Reductive catalytic fractionation   157, 159, 161t, 165
Reductive mechanism   147

Reductive pretreatment   113
Reductive stabilization   156–157, 159
Renewable resource   156, 212
Repolymerization   24, 34, 44, 88, 94–95, 100–101, 108, 128, 137–138
Resin   42t, 146, 176, 182, 193–194, 200
Resinol   2
Resorcinol   106, 203
Resorcinol formaldehyde   203
Resveratrol   222–223
Rhodium   108–109, 157, 162
Rigidity   12, 202, 219
Ring dyer   183
Ring-opening reaction   108, 125
Room sprays   189
Room temperature ionic liquid   147
Ruthenium   106, 108, 157, 162–163, 198

s
Saccharification   72, 159, 216–217
Scented candles   189
Schiff base polymers   165
Secondary pyrolysis   30, 32
Second-generation biofuels   188
Seebeck coefficient   199
Selective delignification   21, 54t
Selective depolymerization   98, 156
Sequential Liquid-Lignin Recovery and Purification technology   178–179
Sequestering agent   188, 203
Sequestrants   220
Sewage sludge   138
Shampoos   189
Shikimate pathway   222
Shower gel   189
Silica   35, 182, 197t, 201
Silica gel   182
Silver nanoparticles   195, 204
Sinapyl alcohol   1, 3f, 4f, 6, 12, 29, 62, 204
Smart Cooking   176
Smelting   203
Soaps   189
Soapstock   138
Soda lignin   7–8, 12, 18–20, 89t, 96t, 174t, 179t, 182, 212–213
Soda pulping   182, 215–216
Sodium borohydride   151

Sodium formate   88
Sodium hydroxide   7, 18, 42, 90–94, 104, 152, 183
Sodium sulfide   7, 18, 216
Softwood lignin   1, 7, 14, 60t, 73, 128, 182
Softwoods   18, 20–22, 191
Soil conditioner   194, 204
Soil stabilisers   203
Soil water holding regulator   194
Solubility   18, 21–22, 39, 90, 97, 102, 192, 196, 199, 220
Solvent exchange   192
Solvolysis   94, 108, 133, 135–136, 138, 140, 156–157, 159
Solvolytic depolymerization   162
Sonication   192
Specialty and commodity chemicals   216
Spent sulfite liquor   18
Spinnability   19, 22
Spruce   106, 164, 179
Stabilizers   18, 203
Steam explosion,   12, 20, 32, 173, 174t, 175, 183, 217
Steam explosion lignin   12, 20–21, 30
Steam explosion process   20, 174t, 175
Steam explosion reactor   175
Stem cell research   192
Stover   58, 63, 72, 175
Straw   57, 182, 215
Streptomyces sp   189
Subcritical water   38, 103t
Sugar crops   8
Sugarcane bagasse   43t, 55t, 56
Sulfamic acid   202
Sulfite pulping   18, 22, 179, 181, 216
Sulfomethylation   202
Sulfonic acid   27, 69, 90, 203, 216
Sulfur dioxide   7, 18, 21, 181
Sulfur free lignin   174t
Sulfur-free soda lignin   7
Sulfuric acid   18, 31, 88, 90, 137, 175, 178–183
SunCarbon lignin   179t, 180–181
Supercapacitors   198–199
Supercritical alcohol   101
Supercritical fluid   40t, 101–102
Supercritical methanol   100–101, 104

Supercritical water   39, 100–101, 197–198
Supercritical water gasification   197–198
Superoxide radical   146, 192
Superplasticizers   201–202
Surface-active properties   203
Surfactants   18, 173
Syngas   138, 173, 188, 219, 220
Syngas products   188
Synthetic polymers   193
Syringaldehyde   69, 128, 136, 139, 189–190, 222
Syringic acid   57, 89t, 128
Syringol   38, 101, 139

t

Tanning   203, 214t
Technical lignin   12, 17, 19–20, 176, 188, 199f
Tetrachloroaluminate anion   96
Tetrachloroaluminates   96
Tetrafluoroborate   96, 146
Tetrahydrofuran   21
Tetrahydrofurfuryl alcohol   153
Thermal depolymerization   29, 30, 37, 138
Thermal stability   29–30, 37, 138
Thermal treatment   22
Thermochemical conversion   30, 104, 219, 221, 223
Thermochemical depolymerization   30, 104, 219, 221, 223
Thermochemical method   197
Thermochemical process   145
Thermochemical techniques   30
Thermolysis   7–8
Thermo-Mechanical Pulp-Bio lignin   179t, 183
Thermoplastic foaming   191
Thermosetting resins   220
Thermostabilization   19
Thiols   20, 67, 71
Tissue engineering   192
Titanium dioxide   108–109, 201
TMP-Bio lignin process   183
Toluene   110t, 165, 189
Total reducing sugars   44
Transition metal   147, 159
Transition metal chlorides   137
Triethylammonium methanesulfonate   146–147

Triflic acid 88
Triterpene 176

u

Urea formaldehyde 203
Urethanes and epoxy resins 188, 194
Urethanization reaction 194
Uv absorber 191

v

Valuable chemicals 145, 163
Value-added chemicals 8, 8*t*, 147, 183, 219
Value-added products 133, 163–165, 173, 212, 219, 222
Vanillic acid 56, 59*t*, 73, 128, 139
Vanillin 38, 57, 61, 69, 125, 127–128, 136, 139, 139*t*, 146, 165, 182, 189, 190, 204, 222
Vanillyl alcohol 137
Vascular plant 1–2, 12, 165
Veratric acid 127
Veratrylglycerol-β-guaiacyl 90
Versatile peroxidase 53, 54*t*, 65, 72, 221
Vinyl ether 90
Vinyl guaiacol 89*t*, 204
Visco-elastic behavior 201

w

Water resistance 201
Wet air 126*t*, 127

Wood 2, 4, 12, 18, 20, 45, 53, 57, 105–106, 108, 138, 140, 153, 159, 162–163, 175–177, 182, 216
Wood extractives 20
Wood flour 96*t*
Woody plant 215

x

Xanathan gum 193
Xylan 161*t*, 203
Xylene 165, 189

y

Yellow pine 96*t*

z

Zeolite 35, 109
Zinc chloride 90, 137
Zinc copper 204
Zingiberene 189
Zirconium (iv) oxide 197

α-aryl ether linkage 87, 90
β-aryl ether 91, 94, 157, 138, 147, 162
β-O-4 bond 91, 94, 137–138, 147, 162
β-O-4 hydrogenolysis 44
β-O-4 lignin linkage 2, 7, 14, 63, 71, 87–88, 90, 97, 105*f*, 109, 126, 128, 151, 155, 159, 162
γ-valerolactone 21

Printed and bound by CPI Group (UK) Ltd, Croydon, CR0 4YY
13/09/2023

08114508-0001